AF388883

Empowering Communities, Beyond Energy Scarcity - BIWAES 2021 Biennial International Workshop Advances in Energy Studies

Empowering Communities, Beyond Energy Scarcity - BIWAES 2021 Biennial International Workshop Advances in Energy Studies

Editors

Sergio Ulgiati
Hans Schnitzer
Remo Santagata

MDPI • Basel • Beijing • Wuhan • Barcelona • Belgrade • Manchester • Tokyo • Cluj • Tianjin

Editors
Sergio Ulgiati
Parthenope University of
Naples
Italy

Hans Schnitzer
Graz University of
Technology
Austria

Remo Santagata
Parthenope University of
Naples
Italy

Editorial Office
MDPI
St. Alban-Anlage 66
4052 Basel, Switzerland

This is a reprint of articles from the Special Issue published online in the open access journal *Energies* (ISSN 1996-1073) (available at: https://www.mdpi.com/journal/energies/special_issues/Advances_Energy_Studies).

For citation purposes, cite each article independently as indicated on the article page online and as indicated below:

LastName, A.A.; LastName, B.B.; LastName, C.C. Article Title. *Journal Name* **Year**, *Volume Number*, Page Range.

ISBN 978-3-0365-4485-4 (Hbk)
ISBN 978-3-0365-4486-1 (PDF)

Contents

Preface to "Empowering Communities, Beyond Energy Scarcity - BIWAES 2021 Biennial International Workshop Advances in Energy Studies"

Energy is a fundamental resource for societal and economic metabolisms; not only do we need energy, but we clearly need to address crucial questions about its use (energy to do what? energy from where?) and appropriate management (top-down vs. bottom-up energy policy making). As is well known, a new kind of energy scarcity is occurring, not only due to limited abundance, but increasingly due to environmental constraints and trade-offs, as well as unequal availability worldwide and market prices. Achieving sustainable economies and shared wellbeing calls for an urgent re-framing of the energy problem toward a balanced mix of solutions. The latter include technological improvements, use of energy resources consistent with their thermodynamic properties, a selection of environmentally friendly sources and carriers, suitable approaches to the monitoring of impacts, efficiency measures with rebound control, lifestyle equity and reduction of energy poverty, decrease in wasteful habits, recognition of environmental limits on a limited planet, and careful management of the energy-water-food-environment nexus.

Sergio Ulgiati, Hans Schnitzer, and Remo Santagata

Editors

 *energies*

Editorial

Empowering Communities, beyond Energy Scarcity †

Sergio Ulgiati [1,2,*], **Hans Schnitzer** [3,4] **and Remo Santagata** [5]

1 Department of Science and Technology, University Parthenope, 80143 Napoli, Italy
2 School of Environment, Beijing Normal University, Beijing 100875, China
3 Process Engineering and Energy Systems, Institute of Process and Particle Engineering,
   Graz University of Technology, 8010 Graz, Austria; hans.schnitzer@tugraz.at
4 StadtLABOR—Innovation for Urban Sustainability, Griesgasse 40, 8020 Graz, Austria
5 Department of Engineering, Parthenope University of Naples, Centro Direzionale, Isola C4,
   80143 Napoli, Italy; remo.santagata@gmail.com
* Correspondence: sergio.ulgiati@uniparthenope.it
† BIWAES 2021—Biennial International Workshop Advances in Energy Studies.

Citation: Ulgiati, S.; Schnitzer, H.; Santagata, R. Empowering Communities, beyond Energy Scarcity. *Energies* **2022**, *15*, 4106. https://doi.org/10.3390/en15114106

Received: 30 May 2022
Accepted: 31 May 2022
Published: 2 June 2022

Publisher's Note: MDPI stays neutral with regard to jurisdictional claims in published maps and institutional affiliations.

"If we talk of promoting development, what have we in mind: goods or people?" [1]

1. A Collaborative Network of Scientists and Social Communities

Since the year 1998, a series of Advances in Energy Studies Workshops (BIWAES) have aimed to sharpen scientific focus and build a critical mass and collaborative network among scientists and social communities researching energy and energy-related wellbeing. The workshop was hosted by different countries (Italy, Brazil, Spain, Austria, India, Sweden). The 2008 workshop (Towards a Holistic Approach Based on Science and Humanity) was held in Graz, Austria [2] The 2021 BIWAES occurred in Graz, as a Special Session of the ERSCP2021, European Roundtable on Sustainable Consumption and Production [3].

This workshop aimed to gather all potential players in the energy field, to share knowledge and practices, regulations and roadmaps, as well as integrating and promoting different ways of looking at energy solutions. If successful, this pattern may finally help society to move beyond fossil fuels, overcome energy scarcity and environmental degradation, and prevent the exclusion of important sources of understanding and knowledge.

2. Interdisciplinary Evaluations

The energy problem cannot be addressed using only thermodynamic or technological terms. As was shown in previous editions of BIWAES, a deeper understanding of trends, solutions and policies can only be achieved by converging the efforts of different disciplinary sectors, so that economic, social, environmental, cultural and psychological expertise can converge into an innovative picture of local and larger communities, towards a shared well–being.

3. Old and New Consumers

After COP 21 in Paris [4], the promotion of international agreements on climate change and societal attention to the sustainable use of energy and resources continued to increase. Energy and environmental security are major problems facing our global economy. The increased growth, although this was recently slowed down by the Covid pandemic, and the demands for welfare and well-being made by developed and developing countries, have placed increased pressure on energy resources. A large fraction of "new consumers" in developing countries, mainly concentrated in megacities, strive to access commodity and energy markets worldwide, thus boosting energy consumption and competition for all kinds of resources.

4. Energy Planning

Fossil fuels contributed heavily to climate change and planetary instability. Their supply is governed by dynamic political, economic and ecological factors, independently of the sometimes-questioned estimates about the remaining storage. However, renewable energies are not exempt from environmental and management problems, which make their use questionable, and they are not yet available everywhere. Not all energies have the same quality and environmental costs. They also differ in their extraction, processing, use, turnover time, and land and water demands. This means that energy planning is a challenge, which reaches beyond the achieved or achievable technological progress. We cannot disregard the fact that all energy sources (both renewable or nonrenewable) have pros and cons; their use affects the environment and quality of life to different but non-negligible extents.

However, energy is a fundamental resource for societal and economic metabolisms; not only do we need energy, but we clearly need to address crucial questions regarding its use (energy to do what? energy from where?) and appropriate management (top-down vs. bottom-up energy policy-making).

5. A New Energy Scarcity

As is well known, a new kind of energy scarcity is occurring, not only due to its limited abundance, but due to environmental constraints and trade-offs, its unequal availability worldwide and the market prices. The latter also affects the spread of renewables and energy-efficiency efforts and programs. The achievement of sustainable economies and shared well-being calls for an urgent re-framing of the energy problem towards a balanced mix of different solutions, including technological improvements, a use of energy resources that is consistent with their thermodynamic properties, a selection of environmentally friendly sources and carriers, suitable approaches to monitoring the impacts, efficiency measures with rebound control, lifestyle equity and reductions in energy poverty, decrease in wasteful habits, recognition of environmental limits in a limited planet, and careful management of the energy–water–food–environment nexus. A deeper understanding of these crucial aspects, including ways to address them in our production and consumption patterns, may help us develop qualitative growth and sustainable lifestyles, beyond the illusion of unlimited energy availability and technological fixes.

6. Empowering Communities

Who is in charge of energy solutions? Scientists and technology experts have provided important contributions within research, business and policy-making frameworks. However, some top-down solutions have not always shown an ability to fully address the needs of communities, nor have they promoted stakeholders' and citizens' participation in tailored solutions for different situations. It may be time to integrate top-down and bottom-up efforts, in order to benefit from community insight and knowledge (from regional, urban, neighborhood and condominium realities, rural organizations, and developing communities worldwide), and find needs and solutions that are visible to local realities and not easily visible to experts and policy-makers.

7. Well-Being

Well-being, at the level of local or larger-scale communities, is not only linked to decreasing fossil energy use and energy scarcity. Instead, well-being is deeply linked to lifestyles, community services, fair relations among ages, social levels, professional categories, and stakeholders, towards the satisfaction of primary needs (access to food, decent housing, suitable mobility, and an appropriate and rewarding job), as well as lifestyles that offer each individual access to quality growth, such as culture, social relations, leisure, recognition, and empowerment. Although we cannot claim that energy is the solution to all world problems, we are well aware that energy is one of the most important developmental drivers. Too many countries suffer from energy poverty, due to insufficient, inadequate, expensive or unequal energy supply. There is energy behind the water supply, food production, urban and extra-urban mobility, housing, health, education, communication, democracy and, ultimately, well-being [5].

Awareness of the planetary limits calls for equity policies based on resource-sharing, education, understanding, happiness, and peace. A radical change in the business-as-usual paradigm is needed to allow for people to develop within the Earth's biophysical limits. The further that business-as-usual is allowed to go, the more difficult it will be to reverse this process.

Funding: This research received no external funding.

Institutional Review Board Statement: Not applicable.

Informed Consent Statement: Not applicable.

Data Availability Statement: Not applicable.

Conflicts of Interest: The authors declare no conflict of interest.

References

1. Schumaker, E.F. *Small is Beautiful: Economics as if People Mattered*; Harper Perennial: New York, NY, USA, 1989.
2. Schnitzer, H.; Ulgiati, S. (Eds.) Towards a Holistic Approach Based on Science and Humanity. In *Proceedings of the 6th International Biennial Workshop "Advances in Energy Studies", Graz, Austria, 29 June–2 July 2008*; Graz University of Technology: Graz, Austria, 2008; p. 606. ISBN 978-3-85125-018-3.
3. ERSPC. 20th European Roundtable on Sustainable Consumption and Production, Graz, Austria. 2021. Available online: www.erscp2021.eu (accessed on 15 May 2022).
4. COP-21. 2015. Available online: https://unfccc.int/process-and-meetings/the-paris-agreement/the-paris-agreement (accessed on 10 May 2022).
5. Ulgiati, S.; Fiorentino, G.; Raugei, M.; Schnitzer, H.; Lega, M. Cleaner production for human and environmental well-being. *J. Clean. Prod.* **2019**, *237*, 117779. [CrossRef]

Article

Influence of Population Income on Energy Consumption for Heating and Its CO_2 Emissions in Cities

Pedro J. Zarco-Periñán [1,*], Irene M. Zarco-Soto [1], Fco. Javier Zarco-Soto [1] and Rafael Sánchez-Durán [2]

[1] Departamento de Ingeniería Eléctrica, Escuela Superior de Ingeniería, Universidad de Sevilla, Camino de los Descubrimientos, s/n, 41092 Sevilla, Spain; imzarco@outlook.com (I.M.Z.-S.); fjzarco@outlook.com (F.J.Z.-S.)
[2] Endesa, Avenida de la Borbolla, 5, 41004 Sevilla, Spain; rafael.sanchez@endesa.es
* Correspondence: pzarco@us.es

Abstract: As a result of the increase in city populations, and the high energy consumption and emissions of buildings, cities in general, and buildings in particular, are the focus of attention for public organizations and utilities. Heating is among the largest consumers of energy in buildings. This study examined the influence of the income of inhabitants on the consumption of energy for heating and the CO_2 emissions in city buildings. The study was carried out using equivalized disposable income as the basis for the analysis and considered the economies of scale of households. The results are shown per inhabitant and household, by independently considering each city. Furthermore, to more clearly identify the influence of the population income, the study was also carried out without considering the influence of the climate. The method was implemented in the case of Spain. For this purpose, Spanish cities with more than 50,000 inhabitants were analyzed. The results show that, both per inhabitant and per household, the higher the income of the inhabitants, the greater the consumption of energy for heating and the greater the emissions in the city. This research aimed to help energy utilities and policy makers make appropriate decisions, namely, planning for the development of facilities that do not produce greenhouse gases, and enacting laws to achieve sustainable economies, respectively. The overall aim is to achieve the objective of mitigating the impact of emissions and the scarcity of energy resources.

Keywords: energy consumption for heating; CO_2 emissions; income; buildings; cities; Spain

Citation: Zarco-Periñán, P.J.; Zarco-Soto, I.M.; Zarco-Soto, F.J.; Sánchez-Durán, R. Influence of Population Income on Energy Consumption for Heating and Its CO_2 Emissions in Cities. *Energies* **2021**, *14*, 4531. https://doi.org/10.3390/en14154531

Academic Editor: Vincenzo Bianco

Received: 4 July 2021
Accepted: 23 July 2021
Published: 27 July 2021

Publisher's Note: MDPI stays neutral with regard to jurisdictional claims in published maps and institutional affiliations.

1. Introduction

1.1. Overview

In 2014, 54% of the world's population lived in cities, and it is projected that by 2050, that number will reach 67%. In Europe and North America, more than 80% of the population will live in cities [1]. Currently, between 60% and 80% of energy is consumed in cities where, in addition, CO_2 emissions account for 75% of the total [2]. From this perspective, buildings are the most important energy consumers in cities, in both the residential sector and the tertiary sector (businesses and activities that provide services but do not produce goods, such as banks, stores, government buildings, etc.). These sectors are responsible for the consumption of 36% of energy and the production of 40% of emissions [3]. The two usual forms of energy consumption in buildings are electrical and thermal in the form of natural gas [4], which is used for heating [5]. Specifically, this form of energy accounts for 45% of energy consumption in the OECD [6]. These factors highlight the importance of city buildings for both policy makers and utilities, in terms of legislation for more efficient consumption of energy and lower production of emissions, and appropriate planning of facilities using renewable energy to mitigate the scarcity of energy resources, respectively.

At the global level, this importance is reflected in the Sustainable Development Goals of the United Nations [7]. Specifically, Goal 11 is exclusive to cities: make cities and human settlements inclusive, safe, resilient, and sustainable. However, the above issues are also reflected in Goals 7 (regarding the use of renewable energies), 12 (regarding sustainable and responsible use), and 13 (regarding the reduction of CO_2 emissions) [2]. At the European level, The European Green Deal was established to make Europe a climate neutral continent by 2050 [8]. In addition, Next Generation EU, a EUR 750 billion recovery plan, has been launched, in which buildings are among its priority fields of action [9]. This plan focuses on The European Green Deal and is also a response to the COVID-19 pandemic, which has mainly affected cities [10].

1.2. Literature Review

The main focus of this research is the consumption of energy for heating in the form of natural gas, and its relationship with population income and CO_2 emissions. For this reason, the following review of the literature was carried out in terms of these three perspectives, the relationship between them, and the novelty of this study in relation to the existing research.

Studies have examined the more efficient use of energy in buildings and the reduction in their emissions. The importance placed on these issues by society is reflected in the increase in the number of these studies. Thus, in the period from 2007 to 2017, their number increased from 9 to 82 [11]. Building envelopes and their influence on energy consumption have been investigated in different types of buildings, namely, non-residential buildings [12]; residential buildings in rural [13] or urban [14] areas; and low-income houses [15]. The conclusion is that by improving the thermal insulation of the building envelope, energy consumption is reduced. The relationship between buildings and the health of occupants has also been studied. Inadequate indoor temperatures imply poor health, and particularly respiratory, cardiovascular, and mental health disorders [16].

Thermal energy in the form of natural gas is commonly used for heating in buildings. Conclusions can thus be drawn about how to reduce natural gas consumption and emissions. The studies carried out on this type of energy have been focused on predicting its daily demand at a global level [17], and at the level of a particular sector, such as residential [18] or residential and commercial [19] sectors. Artificial neural networks [20] and learning methods [21] are among the methods most commonly used to undertake demand prediction.

Gross domestic product (GDP), rather than the income of the population, is usually included in the variables used to make these predictions. Because GDP is a more general variable than income, it does not allow analysis at a more specific level of detail. However, the income variable has been used to analyze the detail of the thermal conditions of the low-income segment in some countries [22].

Regarding the use of GDP, certain studies have used the GDP variable to predict demand, although it is usually used with others, such as price [23], heating degree days [24], population [25], and urbanization [26]. Other studies analyzed the influence of different variables on each other, such as gas consumption and GDP. The conclusion they reached is that the elasticity of gas demand is very low; that is, consumers do not respond to price changes by adapting their consumption or using other sources of energy [27].

Regarding CO_2 emissions, studies have examined the emissions produced in buildings in general, without specifically examining those that come from heating. Studies that have taken natural gas into account as an energy source have analyzed the emissions produced in distributed generation projects [28], or the differences in emissions when district heating systems or heat pumps are used for heating in buildings [29].

However, as was the case with consumption, most studies that analyze emissions have focused on the influence of different variables on each other. Among these, rather than income, GDP has again been used, but in this case, it has been related to energy consumption in general [30–32], and few studies have related it to gas consumption in particular [33]. Population is another of the commonly used variables. In two studies comprising a group of 83 countries [34] and the OECD countries [35], the results show that the higher the income, the greater the emissions.

The geographical scope of the studies is relatively wide, and does not focus on local detail. At the country level, household emissions in Ireland [36], and France and the USA [37], have been investigated. The conclusion of both studies is that the higher the income, the greater the emissions. Other studies at the national level have been carried out for China [38] and a group of 170 countries [39]. However, analysis of emissions at the city level has only been carried out for 10 cities considering technical and geophysical factors [40]. In addition, the complete CO_2 emissions of the households of four Chinese cities, in both their urban and rural areas, have been analyzed. The conclusions indicate that emissions are higher in urban areas than in rural areas [41].

The above review indicates that, to the best of the knowledge of the authors, the previous studies have not addressed the analysis of energy consumption for heating and its CO_2 emissions in cities using the approach undertaken in the current research: at the income level of the population, considering the cities independently, and eliminating the influence of the climate. Hence, the authors consider this study to be novel.

1.3. Aim of the Research

The importance for energy consumption and CO_2 emissions of cities in general, their buildings in particular and, more specifically, thermal energy for heating, is highlighted in this research. It is necessary to know the starting point to allow governments and utilities to take the appropriate measures to reduce energy consumption and emissions, and plan the necessary infrastructure to achieve this. For this purpose, cities must be considered to be independent elements of study, rather than being treated with a single criterion for the application of the same solution to all cities. For this reason, this study focused on the energy consumption related to heating in city buildings and the emissions they produce. To allow analysis and comparison of the cities as independent elements, the inhabitants and households are used as basic units, and the energy consumption for heating of all the buildings in the city is assigned to these units. For this purpose, the consumption corresponding to non-residential buildings is also distributed among the inhabitants and households. This is because the purpose of non-residential buildings is to fulfill the needs of the city's residents, and the number of these buildings is proportional to the number of inhabitants.

The methodology used in the study is based on other classical approaches that have been used recently [42–45], and on others that are based on the creation of synthetic populations [46]. These approaches use aggregated public data to represent the population of each city in a simplified manner. The practical application of the method was implemented for the 145 Spanish cities, considered individually, with more than 50,000 inhabitants.

The main contributions of this paper are to: analyze energy consumption for heating and related emissions in buildings; study energy consumption and related emissions based on the income of building inhabitants; analyze the results at the inhabitant and household level; consider all of the cities of a country separately rather than in aggregate form; and eliminate the influence of the climate on the results. To the best of the authors' knowledge, a similar investigation has not been carried out previously.

The paper is structured as follows: Section 2 presents the proposed methodology; the application of the method to the case of Spain, in which the scope is defined as cities with more than 50,000 inhabitants, is shown in Section 3; in Section 4 the results are presented and discussed; finally, Section 5 summarizes the findings with respect to the study's aims.

2. Method

The sequence followed in the proposed method was: first, define the study area; second, define the criteria for selecting the cities under study; third, classify cities according to the income of their inhabitants; fourth, determine the consumption of thermal origin of the buildings, per inhabitant and per household (it should be recalled that energy for heating usually has a thermal origin in the form of natural gas); fifth, eliminate the influence of the climate on consumption; and sixth, calculate the CO_2 emissions produced by this consumption. Because the climate is a variable that notably influences the consumption of heating, this influence was eliminated. In addition, other variables can also influence energy consumption: population density, income of its inhabitants, characteristics of the house, age of buildings, etc. However, their elimination is beyond the scope of this study. All of the data used was publicly accessible and, where possible, of governmental origin. The aim was that the obtained results would provide individualized information for each city. Thus, the information relating to the energy consumed by each city allowed the determination of the power demand for each city, and thus the corresponding level of emissions and the requirement for energy from renewable sources to reduce the city's emissions. Figure 1 presents the methodological approach.

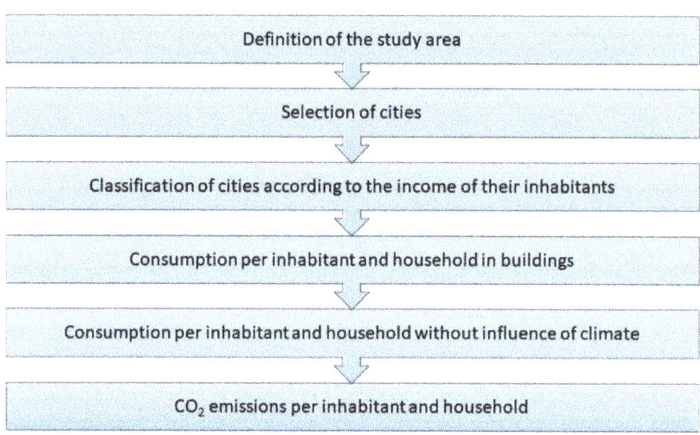

Figure 1. Methodological approach.

2.1. Classification of Cities by Income of Their Inhabitants

First, the geographical area of interest was selected, which could be as large as desired. The cities to be studied and that met a certain criterion of interest were chosen. This criterion could include the number of inhabitants, the possible saturation of its infrastructure, etc. Next, each city was assigned a certain income value of its inhabitants, thus allowing a segmentation of that income. This was based on different criteria, such as the National Minimum Wage (NMW) or a relationship to it.

To perform the analysis, the main statistical data of thermal consumption for each group of cities were studied:

Mean:

$$\overline{E}_i = \frac{\sum_j E_{ij}}{n_i} \tag{1}$$

Standard deviation:

$$s_i = \sqrt{\frac{\sum_{ij}\left(E_{ij} - \overline{E}_i\right)^2}{(n_i - 1)}} \tag{2}$$

Median:

$$Median_i = \left[\frac{n_i+1}{2}\right] th \text{ term if the total number of the elements is an odd number,}$$
$$\text{otherwise } Median_i = \frac{\left(\frac{n_i}{2}\right)th \text{ term} + \left(\frac{n_i}{2}+1\right)th \text{ term}}{2} \tag{3}$$

where n_i is the number of cities that belong to group i; $\overline{E}_i$ is the mean energy consumed in group i; E_{ij} is the energy consumption of city j, which is in group i; s_i is the standard deviation of the energy consumed in the cities of group i. The consumption of thermal energy and cities' consumption were listed in ascending order to calculate the median.

Furthermore, to analyze the variations in consumption, an index was defined. A similar index for monthly electric energy consumption was defined in [47]. The income variation index (*IVI*) was defined as follows:

$$IVI_i = \overline{E}_i / \overline{E} \tag{4}$$

where IVI_i is the index of the group of cities that have size i, $\overline{E}_i$ is the energy consumption mean value of group i, and $\overline{E}$ is the mean energy consumption of all cities (of all groups). This index allows visualization of the variations of each group of cities, and identification of those with the highest and lowest consumption.

A common reference was selected to make the comparison between the city groups. For this purpose, different possible criteria may be used. The most basic is the income per inhabitant. However, equivalized disposable income is another possible criterion. Because this research was carried out both per inhabitant and per household, in the opinion of the authors, the latter criterion is more appropriate and was therefore used in the analysis.

2.2. Equivalized Disposable Income

Households are characterized by a certain economy of scale depending on the number of members and the ages of the people who form it. This is considered in the concept of equivalized disposable income, which also considers total household income after taxes. Each household was assigned an equivalent size, which is the sum of the weights of all of the members of a given household. The weight of each member was calculated using the modified OECD equivalence scale. This scale attributes the following weights to each member: 1 to the first adult, 0.5 to the other adults in the household, and 0.3 to those under 14 years of age. Finally, equivalized disposable income was calculated by dividing the household's total income from all sources by its equivalent size. Using the equivalized disposable income of each household, it is possible to obtain that corresponding to the city. Thus:

$$ES = n_i + 0.5 \, n_j + 0.5 \, n_k \tag{5}$$

$$EDI = \frac{\sum_i Income}{ES} \tag{6}$$

where ES is the equivalent size, n_i is equal to 1 for the first adult in the household, n_j is the total number of people over 14 years of age minus 1, n_k is the number of people under 14 years of age living in the household, and EDI is the equivalent disposable income.

Therefore, it is necessary to have access to data on inhabitants per household and their ages, and on disposable income per inhabitant and city. Note that, hereinafter, the word income is used to refer to the equivalized disposable income.

2.3. Thermal Energy Consumption

Energy consumption for heating in homes is produced almost exclusively by means of thermal energy in the form of natural gas. The pressure at these supply points is equal to or less than 4 bar. The official data that are usually published for this supply pressure do not distinguish between whether these consumption points are residential, commercial, or administrative office buildings. In addition, stores and administrative offices exist in a city to fulfill the needs of its inhabitants. For this reason, it was decided to carry out the study by distributing the energy consumed in these non-residential buildings among all citizens. Therefore, the allocated energy consumption includes both that consumed in households and that of stores and administrative offices. All information from public databases was processed to obtain the necessary data for the investigation.

2.4. Elimination of the Influence of Climate

Climate is a parameter that significantly influences energy consumption in cities [48]. Energy consumption varies depending on the climate [49]. To better analyze the influence of income on consumption for heating, the influence of climate can be eliminated. For this purpose, each city must be assigned a climate, for which the city must be geographically located on the climate map of the study area. Subsequently, the mean consumption of the climatic zone to which it belongs must be identified.

The correction is made by a factor [43] as follows:

$$K_{ci} = \overline{E}_c / \overline{E}_{ci} \tag{7}$$

where $\overline{E}_{ci}$ is the mean energy consumed in the climate zone i; $\overline{E}_c$ is the mean energy consumed in all cities studied; and K_{ci} is the correction factor by which each city will be affected according to its climate. Finally, the energy consumption of the city must be corrected. The correction factor is applied to the energy consumed by each city according to the climatic zone in which it is located.

2.5. CO₂ Emissions

CO_2 emissions are obtained based on energy consumption in buildings. Therefore, based on the information obtained on energy consumed, it is possible to determine the emissions.

3. Application of the Method to the Case of Spain

The case of Spain is presented as an application of the proposed method. As a study area, Spain as a whole was considered. The data used were those corresponding to the year 2016. The study cities were those with more than 50,000 inhabitants. These represent more than 50% of the Spanish population [50]. Knowledge of the consumption of thermal energy and the emissions they produce will allow correct planning of infrastructure to cover future needs, in addition to promoting ad hoc measures to reduce energy consumption and emissions in each city.

3.1. Classification of Study Cities

In Spain there are 145 cities with more than 50,000 inhabitants. For each of these, the income of its inhabitants, its thermal energy consumption, and its CO_2 emissions were considered. The cities were separated into five groups based on the mean income value of their inhabitants. This value was assigned to each of the cities. NMW was chosen as the basis for segmentation and its value in 2016 amounted to EUR 7429.97 per year [51]. Initially, a division into groups based on a differentiation at 0.5 NMW was studied. However, this resulted in a highly unbalanced number of cities per group. The number of cities with income between 2 and 2.5 NMW was much higher than the remainder. To achieve a more balanced number of cities per group, groups with different multiples of the NMW were established. Thus, the difference between some groups is 0.5 NMW and in others it is 1 NMW.

Cities with incomes less than 2 times the NMW form Group 1; those between 2 and 2.5 times the NMW form Group 2; Group 3 comprises those with income between 2.5 and 3 times the NMW; Group 4 includes those whose inhabitants have incomes between 3 and 4 NMW; finally, the cities with incomes greater than 4 NMW make up Group 5. The cities that make up each group, arranged alphabetically, are shown in Table 1.

Normally, the information on the average income of the inhabitants is provided in the databases, and is the situation in Spain. Alternatively, if information corresponding to the median income was available, rather the average, this could be used instead.

Table 1. Classification of cities according to their equivalized disposable income.

Equivalized Disposable Income	Cities
Group 1: income less than 2 times the NMW	Alcalá de Guadaíra, Alcoy/Alcoi, Arona, Arrecife, Benalmádena, Benidorm, Chiclana de la Frontera, Dos Hermanas, Ejido (El), Elche/Elx, Elda, Estepona, Fuengirola, Gandía, Jerez de la Frontera, Linares, Línea de la Concepción (La), Lorca, Marbella, Mijas, Motril, Orihuela, Parla, Puerto de Santa María, Roquetas de Mar, San Bartolomé de Tirajana, San Fernando, San Vicente del Raspeig, Sanlúcar de Barrameda, Santa Coloma de Gramenet, Santa Lucía de Tirajana, Talavera de la Reina, Telde, Torremolinos, Torrent, Torrevieja, Utrera, Vélez-Málaga,
Group 2: income between 2 and 2.5 times the NMW	Albacete, Alcalá de Henares, Alcorcón, Algeciras, Alicante/Alacant, Almería, Aranjuez, Arganda del Rey, Ávila, Avilés, Badajoz, Badalona, Cáceres, Cádiz, Cartagena, Castellón de la Plana, Ceuta, Ciudad Real, Collado Villalba, Córdoba, Cornellà de Llobregat, Coslada, Cuenca, Ferrol, Fuenlabrada, Getafe, Gijón, Granada, Guadalajara, Huelva, Huesca, Jaén, Las Palmas, Leganés, L'Hospitalet de Llobregat, Lleida, Logroño, Lugo, Málaga, Manresa, Mataró, Melilla, Mérida, Molina de Segura, Mollet del Vallès, Móstoles, Murcia, Ourense, Palencia, Palma de Mallorca, Paterna, Pinto, Ponferrada, Pontevedra, Prat de Llobregat (El), Reus, Rubí, Sabadell, Sagunto/Sagunt, Salamanca, San Cristóbal de la Laguna, Sant Boi de Llobregat, Santa Cruz de Tenerife, Segovia, Sevilla, Siero, Terrassa, Torrejón de Ardoz, Torrelavega, Valdemoro, Valencia, Vigo, Viladecans, Vilanova i la Geltrú, Vila-Real, Zamora
Group 3: income between 2.5 and 3 times the NMW	A Coruña, Barakaldo, Burgos, Cerdanyola del Vallès, Girona, Granollers, Irún, León, Oviedo, Pamplona/Iruña, Rivas-Vaciamadrid, San Sebastián de los Reyes, Santander, Santiago de Compostela, Tarragona, Toledo, Valladolid, Zaragoza
Group 4: income between 3 and 4 times the NMW	Alcobendas, Barcelona, Bilbao, Castelldefels, Getxo, Madrid, San Sebastián/Donostia, Vitoria/Gasteiz
Group 5: income greater than 4 times the NMW	Boadilla del Monte, Majadahonda, Pozuelo de Alarcón, Rozas de Madrid (Las), Sant Cugat del Vallès

3.2. Thermal Energy Consumption

Data for the study were obtained from the Ministry of Economic Affairs and Digital Transformation. Those corresponding to population and cities were taken from the Spanish National Statistics Institute [52], and those for consumption were taken from the National Commission on Markets and Competition [53].

3.3. CO_2 Emissions

Directive 2010/31/UE of the European Parliament and the Council of 19 May 2010 was issued in 2010 by the European Union. This directive establishes energy performance of buildings, based on which, Spain established the emission factor for natural gas at 0.252 t CO_2/MWh [54].

4. Results and Discussion

4.1. Sample of Study

The study area considered comprised Spain and, separately, the 145 cities with more than 50,000 inhabitants. The consumption of energy for heating in buildings was studied. The data obtained from public sources were processed to obtain the conclusions of the investigation. Five groups of cities were defined based on the NMW.

The number of cities in each group is presented in Figure 2. Almost 80% of the cities have incomes lower than 2.5 times the NMW, of which the cities with NMW between 2 and 2.5 are the most numerous. The number of cities decreases as the NMW of their inhabitants increases, with those corresponding to incomes greater than 4 NMW the least numerous. Regarding the population of cities, almost half of the inhabitants have incomes between 2 and 2.5 times the NMW (Group 2) and almost 25% have incomes between 3 and 4 times the NMW (Group 4), despite accounting for only 5% of cities. Cities with fewer inhabitants are those with incomes greater than 4 NMW (Group 5), for which the number of inhabitants is less than 2% (Figure 3). The main statistical data of population and households are shown in Table 2. Both have a similar behavior.

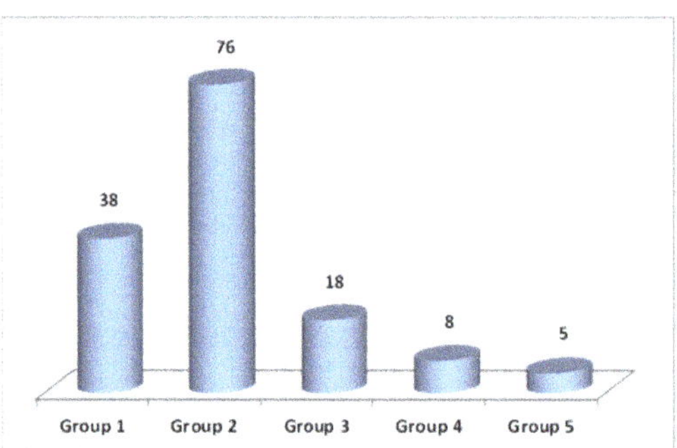

Figure 2. Number of cities of each group.

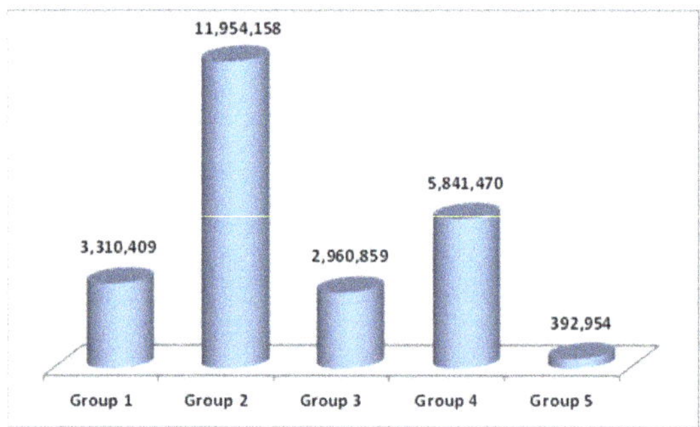

Figure 3. Population of each group of cities.

Table 2. Statistical data by population and household of each group of cities.

Equivalized Disposable Income	POPULATION						NUMBER OF HOUSEHOLDS					
	Total	Mean	Std. Dev.	Median	Maximum	Minimum	Total	Mean	Std. Dev.	Median	Maximum	Minimum
Group 1	3,310,409	87,116	38,473	76,624	228,675	52,620	1,220,128	32,109	13,571	29,249	83,182	18,927
Group 2	11,954,158	157,292	140,553	104,380	787,808	50,334	4,517,519	59,441	53,452	41,936	312,339	17,901
Group 3	2,960,859	164,492	142,846	112,815	664,938	57,723	1,188,755	66,042	59,001	46,331	269,347	21,470
Group 4	5,841,470	730,184	1,116,856	216,673	3,182,981	65,954	2,345,167	293,146	445,649	90,617	1,262,282	23,811
Group 5	392,954	78,591	17,530	85,605	95,071	51,463	122,900	24,580	5963	26,291	29,937	15,434

4.2. Energy Consumption per Group

Statistical data of the energy consumed in cities, in MWh per year, are shown in Table 3. Group 4 has the highest consumption, although it has half the population of Group 2, which has the next highest, but similar, consumption. These are followed by Group 3 which, although it has a population similar to that of Group 1, has a consumption that is almost five-fold higher. The final group is that with incomes above 4 NMW, which has a similar consumption, despite having a population 10-fold lower than that of Group 1. The total consumption of each group, in GWh per year, and the of the mean groups, are shown in Figure 4.

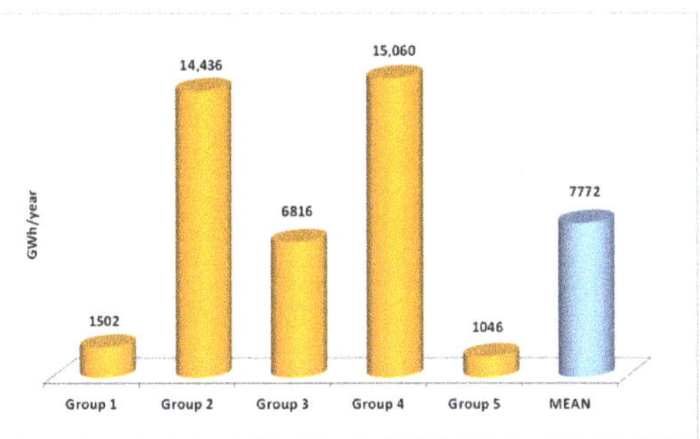

Figure 4. Mean energy consumption by group of cities.

Table 3. Statistical data of consumption of the groups.

Equivalized Disposable Income	MWh/Year					
	Total	Mean	Std. Dev.	Median	Maximum	Minimum
Group 1	1,501,718	39,519	67,369	19,955	354,793	0
Group 2	14,435,580	189,942	171,343	140,476	643,146	0
Group 3	6,815,922	378,662	400,478	234,313	1,627,614	78,422
Group 4	15,059,610	1,882,451	3,062,082	550,718	8,969,965	140,304
Group 5	1,045,547	209,109	47,470	200,928	267,920	145,028

4.3. Energy Consumption per Household

Table 4 presents the statistical data of the consumption per household of each group, in MWh per year, and Figure 5 shows the mean values and the mean consumption of all the groups. Consumption increases as income increases, and the group of cities with incomes greater than 4 times the NMW has the highest consumption. With the exception of the group of cities with incomes below 2 NMW, the groups have a higher value than the median. The difference between the consumption of the group with less than 2 NMW and that of the remainder is marked. Thus, the next highest group, Group 2, consumes almost four-fold that of Group 1, and that with the highest consumption, Group 5, consumes more than seven-fold that of Group 1. The consumption of the highest income group is more than 35% greater than that of the next highest.

Table 4. Statistical data of household consumption.

Equivalized Disposable Income	MWh/Year				
	Mean	Std. Dev.	Median	Maximum	Minimum
Group 1	1.11	1.57	0.69	8.29	0.00
Group 2	3.89	2.72	4.36	8.43	0.00
Group 3	5.53	2.15	5.67	8.65	1.94
Group 4	6.29	1.63	5.79	8.80	4.44
Group 5	8.46	1.17	8.99	9.16	6.38

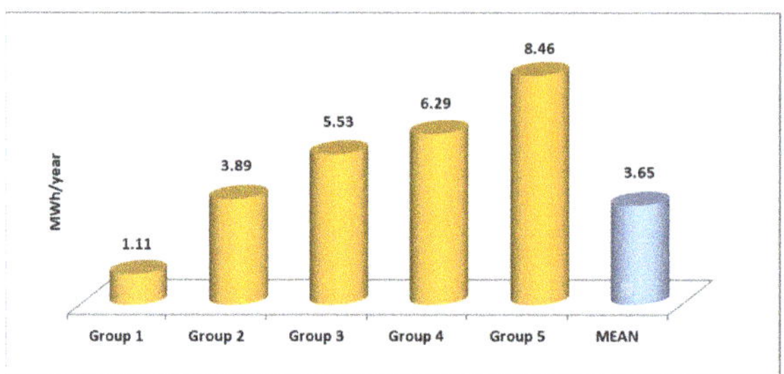

Figure 5. Mean energy consumption per household by group of cities.

Consumptions was analyzed using the *IVI* index defined in Section 2.1. Figure 6 presents the variation of the index in the groups. The higher the income in the cities, the higher the index and, therefore, the consumption. The lowest value of the index is 0.3 in the group with incomes less than 2 times the NMW; that is, consumption in these cities is 70% lower than the average consumption. On the contrary, cities with incomes greater than 4 times the NMW have an index value of 2.32; that is, their consumption is 132% higher than the average. Therefore, there is a consumption difference of more than 200% between the cities with the highest and lowest consumption values.

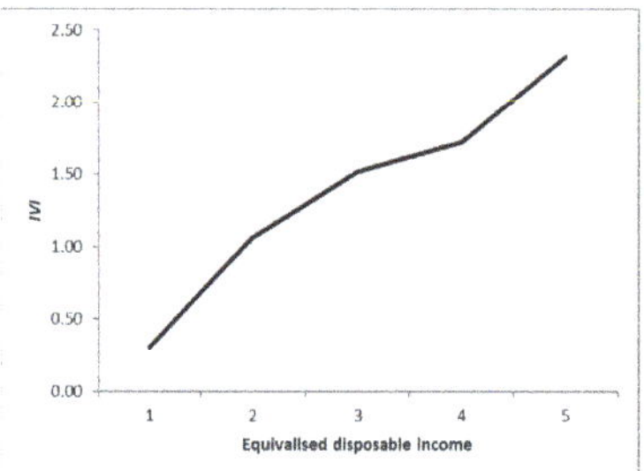

Figure 6. Variation of the *IVI* index for energy consumption per household by group of cities.

In conclusion, the higher the income, the higher the consumption per household.

4.4. Energy Consumption per Inhabitant

To calculate consumption per inhabitant, the value of consumption per household in each city was used. Furthermore, the number of inhabitants per household in each city was determined. Using these data, the information corresponding to energy consumption per inhabitant was calculated.

The main statistical data of the consumption per inhabitant of each group, in MWh per year, are shown in Table 5, and the mean values and the mean consumption of all the groups are presented in Figure 7. As in the case of energy consumed per household, the higher the income in the cities, the higher the consumption. Similarly, only the consumption of cities with incomes less than 2 NMW is less than the mean consumption. In this case, the greatest difference between groups occurs between Groups 1 and 2; consumption is three-fold greater in Group 2. Consumption of the group with the higher income is six-fold greater. Therefore, the differences between the groups are somewhat smaller than in consumption per household.

Table 5. Statistical data of inhabitant consumption.

Equivalized Disposable Income	MWh/Year				
	Mean	Std. Dev.	Median	Maximum	Minimum
Group 1	0.41	0.55	0.26	2.82	0.00
Group 2	1.48	1.00	1.76	3.28	0.00
Group 3	2.18	0.82	2.13	3.64	0.81
Group 4	2.48	0.64	2.28	3.76	1.88
Group 5	2.68	0.31	2.82	2.82	21.3

Analyzing the variations in consumption using the *IVI* index (Figure 8), it is once again found that, as income increases in city groups, consumption increases. In this case, the increase with respect to the previous group of cities is less pronounced. The index in Group 1 is 0.3, equal to that in the case of households, whereas in Group 5 it is 1.94 and, therefore, lower than that in the case of households, which is 2.32.

Therefore, the higher the income, the higher the consumption per inhabitant.

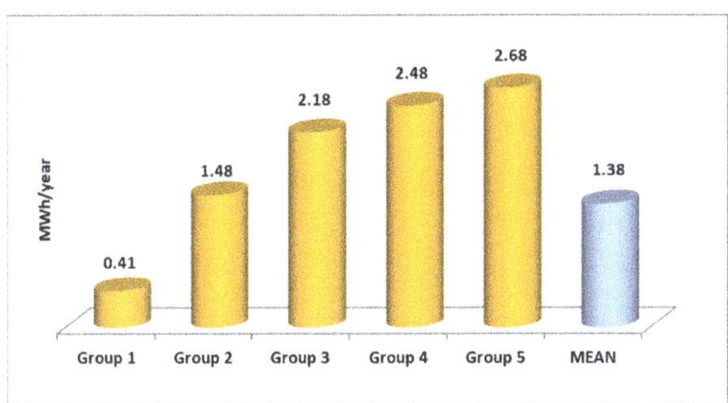

Figure 7. Mean energy consumption per inhabitant by group of cities.

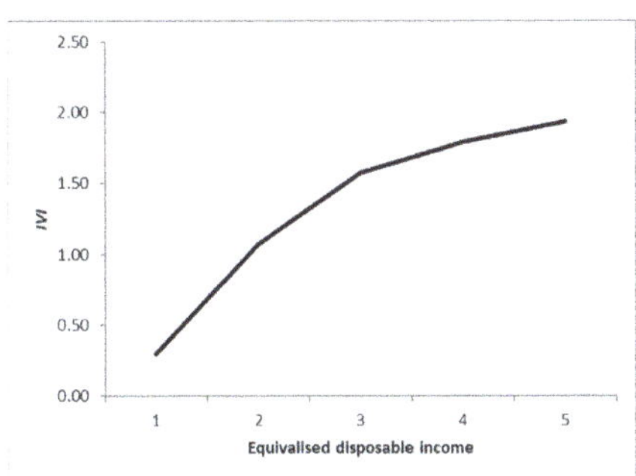

Figure 8. Variation of the *IVI* index for energy consumption per inhabitant by group of cities.

4.5. Energy Consumption without the Influence of Climate

The climate influences the energy consumption in each city. To analyze the influence of the income of the inhabitants on the energy consumption, without being masked by another important variable, the influence of the climate was eliminated. For this, the 145 analyzed cities were located on the climate map of Spain. Thus, each city was assigned a climate zone. Using the research carried out in [49], the consumption per household (Figure 9) and per inhabitant (Figure 10) of each city as a function of its climate were identified.

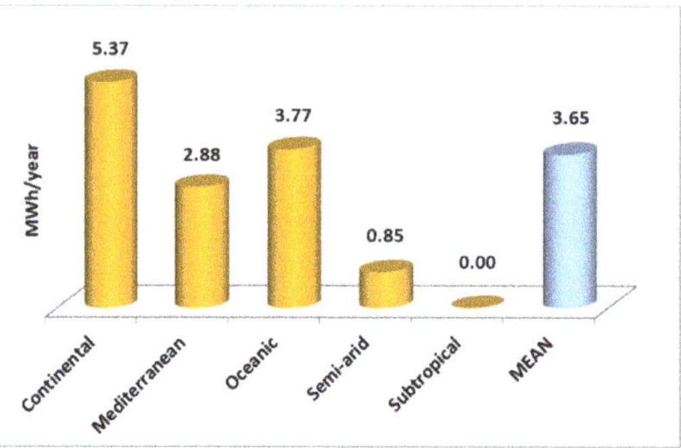

Figure 9. Mean energy consumption per inhabitant by climate zone.

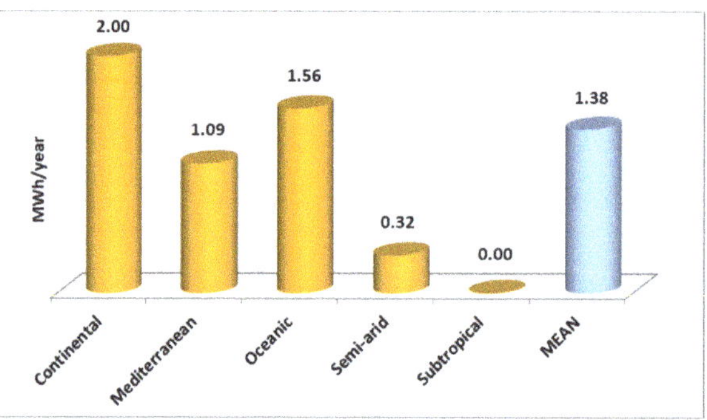

Figure 10. Mean energy consumption per household by climate zone.

Considering the consumption of the cities calculated in Sections 4.3 and 4.4, and the consumption based on the climate of each city, the consumption of the cities per household and inhabitant was obtained without the influence of the climate. Thus, Table 6 and Figure 11, and Table 7 and Figure 12, show the main statistical parameters and mean consumption, according to the income of each group, per household and inhabitant, respectively.

Table 6. Statistical data of household consumption without the influence of climate.

	MWh/Year				
Equivalized Disposable Income	**Mean**	**Std. Dev.**	**Median**	**Maximum**	**Minimum**
Group 1	1.59	1.76	0.80	7.39	0.00
Group 2	3.67	2.35	3.48	7.47	0.00
Group 3	4.63	1.62	4.60	7.23	1.87
Group 4	5.62	0.98	5.68	7.33	4.30
Group 5	6.50	0.89	6.11	8.09	5.96

The conclusions regarding the consumption per household obtained by eliminating the influence of the climate are similar to those obtained previously: the higher the income, the higher the consumption. However, the differences, although large, are not as pronounced, and are about half the magnitude of the previous values. Thus, the group with the highest income has a consumption that is more than three-fold higher than the group with the lowest income. In addition, the consumption of the group with incomes above 4 NMW is 15% higher than that of the next highest. A similar result is found with the lowest income group compared to the next highest, for which the difference is more than double.

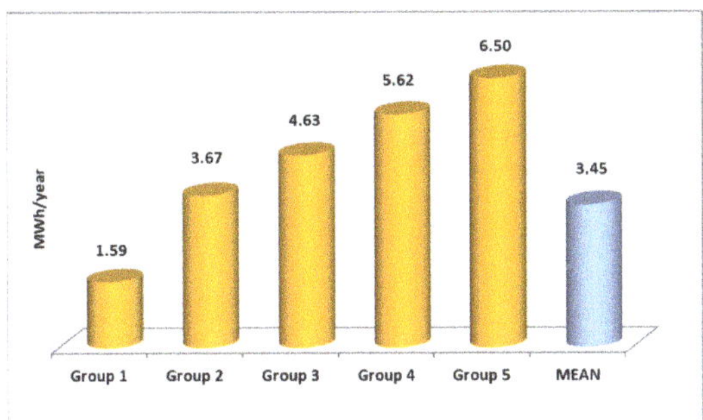

Figure 11. Mean energy consumption per household by group of cities without the influence of climate.

Table 7. Statistical data of inhabitant consumption without the influence of climate.

	MWh/Year				
Equivalized Disposable Income	**Mean**	**Std. Dev.**	**Median**	**Maximum**	**Minimum**
Group 1	0.60	0.64	0.33	2.69	0.00
Group 2	1.39	0.87	1.48	2.69	0.00
Group 3	1.79	0.62	1.81	2.69	0.72
Group 4	2.17	0.44	2.05	2.69	1.66
Group 5	2.09	0.33	1.94	2.69	1.94

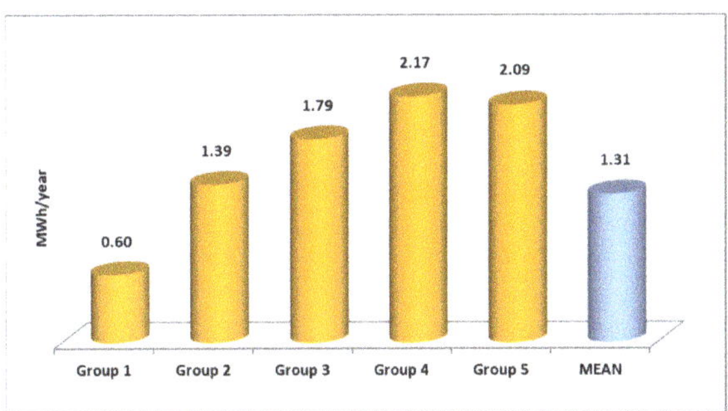

Figure 12. Mean energy consumption per inhabitant by group of cities without the influence of climate.

This also occurs when analyzing the results obtained per inhabitant by eliminating the influence of the climate: higher income implies higher consumption. The differences between the different groups of cities are approximately half those that existed without eliminating the influence of the climate. However, a difference is observed in the groups with incomes greater than 3 times the NMW. In this case, their consumption values are very similar, and those of Group 4 are only 3% higher than those of Group 5.

4.6. CO_2 Emissions

In accordance with the method presented in Section 3.3, Spain established the emission factor for natural gas at 0.252 t CO_2/MWh. Therefore, once the demand for this type of energy is known, its emissions can be obtained. In this section, the emissions produced by heating in cities are analyzed according to the income of their inhabitants. As undertaken previously, emissions are analyzed with and without the influence of the climate.

Figures 13 and 14 present the CO_2 emissions by households and inhabitants, respectively. In both cases, a similar behavior is observed: the higher the income, the higher the emissions. Regarding household emissions, only cities with incomes less than 2 times the NMW have emissions below the mean, and are 70% lower. On the contrary, cities with incomes above 4 NMW produce the most emissions, which are almost 90% higher than the mean. In addition, the jumps that occur between Groups 1 and 2, and between Groups 4 and 5, are greater than that in the groups who have incomes between 2 and 4 times the NMW.

Emissions per person display a similar behavior to those of households. However, the difference in emissions between Groups 1 and 2 is significantly more pronounced, and the emissions of the latter are almost four-fold higher than those of the former. By comparison, the increase in emissions among higher income groups is less pronounced than in the case of households. Regarding their comparison with respect to mean emissions, the group with the lowest emissions, Group 1, and the group with the largest, Group 5, have a behavior similar to that of households.

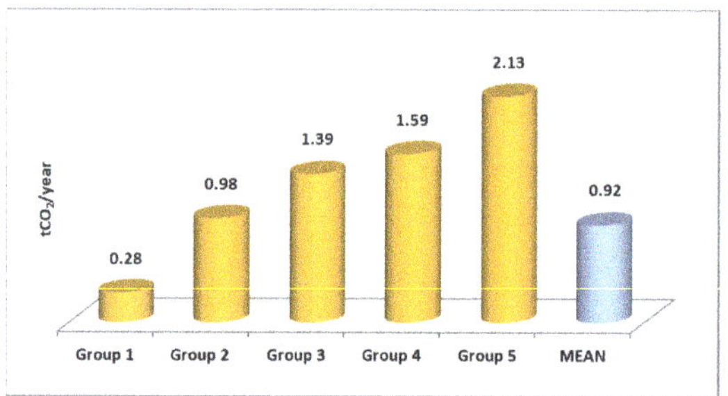

Figure 13. Mean CO_2 emissions per household by group of cities.

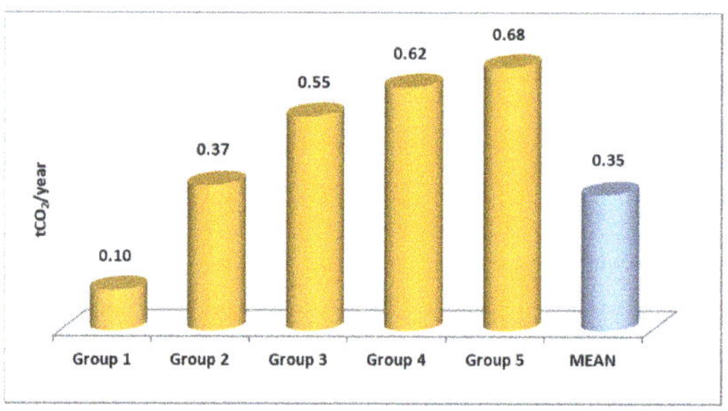

Figure 14. Mean CO_2 emissions per inhabitant by group of cities.

Emissions per household and inhabitant, after eliminating the influence of the climate, are reflected in Figures 15 and 16, respectively. In this case, a similar behavior is once again evident: cities with higher incomes produce higher emissions. In both cases, only the cities with incomes below 2 NMW produce emissions lower than the mean, and furthermore, these cities have a greater difference with respect to the emissions of the upper group of cities. In the case of emissions per household, the growth between groups is approximately linear. In the case of emissions per inhabitant, growth among groups with incomes between 2 and 4 NMW is also approximately linear. However, cities with more than 4 NMW produce slightly fewer emissions than those in the group of cities between 3 and 4 times the NMW, although the difference is 3%.

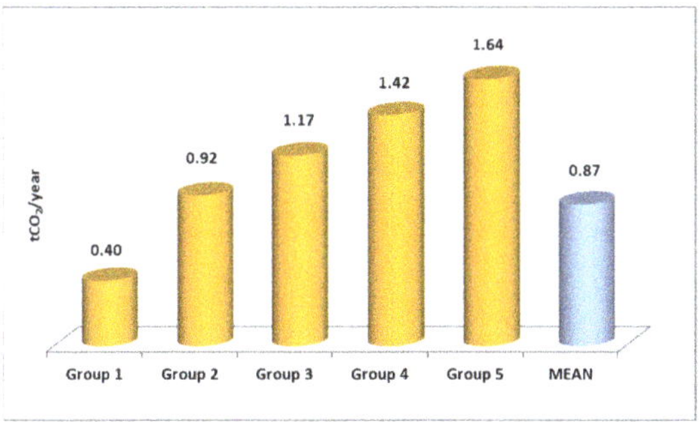

Figure 15. Mean CO_2 emissions per household by group of cities without the influence of climate.

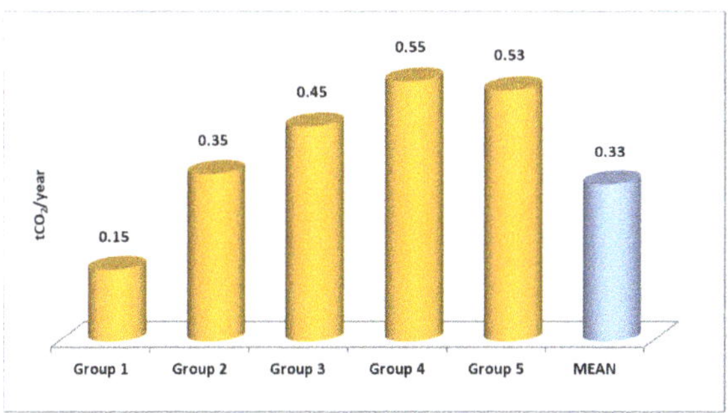

Figure 16. Mean CO_2 emissions per inhabitant by group of cities without the influence of climate.

5. Conclusions

More than 50% of the world's population lives in cities, and that share is expected to exceed two-thirds by 2050. In certain areas, the urbanized proportion of the population is even larger. Furthermore, buildings consume more than 35% of energy and produce 40% of CO_2 emissions. In particular, heating is among the uses that consumes the most energy, accounting for almost 50%, and produces the most emissions. Hence, cities in general, and buildings in particular, have significant importance for both public administration and utilities. This is even more so due to the COVID-19 pandemic, by which cities have been the most affected. Cities and buildings are important to public administrators because they are responsible for legislating to achieve sustainable development goals, and to utilities because they must plan infrastructure for the proper functioning of the cities.

Because of the importance of heating in buildings, this paper presents a method to analyze the influence of the income of city inhabitants on consumption and its related emissions. The study was carried out at the city level and was based on equivalized disposable income, which was used to consider the economies of scale of households. The method selected cities in a geographic area and grouped them based on the national minimum wage. The study was carried out using households and inhabitants as the basic unit. The energy consumed in all of the buildings of each city was distributed among all of its inhabitants and all of its households. In addition, to more clearly analyze the influence of income, the influence of climate was removed. For this purpose, it was necessary to locate each city on a climate map of the study area and thus assign it to a climate zone. To facilitate the analysis, an index was introduced.

The method was applied to 145 Spanish cities with more than 50,000 inhabitants. The inhabitants of the cities and the members of the households were identified, in addition to the heating energy consumed in all of the buildings. The results show that the higher the income, the higher the consumption for heating and the higher the emissions, at both the household and the inhabitant levels. Subsequently, the climate of each city was identified, and the study was carried out by eliminating the influence of the climate on the city. The results obtained again confirm the same finding: the higher the income, the higher the consumption and the higher the emissions.

Author Contributions: Conceptualization, P.J.Z.-P., I.M.Z.-S. and R.S.-D.; methodology, P.J.Z.-P. and I.M.Z.-S.; validation, P.J.Z.-P., I.M.Z.-S. and F.J.Z.-S.; formal analysis, P.J.Z.-P., I.M.Z.-S. and F.J.Z.-S.; investigation, P.J.Z.-P. and I.M.Z.-S.; data curation, I.M.Z.-S. and F.J.Z.-S.; writing—original draft preparation, P.J.Z.-P. and I.M.Z.-S.; writing—review and editing, I.M.Z.-S., F.J.Z.-S. and P.J.Z.-P.; visualization, I.M.Z.-S.; supervision, P.J.Z.-P. and R.S.-D. All authors have read and agreed to the published version of the manuscript.

Funding: This research received no external funding.

Institutional Review Board Statement: Not applicable.

Informed Consent Statement: Not applicable.

Data Availability Statement: Not applicable.

Acknowledgments: The authors would like to thank eCitySevilla project for providing facilities to conduct the research.

Conflicts of Interest: The authors declare no conflict of interest.

References

1. Department of Economic and Social Affairs. *2014 Demographic Yearbook*, 64th ed.; United Nations: New York, NY, USA, 2015.
2. United Nations. Sustainable Development Goals. Available online: https://www.un.org/sustainabledevelopment/ (accessed on 13 May 2020).
3. International Energy Agency. *2018 Global Status Report: Towards a Zero-Emission, Efficient and Resilient Buildings and Construction Sector*; United Nations Environment Programme: Nairobi, Kenya, 2018.
4. Shahrokni, H.; Levihn, F.; Brandt, N. Big meter data analysis of the energy efficiency potential in Stockholm's building stock. *Energy Build.* **2014**, *78*, 153–164. [CrossRef]
5. Lukic, N.; Nikolic, N.; Timotijevic, S.; Tasic, S. Influence of an unheated apartment on the heating consumption of residential building considering current regulations—Case of Serbia. *Energy Build.* **2017**, *155*, 16–24. [CrossRef]
6. IEA. *Energy Technology Perspective 2017, Catalysing Energy Technology Transformations*; International Energy Agency: Paris, France, 2017.
7. United Nations Educational, Scientific and Cultural Organization (UNESCO). Available online: https://en.unesco.org/sdgs (accessed on 29 June 2020).
8. COM(2019) 640 Final. *The European Green Deal*; European Commission: Brussels, Belgium, 2019.
9. European Parliament News. Available online: https://www.europarl.europa.eu/news/en/headlines/society/20200618STO81513/green-deal-key-to-a-climate-neutral-and-sustainable-eu (accessed on 15 August 2020).
10. United Nation. Department of Economic and Social Affairs. Available online: https://sdgs.un.org/goals/goal11 (accessed on 14 September 2020).
11. Wang, L.; Long, R.; Chen, H.; Li, W.; Yang, J. A review of studies on urban energy performance evaluation. *Environ. Sci. Pollut. Res.* **2019**, *26*, 3243–3261. [CrossRef]
12. D'Agostino, D.; Zangheri, P.; Castellazzi, L. Towards nearly zero energy buildings in Europe: A focus on retrofit in non-residential buildings. *Energies* **2020**, *10*, 117. [CrossRef]
13. Cui, Y.; Sun, N.; Cai, H.; Li, S. Indoor temperatura improvement and energy-saving renovations in rural houses of China's cold region—A case study of Shandong province. *Energies* **2020**, *13*, 870. [CrossRef]
14. Ala-Kotila, P.; Vainio, T.; Laamanen, J. The influence of building renovations on indoor comfort—A field test in an apartment building. *Energies* **2020**, *13*, 4958. [CrossRef]
15. Hashemi, A. Climate resilient low-income tropical housing. *Energies* **2020**, *9*, 468. [CrossRef]
16. Lima, F.; Ferreira, P.; Leal, V. A review of the relation between household indoor temperature and health outcomes. *Energies* **2020**, *13*, 2881. [CrossRef]
17. Wang, D.; Liu, Y.; Wu, Z.; Fu, H.; Shi, Y.; Guo, H. Scenario analysis of natural gas consumption in China based on wavelet neural network optimized by particle swarm optimization algorithm. *Energies* **2018**, *11*, 825. [CrossRef]
18. Scarpa, F.; Bianco, V. Assessing the quality of natural gas consumption forecasting: An application to the Italian residential sector. *Energies* **2017**, *10*, 1879. [CrossRef]
19. Tavakoli, E.; Montazerin, N. Stochastic analysis of natural gas consumption in residential and commercial buildings. *Energy Build.* **2011**, *43*, 2289–2297. [CrossRef]
20. Akpinar, M.; Adak, M.F.; Yumusak, N. Day-ahead natural gas demand forecasting using optimized ABC-based neural network with sliding window technique: The case study of regional basis in Turkey. *Energies* **2017**, *10*, 781. [CrossRef]

21. De, G.; Gao, W. Forecasting China's natural gas consumption based on AdaBoost—Particle swarm optimization—Extreme learning machine integrate learning method. *Energies* **2018**, *11*, 2938. [CrossRef]
22. Flores-Larsen, S.; Filippín, C. Energy efficiency, thermal resilience, and health during extreme heat events in low-income housing in Argentina. *Energy Build.* **2021**, *231*, 110576. [CrossRef]
23. Erdogdu, E. Natural gas demand in Turkey. *Appl. Energy* **2010**, *87*, 211–219. [CrossRef]
24. Bianco, V.; Scarpa, F.; Tagliafico, L.A. Analysis and future outlook of natural gas consumption in the Italian residential sector. *Energy Conv. Manag.* **2014**, *87*, 754–764. [CrossRef]
25. Behrouznia, A.; Saberi, M.; Azadeh, A.; Asadzadeh, S.M.; Pazhoheshfar, P. An adaptive network based fuzzy inference system-fuzzy data envelopment analysis for gas consumption forecasting and analysis: The case of South America. In Proceedings of the 2010 International Conference on Intelligent and Advanced Systems, Kuala Lumpur, Malaysia, 15–17 June 2010; pp. 1–6. [CrossRef]
26. Li, J.; Dong, X.; Shangguan, J.; Hook, M. Forecasting the growth of China's natural gas consumption. *Energy* **2011**, *36*, 1380–1385. [CrossRef]
27. Wang, T.; Lin, B. China's natural gas consumption and subsidies—From a sector perspective. *Energy Policy* **2014**, *65*, 541–551. [CrossRef]
28. Liu, H.; Zhou, S.; Peng, T.; Ou, X. Life cycle energy consumption and greenhouse gas emissions analysis of natural gas- based distributed generation projects in China. *Energies* **2017**, *10*, 1515. [CrossRef]
29. Gustafsson, M.; Thygesen, R.; Karlsson, B.; Ödlund, L. Rev-changes in primary energy use and CO_2 emissions—An impact assessment for a building with focus on the Swedish proposal for nearly zero energy buildings. *Energies* **2017**, *10*, 978. [CrossRef]
30. Ouyang, X.; Lin, B. Carbon dioxide (CO_2) emissions during urbanization: A comparative study between China and Japan. *J. Clean Prod.* **2017**, *143*, 356–368. [CrossRef]
31. Alam, M.M.; Murad, M.W.; Noman, A.H.M.; Ozturk, I. Relationships among carbon emissions, economic growth, energy consumption and population growth: Testing Environmental Kuznets Curve hypothesis for Brazil, China, India and Indonesia. *Ecol. Indic.* **2016**, *70*, 466–479. [CrossRef]
32. Sharma, S.S. Determinants of carbon dioxide emissions: Empirical evidence from 69 countries. *Appl. Energy* **2011**, *88*, 376–382. [CrossRef]
33. Solarin, S.A.; Lean, H.H. Natural gas consumption, income, urbanization, and CO_2 emissions in China and India. *Environ. Sci. Pollut. Res.* **2016**, *23*, 18753–18765. [CrossRef] [PubMed]
34. Sohag, K.; Mamun, M.d.a.; Uddin, G.S.; Ahmed, A.M. Sectoral output, energy use, and CO_2 emission in middle-income countries. *Environ. Sci. Pollut. Res.* **2017**, *24*, 9754–9764. [CrossRef] [PubMed]
35. Chen, J.; Wang, P.; Cui, L.; Huang, S.; Song, M. Decomposition and decoupling analysis of CO_2 emissions in OECD. *Appl. Energy* **2018**, *231*, 937–950. [CrossRef]
36. Lyons, S.; Pentecost, A.; Tol, R.S.J. Socioeconomic distribution of emissions and resource use in Ireland. *J. Environ. Manag.* **2012**, *112*, 186–198. [CrossRef] [PubMed]
37. Chancel, L. Are younger generations higher carbon emitters than their elders? Inequalities, generations and CO_2 emissions in France and in the USA. *Ecol. Econ.* **2014**, *100*, 195–207. [CrossRef]
38. Golley, J.; Meng, X. Income inequality and carbon dioxide emissions: The case of Chinese urban households. *Energy Econ.* **2012**, *34*, 1864–1872. [CrossRef]
39. Wang, S.; Guangdong, L.; Fang, C. Urbanization, economic growth, energy consumption, and CO_2 emissions: Empirical evidence form countries with different income level. *Renew. Sustain. Energy Rev.* **2018**, *81*, 2144–2159. [CrossRef]
40. Kennedy, C.; Steinberger, J.; Gasson, B.; Hansen, Y.; Hillman, T.; Havránek, M.; Pataki, D.; Phdunngsilp, A.; Ramaswami, A.; Mendez, G.V. Greenhouse gas emissions from global cities. *Environ. Sci. Technol.* **2009**, *43*, 7297–7302. [CrossRef]
41. Huang, R.; Zhang, S.; Liu, C. Comparing urban and rural household CO_2 emissions—Case from China's four megacities: Beijing, Tianjin, Shanghai, and Chongqing. *Energies* **2018**, *11*, 1257. [CrossRef]
42. Jiang, L.; Xing, R.; Chen, X.; Xue, B. A survey-based investigation of greenhouse gas and pollutant emissions from household energy consumption in the Qinghai-Tibet Plateau of China. *Energy Build.* **2021**, *235*, 110753. [CrossRef]
43. Zarco-Soto, I.M.; Zarco-Periñán, P.J.; Sánchez-Durán, R. Influence of cities population size on their energy consumption and CO_2 emissions: The case of Spain. *Environ. Sci. Pollut. Res.* **2021**, *28*, 28146–28167. [CrossRef]
44. Urquizo, J.; Calderón, C.; James, P. Metrics of urban morphology and their impact on energy consumption: A case study in the United Kingdom. *Energy Res. Soc. Sci.* **2017**, *32*, 193–206. [CrossRef]
45. Hekkenberg, M.; Benders, R.M.J.; Moll, H.C.; Schoot, A.J.M. Indications for a changing electricity demand pattern: The temperature dependence of electricity demand in the Netherlands. *Energy Policy* **2009**, *37*, 1542–1551. [CrossRef]
46. Nageli, C.; Camarasa, C.; Jakob, M.; Catenazzi, G.; Ostermeyer, Y. Synthetic building stocks as a way to assess the energy demand and greenhouse gas emissions of national building stocks. *Energy Build.* **2018**, *173*, 443–460. [CrossRef]
47. Valor, E.; Meneu, V.; Caselles, V. Daily air temperature and electricity load in Spain. *Appl. Meteorol.* **2001**, *40*, 1413–1421. [CrossRef]
48. Li, D.H.W.; Yang, L.; Lam, J.C. Impact of climate change on energy use in the built environment in different climate zones—A review. *Energy* **2012**, *42*, 103–112. [CrossRef]
49. Zarco-Soto, I.M.; Zarco-Periñán, P.J.; Sánchez-Durán, R. Influence of climate on energy consumption and CO_2 emissions: The case of Spain. *Environ. Sci. Pollut. Res.* **2020**, *27*, 15645–15662. [CrossRef] [PubMed]

50. Instituto Nacional de Estadística, Cifras de Población, Ministerio de Asuntos Económicos y Transformación Digital. Available online: http://www.ine.es/dyngs/INEbase/es/operacion.htm?c=Estadistica_C&cid=1254736176951&menu=ultiDatos&idp=1254735572981 (accessed on 12 October 2020).
51. Ministerio de Empleo y Seguridad Social. *Real Decreto 1171/2015, de 29 de Diciembre, por el que se fija el Salario Mínimo Interprofesional para 2016*; Boletín Oficial del Estado: Madrid, Spain, 2015.
52. Instituto Nacional de Estadística, Demografía y Población, Ministerio de Asuntos Económicos y Transformación Digital. Available online: http://www.ine.es/ss/Satellite?L=es_ES&c=Page&cid=1254735910183&p=1254735910183&pagename=INE%2FINELayout (accessed on 1 November 2020).
53. Comisión Nacional de los Mercados y la Competencia. *Informe de Supervisión del Mercado de Gas Natural en España*; Ministerio de Economía, Industria y Competitividad: Madrid, Spain, 2017.
54. Ministerio de Industria, Energía y Turismo & Ministerio de Fomento, Factores de Emisión de CO_2 y Coeficientes de paso a Energía Primaria de Diferentes Fuentes de Energía Final Consumidas en el Sector de Edificios de España. Available online: https://energia.gob.es/desarrollo/EficienciaEnergetica/RITE/Reconocidos/Paginas/IndexDocumentosReconocidos.aspx (accessed on 19 March 2020).

Article

The Role of Discounting in Energy Policy Investments

Gabriella Maselli and Antonio Nesticò *

Department of Civil Engineering, University of Salerno, Via Giovanni Paolo II, 132, 84084 Fisciano, Italy; gmaselli@unisa.it
* Correspondence: anestico@unisa.it; Tel.: +39-089964318

Abstract: For informing future energy policy decisions, it is essential to choose the correct social discount rate (SDR) for ex-ante economic evaluations. Generally, costs and benefits—both economic and environmental—are weighted through a single constant discount rate. This leads to excessive discounting of the present value of cash flows progressively more distant over time. Evaluating energy projects through constant discount rates would mean underestimating their environmental externalities. This study intends to characterize environmental–economic discounting models calibrated for energy investments, distinguishing between intra- and inter-generational projects. In both cases, the idea is to use two discounting rates: an economic rate to assess financial components and an ecological rate to weight environmental effects. For intra-generational projects, the dual discount rates are assumed to be constant over time. For inter-generational projects, the model is time-declining to give greater weight to environmental damages and benefits in the long-term. Our discounting approaches are based on Ramsey's growth model and Gollier's ecological discounting model; the latter is expressed as a function of an index capable of describing the performance of a country's energy systems. With regards to the models we propose, the novelty lies in the calibration of the "environmental quality" parameter. Regarding the model for long-term projects, another innovation concerns the analysis of risk components linked to economic variables; the growth rate of consumption is modelled as a stochastic variable. The defined models were implemented to determine discount rates for both Italy and China. In both cases, the estimated discount rates are lower than those suggested by governments. This means that the use of dual discounting approaches can guide policymakers towards sustainable investment in line with UN climate neutrality objectives.

Keywords: energy policy investments; cost-benefit analysis; social discount rate; dual discounting; energy transition index

Citation: Maselli, G.; Nesticò, A. The Role of Discounting in Energy Policy Investments. *Energies* **2021**, *14*, 6055. https://doi.org/10.3390/en14196055

Academic Editors: Sergio Ulgiati, Hans Schnitzer and Remo Santagata

Received: 16 August 2021
Accepted: 18 September 2021
Published: 23 September 2021

Publisher's Note: MDPI stays neutral with regard to jurisdictional claims in published maps and institutional affiliations.

1. Introduction

Nowadays, energy policies are a key governmental instrument for achieving economic, environmental, and social objectives, encouraging sustainable development, providing environmental protection, and containing greenhouse gas emissions (GHG) [1]. In this respect, the path to energy transition—increasingly advocated for by governments—is driven by investment programmes whose effects often manifest themselves in the long term; these include energy infrastructure and the pricing of environmental externalities such as carbon emissions [2]. Thus, choosing more sustainable investments means making intertemporal decisions. Such choices involve trade-offs between benefits and costs that occur at different times [3]. It follows that a critical issue in environmental and resource economics is the choice of social discount rate (SDR), as it significantly influences the outcome of a cost–benefit tests [4,5]. A social discount rate reflects a society's relative assessment of well-being today versus well-being in the future [6]. The SDR allows the costs and benefits that an investment generates over time to be weighted to make them economically comparable. It is therefore a fundamental parameter for being able to express an opinion on the economic performance of an investment project whenever the analysis is conducted from the point of view of a public operator or of the community [7].

Choosing an appropriate social discount rate is crucial for cost–benefit analysis. Choosing too high a social discount rate could preclude the realization of many desirable public projects for society, in terms of extra-financial repercussions. Conversely, setting an SDR that is too low would risk steering investment decisions towards economically inefficient investments. Furthermore, a relatively high social discount rate ends up giving less weight to the benefit and cost streams that occur in progressively more distant times, favouring projects with benefits that occur at the beginning of the analysis period [8].

The choice of social discount rate affects both the ex-ante decision that allows the testing of whether a specific public sector project deserves funding, and the ex-post evaluation of its performance [9].

The issue of discounting is also crucial for energy efficiency projects. In this case, investors must weight higher initial costs against future energy savings [10]. There are two aspects of energy projects that need to be addressed: Firstly, these are investments that have multiple extra-financial effects on the community, so their effectiveness is more of a social nature rather than a specifically financial one. Secondly, the time perspective is very long for some initiatives [11]; see, among others, the European Green Deal projects, with targets for 2050 [12], or energy transition programmes to curb global warming, whose effects last for centuries [13].

To guide the decision-making process towards efficient investments that respect the defined programmatic guidelines, it is necessary to attribute a greater 'value' to the extra-financial effects that the intervention initiatives generate on the community in the analysis. According to Kula and Evans [14], in a moment of strong environmental stress like the one we are experiencing, environmental effects should be discounted separately and differently from economic impacts. In particular, the challenge today is to fix the discount rate for environmental effects at a rate that reaches either a natural capital depletion rate that maximises the utility of consumption of current and future generations, or the preservation of natural capital. One cannot assume a common discount rate for both natural and man-made capital, since natural capital is finite, while man-made capital is unlimited. So, there should be two discount rates. On the contrary, the two discount rates can only coincide if the demand for ecosystem goods and services does not exceed the ecosystem's regenerative ability.

The aim of this paper is to propose an innovative economic–environmental (or dual) discounting approach in which environmental externalities are weighted at a different and lower rate than that used for strictly financial cash-flows. This is possible because the social welfare function (SWF), from which the social discount rate derives, is no longer only a function of consumption—and therefore of economic parameters—but also of environmental quality. With this research, we want to define a dual discounting specification for energy projects. Specialising the discounting rate according to the investment sector can lead to a fairer and more equitable allocation of resources [11,15]; specifically, to consider the performance of the energy systems of individual countries, the variable "environmental quality" is defined as a function of the Energy Transition Index (ETI) [16].

In addition, we distinguish between intra-generational energy projects (or those with short-term effects) and inter-generational energy projects (or those with long-term effects). In the first case, we define a dual discounting approach based on time-constant environmental and economic discount rates. In the second case, both discount rates—environmental and economic—are based on a time-declining structure. The use of constant discount rates for projects with long-term implications would end up excessively contracting the present value of progressively more distant costs and benefits over time.

This paper is divided into the following four sections: Section Two first proposes a review of the relevant literature. Section Three defines the theoretical framework of the two environmental–economic discounting models. In Section Four, we implement the models defined to estimate constant and declining discount rates, with reference to both the Italian and Chinese economies. Section Five concludes and discusses energy policy implications.

2. Literature Review

The social discount rate (SDR) plays a critical role in cost–benefit analysis (CBA). The SDR allows the comparison of socio-economic costs and benefits—expressed in monetary terms—in order to make a judgement on the efficiency of a project, programme, or policy [17]. This judgement is summarised by performance indicators such as the economic net present value (ENPV). This indicator is a measurement of an investment's marginal utility for 'present' society [18,19]:

$$\text{ENPV} = \frac{B_t - C_t}{(1 + SDR)^t} \tag{1}$$

In which B_t and C_t represent, respectively, the benefits and costs arising at time t; $1/(1 + SDR)$ is the discount factor. (1) shows that as the discount rate increases, the present value of net benefits decreases, as they become more distant from the time of valuation.

The effect of the contraction of the present value of cash flows is a crucial issue when the objects of analysis are long-lived projects, whose effects extend for at least 30–40 years and therefore involve more than one generation [6]. In the valuation of intergenerational projects, such as those with environmental impacts, the choice of appropriate discount rate involves the additional challenge of taking intergenerational equity into account [9,20].

This is one of the main reasons why there is still no consensus on the discount rate to be used in valuations. The issue becomes even more complex when environmental effects make large contributions and mainly occur in the long run.

The literature review shows that the most widely used approach to estimate the discount rate is the social rate of time preference (SRTP) [21,22]. According to this approach, the social welfare function (SWF) depends on the utility $U(c)$ of income or consumption c alone. In the formula:

$$\text{SWF} = \int_{t=0}^{\infty} U(c_t)\, e^{-\rho t} dt \tag{2}$$

SWF is dependent on the following parameters: $U(c_t)$, which represents the utility that society derives from public and private per capita consumption at time t; $e^{-\rho t}$ is the discount factor that allows the incremental utility resulting from an additional unit of consumption at time t to be weighted; ρ represents the rate at which future utility is discounted. This last parameter is also called the pure rate of time preference. In order to determine the discount rate that society should apply to incremental consumption, it is first necessary to estimate the discount factor by maximising the SWF. If W denotes the integral of equation 2, then the derivative of W with respect to consumption in period t represents the discount factor and can be interpreted as the social present value of an incremental unit of consumption in period t [21]. The social discount rate is equal to the proportional rate of decrease in this discount factor over time. In other words, this parameter—also called SRTP—is the rate at which the value of a small increment of consumption falls as time changes. It is shown that the SRTP is a function of two components [9]. The first is ρ, the pure time preference rate (or the utility discount rate). ρ reflects the importance that society attaches to the welfare of the current generation relative to the welfare of the future generation. The second contribution is the product of the elasticity of the marginal utility of consumption η and the growth rate of per capita consumption g. This product shows that an additional unit of consumption for a future generation has a lower utility value than an incremental unit of consumption for the current generation [8]. The formula:

$$\text{SRTP} = \rho + \eta \times g \tag{3}$$

(3), also known as the Ramsey formula, depends only on economic parameters and is time-constant, i.e., it leads to estimating a constant discount rate throughout the analysis period. Therefore, according to some authors, this approach fails to properly consider environmental externalities, which often occur in the long term. In this regard, Emmerling et al. [23] argue that the climate goals of the Paris Agreement (2015) can only be achieved by employing very low discount rates, such as the one estimated by Stern [24]. Similarly, van den Bijgaart et al. [25] and van der Ploeg and Rezai [26], using analytical integrated assessment models (IAMs), reveal that the discount rate is a key factor in the social cost of carbon. Gollier [27] proposes an extension of the Ramsey formula for projects with long-term effects, e.g., investments for climate change that reduce greenhouse gas emissions. The assumption is that the consumption level in SWF is uncertain and that fluctuations in consumption growth are distributed independently and normally. According to these assumptions, (3) becomes:

$$\text{SRTP} = \rho + \eta \times \mu_g - 0.5 \times \eta^2 \times \sigma_g^2 \tag{4}$$

where μ_g and σ_g^2 are respectively the consumption growth rate mean and variance. $0.5 \times \eta^2 \times \sigma_g^2$ is the precautionary term and indicates the planner's intention to save more now in favour of future benefits. This term, called "precautionary", summarizes the uncertainty of the growth rate of consumption and determines a reduction in the value of the discount rate [18,27]. Luo et al. [3] demonstrate that non-diversifiable idiosyncratic risk reduces the discount rate and increases the present value of the uncertain future benefits of projects.

Other scholars suggest the use of dual discounting approaches, whereby environmental components are weighted at a lower "ecological" rate than the "economic" rate, which is useful for assessing strictly financial costs and revenues [14,28–31]. This means that the economic net present value (ENPV) is given by the sum of two rates:

$$\text{ENPV} = \sum_{t=0}^{n} \frac{F_t}{(1+r_c)^t} + \sum_{t=0}^{n} \frac{E_t}{(1+r_q)^t} \tag{5}$$

where: F_t and E_t indicate, respectively, the annual economic cash flows and net environmental benefits at time t; r_c represents the consumption discount rate (or economic discount rate); r_q is the environmental quality discount rate (or more simply environmental discount rate), with $r_q < r_c$. In other words, the environmental and social damages and benefits generated by the project, after being transformed into monetary terms, are discounted using r_q. While the economic benefits and costs are assessed through the r_c [18,20,27]. The formulas for estimating r_c and r_q are derived in the following section, via Formulas (7) and (8).

Another branch of the literature proposes the use of time-declining discount rates to give more weight to distant project effects than is the case when using time-constant discount rates [32–36]. Two methods are used to estimate the declining discount rate (DDR): The expected net present value approach and the consumption-based approach. For both, the theoretical assumption is to include an uncertainty factor in the time-structure of the discount rate. In the ENPV approach, the same discount rate is modelled as an uncertain parameter, while in the consumption-based approach, the uncertainty concerns the growth rate of consumption which appears in the Ramsey formula.

With reference to the first approach, Weitzman [6] shows that estimating ENPV with an uncertain but constant discount rate is equivalent to computing net present value (NPV) with a certain but decreasing "certainty equivalent" until it reaches the minimum possible value at time $t = \infty$. Thus, if the discount rate is modelled as a stochastic variable, we can first estimate the certainty equivalent discount factor, then the corresponding certainty equivalent discount rate, understood as the exchange rate of the expected discount factor or rate of progression from t to $t + 1$.

According to Gollier's consumption-based approach [18,27,29], the absence of a sufficiently large dataset covering the growth process of the economy in the long run implies that parameters μ and σ of (3) can be treated as uncertain. It is then assumed that the consumption log follows a Brownian motion with trend $\mu(\theta)$ and volatility $\sigma(\theta)$. These values depend on parameter θ, which is uncertain at time 0. These assumptions allow us to transform (3) into a time-declining function.

Weitzman's [6,32] findings guided the UK and France to adopt discount rates with a declining structure for projects with long-term consequences [37,38]. The U.S. Environmental Protection Agency [39] has also followed suit.

Finally, recent studies analyse the need to use a specific discount rate for environmental sectors and services. Baumgärtner et al. [31] show that ecosystem services should be discounted at significantly lower rates than those used to weight consumer goods. Vazquez-Lavín et al. [40], with reference to projects aimed at preserving biodiversity in marine protected areas in Chile, estimate a declining SDR for eco-system services. Muñoz-Torrecillas et al. [41] estimate an SDR to be employed in the appraisal of afforestation projects in the United States.

With specific reference to the energy sector, Steinbach and Staniaszek [42], Kubiak [10], and Poudineh and Penyalver [2] offer a review of social discount rates for energy transition policies and their implications for decision-making. Foltyn-Zarychta et al. [11] consider employing a lower discount rate than that suggested by the government, as energy policy planning horizons are generally very long. The US Department of Energy (DOE) evaluates a rate of 3% for energy conservation and RES projects. The estimate is based on long-term Treasury bonds, averaged over a 12-year period [42].

The following Table 1 summarises the main literature studies concerning approaches to estimating the discount rate.

Table 1. Literature review on the social discount rate.

Literature Branch	References
Constant and single discounting	[8,21,43–50]
Declining discounting	[3–7,17,20,32–36,51]
Dual discounting	[14,18,28,30]
Specific discount rate per investment sector/area of intervention	Energy systems [11,42] Application for different investment sectors [15] GHG emissions [23] Ecosystem Services [31,40] Afforestation Projects [41]

Considering the framework outlined, this research intends to characterise new approaches for estimating SDR for use in economic evaluations of energy interventions and policies. As the literature review shows, there is a lack of studies proposing both constant and declining dual models specifically for the economic evaluation of energy projects. Thus, building on the existing literature, we define a new discounting model in which environmental quality is described as a function of an energy transition index. Specifically, we define: (i) a constant-dual discounting model for intra-generational energy projects, whose effects can be assessed over a thirty-year period. In this case we define an environmental and an economic discount rate, which are constant over time; (ii) a declining-dual discounting model for inter-generational investments, i.e., those with appreciable effects over the long run. In the second case, however, we define an environmental discount and an economic discount, both with a declining structure over time; this is possible because we take macroeconomic risks into account in the modelling.

3. Modelling the Social Discount Rate for Energy Policies

In this section, we characterise discounting models that can fairly account for the environmental impacts of energy policies, both short- and long-term. Section 3.1 focuses on the model for estimating discount rates for intra-generational energy projects, i.e., investments whose impacts occur over a period of at most thirty years. Section 3.2 defines the discounting model for energy projects with long-term effects for which inter-generational equity issues need to be considered.

Both models are based on the use of discount rates, that are lower for discounting environmental externalities than rates which weight only the strictly economic components.

This is because the mathematical structure of the discount rate is a function not only of consumption, but also of environmental quality. The latter is, for the first time, expressed as a function of the Energy Transition Index (ETI), to orient decision-making towards investments increasingly in line with climate neutrality goals.

The model for energy intra-generational projects proposes the use of time-constant rates. This is legitimate as the contraction effects on the present value of cash flows are acceptable for time intervals limited to 20–30 years. Instead, in the case of investments with long-run effects, inter-generational equity issues are addressed by using rates with a declining structure over time. Otherwise, long-term environmental damage and benefits would be underestimated, or not considered at all in the analysis.

3.1. A New Discounting Model for Energy Intra-Generational Projects

Our approach to discount the effects of intra-generational projects in the energy field is based on Ramsey's growth model [47] and Gollier's ecological discounting model [29].

Gollier [29,51] proposes discounting the environmental components of investment at a different and lower r_q than the r_c needed to weight the strictly financial effects. To derive useful rates discounting different costs and benefits at different time horizons, it is necessary to consider a representative agent consuming two goods whose availability evolves stochastically over time. This is possible by extending Ramsey's rule (Equation (3)—taking into account the degree of substitutability between the two goods and the uncertainty surrounding economic and environmental growth. The rate at which environmental impacts should be discounted is in general different from the rate at which monetary benefits should be discounted. It is shown that, under Cobb–Douglas certainty and preferences, the difference between the economic and ecological discount rates is equal to the difference between the economic and ecological growth rates.

More specifically, it is assumed that the utility function U_t also depends on environmental quality q_t as well as consumption c_t, i.e., $U_t = U(c_t, q_t)$. In addition, since the environment tends to deteriorate over time, an incremental improvement in environmental quality will be more valuable to future generations than to current ones. Assuming again that c_t is a partial substitute for environmental quality, economic growth has a positive impact on the ecological discount rate, potentially offsetting the effect of environmental deterioration. If the substitutability is limited, the effect of environmental deterioration dominates economic growth. This leads to a low ecological discount rate that allows environmental assets to be preserved.

Based on the assumptions introduced, the inter-temporal SWF becomes the sum of the utilities derived from both consumption c_t and environmental quality q_t:

$$SWF = \int_{t=0}^{\infty} U(c_t, q_t) \cdot e^{-\rho t} dt \tag{6}$$

To derive the economic discount rate and the environmental discount rate, we assume that environmental quality is a deterministic function of economic performance: $q_t = f(c_t)$. Common sense implies that environmental quality is a decreasing function of GDP per capita, but this is much debated in scientific circles. For this reason, it is permissible to assume the following monotone relationship $q_t = c_t^{\rho}$, where ρ can be either positive or negative. If we assume that q_t follows a geometric Brownian motion, we obtain an analytical solution for r_c and r_q. Without going into the analytical demonstration of the formulae, for which we refer to Gollier [29], it should be noted that deriving $U(c_t, q_t)$ with respect to consumption c_t, we have the function describing the economic discount rate r_c:

$$r_c = \rho + [\eta_1 + \delta \cdot (\eta_2 - 1)] \cdot [g_1 - 0.5 \cdot (1 + \eta_1 + \delta \cdot (\eta_2 - 1)] \cdot \sigma_{11} \tag{7}$$

Deriving $U(c_t, q_t)$ with respect to environmental quality q_t, we obtain the ecological discount rate function r_q :

$$r_q = \rho + [(\delta \cdot \eta_2 + \eta_1 - 1)] \cdot [g_1 - 0.5 \cdot (\delta \cdot \eta_2 + \eta_1)] \cdot \sigma_{11} \tag{8}$$

(7) and (8) show how r_c and r_q depend on: (i) socio-economic parameters, such as the time preference rate ρ, risk aversion to income inequality η_1, the growth rate of consumption g_1, the uncertainty of the consumption growth rate σ_{11} in terms of the mean square deviation of the variable; (ii) environmental variables, such as the degree of environmental risk aversion η_2 and the elasticity δ of environmental quality to changes in the growth rate of consumption g_1. The estimation of each parameter is detailed at the end of this section.

The aim of this research is to propose discount rates that adequately account for the costs and benefits of energy investments. The main novelty of the model is therefore the modelling of environmental quality q_t, which for the first time is defined as a function of the Energy Transition Index (ETI). The index, estimated by the World Economic Forum (WEF), provides a framework to compare and support countries in their energy transition needs, considering their current energy system performance and the readiness of their macroeconomic, social, and regulatory environment for transition. The index, which summarises 40 different indicators, is currently available for 114 countries. The scores show that while 92 countries have risen their score over the last 10 years, only 10% of countries have been able to reach consistent gains, which are necessary to achieve climate targets for the next decade.

According to the World Economic Forum report 'Fostering Effective Energy Transition [16], even as countries continue in their progress in clean energy transition, it becomes necessary to embed the transition in economic, political, and social practices to ensure irreversible progress. For this reason, it is essential to introduce a variable into the mathematical structure of the discount rates that sees the progress of countries on the path to energy transition. This introduces an acceptance criterion that can guide decision-making towards those projects that are in keeping with climate neutrality goals to be achieved by 2030 and 2050.

Defining q_t = f(ETI), we can derive the value of the elasticity δ of environmental quality to changes in the growth rate of consumption as follows. Let c_1 be the GDP per capita of a country and c_2 the relative ETI. The slope of the regression line that correlates the two parameters GDP per capita and ETI corresponds to the value of δ. It follows that a different definition of environmental quality may allow the model to be adapted to the assessment of project categories other than energy projects.

In the following, the approaches to estimate the parameters that make up (7) and (8) are defined.

With reference to socio-economic variables, the time preference rate ρ is the sum of (i) l, which coincides with the average mortality rate for a country—this is because individuals tend to discount future utility according to the probability of being alive at the time of the decision; and (ii) r, or the pure time preference rate. This parameter reflects the irrational behaviour of individuals in making choices about the distribution of resources over time and is generally between 0 and 0.5% [49,50].

The elasticity η_1 of the marginal utility of consumption represents the percentage change in marginal utility resulting from a unit change in consumption [51]. It is a measure of risk aversion to income inequality, and it is estimated using the formula proposed by both Stern [52] and Cowell et al. [53]:

$$\eta_1 = \frac{\log\left(1 - t\right)}{\log\left(1 - \frac{T}{Y}\right)} \tag{9}$$

(9) is a function of t, the marginal tax rate, and T/Y, the average tax rate.

The growth rate of consumption g_1 expresses the degree of wealth in society and it is generally at the average growth rate of a country's GDP per capita [46,48].

Finally, a further environmental parameter is η_2, which represents the degree of environmental risk aversion. It can be expressed as a function of the consumption expenditure η^* to be allocated to environmental quality, considering that $10\% < \eta^* < 50\%$ [29,54,55]:

$$\eta^* = \frac{\eta_2 - 1}{\eta_1 + \eta_2 - 2} \tag{10}$$

3.2. A New Discounting Model for Energy Inter-Generational Projects

To provide the "right" weighting for the environmental effects of energy projects and policies in the long run, a dual and diminishing discounting approach is proposed. In other words, the structure of the two functions of the discount rate, economic and environmental, defined in the previous section begins decreasing over time.

This can be done by considering macroeconomic risk, i.e., we assume that the growth rate of consumption g_1 is a risky variable. To do this, we must first analyse the variable's trend over time, then define the probability distribution that best approximates the historical data. From the probability distribution of g_1 thus obtained, we derive the probability distributions of the unknowns r_c and r_q. From these parameters we then derive the values of the economic and environmental discount rates for each of the n years of the analysis period.

The next step is to move from the two uncertain and constant discount rates r_c and r_q, which coincide with the expected value of the probability distributions obtained, to certain but decreasing rates with a 'certainty equivalent'. This is possible by using the expected net present value (ENPV) approach, according to which, assessing the ENPV with an uncertain but constant discount rate is correspondent to evaluating the NPV with a certain rate, but diminishing with a 'certainty equivalent' until it has the minimum value at time $t = \infty$ [33]. In order to move from the uncertain and constant discount rate to the certain but decreasing discount rate with a 'certainty equivalent', it is first necessary to assess the economic discount factors $E_c(P_t)$ and environmental discount factors $E_q(P_t)$, and then r_{ct} and r_{qt}:

$$r_{ct} = \frac{E_c(P_t)}{E_c(P_t + 1)} - 1 \tag{11}$$

$$r_{qt} = \frac{E_q(P_t)}{E_q(P_t+1)} - 1 \tag{12}$$

In (11) $E_c(P_t)$ is calculated using the following formula:

$$E_c(P_t) = E_c\left[\sum_{i=1}^{m} p_{rci} \cdot e^{(-r_{ci} t)}\right] \tag{13}$$

where r_{ci} is the value of the i-th economic discount rate, resulting from the probability distribution of r_c; p_{ci} = probability of the i-th value of r_c; m = intervals of discretization of probability distributions r_c and r_q;

In (12) $E_q(P_t)$:

$$E_q(P_t) = E_q\left[\sum_{i=1}^{m} p_{rqi} \cdot e^{(-r_{qi} t)}\right] \tag{14}$$

In which r_{qi} is the value of the i-th environmental discount rate, deriving from the probability distribution of r_q; p_{qi} = probability of the i-th value of r_q.

4. Application: Estimation of SDRs for Italy and China

The approaches described in Sections 3.1 and 3.2 are implemented below to estimate discount rates for intra- and inter-generational energy projects for two very different economies: Italy and China. This is to demonstrate how: (i) the model can be applied to any territorial context; (ii) different social, economic, and environmental conditions lead to significantly dissimilar results.

4.1. Estimation of Constant and Dual Discount Rates for Italy and China

In the following we detail the estimation of the socio-economic and environmental parameters in (7) and (8).

The time preference rate ρ is a function of the mortality-based discount rate l and the pure time preference rate r. The first parameter, l, corresponds with the time-averaged mortality rate of the country. Since this rate undergoes small variations over time, it is considered correct to consider data from the last 30 years. l is estimated using mortality rates for the period 1991–2020 given by ISTAT for Italy and by the World Bank for China. Table 2 below shows the result of the calculations.

Table 2. Mortality rates for Italy and China over the 30-year period 1991–2020.

Year	1991	1992	1993	1994	1995	1996	1997	1998	1999	2000	Decade average rate
Death rate Italy (%)	0.97	0.96	0.97	0.97	0.98	0.97	0.98	1.00	0.98	0.98	0.98
Death rate China (%)	0.67	0.66	0.66	0.64	0.66	0.66	0.65	0.65	0.65	0.65	0.66
Year	2001	2002	2003	2004	2005	2006	2007	2008	2009	2010	Decade average rate
Death rate Italy (%)	0.96	0.98	1.02	0.95	0.98	0.96	0.98	0.99	1.00	0.99	0.98
Death rate China (%)	0.64	0.64	0.64	0.64	0.65	0.68	0.69	0.71	0.71	0.67	0.67
Year	2011	2012	2013	2014	2015	2016	2017	2018	2019	2020	Decade average rate
Death rate Italy (%)	1.00	1.03	1.00	0.98	1.07	1.00	1.04	1.05	1.05	1.06	1.03
Death rate China (%)	0.71	0.72	0.72	0.72	0.71	0.71	0.71	0.71	0.73	0.74	0.62
Thirty-year average rate Italy (%)											1.00
Thirty-year average rate China (%)											0.68

The result for Italy is $l = 1.00\%$, in line with the estimations obtained by Percoco [46] and Florio and Sirtori [48]. For China $l = 0.68\%$. This lower value compared to Italy is the effect of lower mortality rates over the 30-year period.

The pure time preference rate r is positive and reflects the irrational behaviour of individuals in making choices about the distribution of resources over time. As suggested by both Pearce and Ulph [49] and Evans and Kula [50], $0 < r < 0.5\%$ and is assumed to be 0.3%. It follows that:

$$\rho_{Italy} = 1.00\% + 0.3\% = 1.30\%;$$

$$\rho_{China} = 0.68\% + 0.3\% = 0.98\%.$$

By implementing (9) we calculate the elasticity η_1 of the marginal utility of consumption. Using the data of the marginal t and average T/Y individual income tax rates given by the Organization for Economic Cooperation and Development Countries (OECD), we assess $\log(1 - t)$, $\log(1 - T/Y)$, and the corresponding ratio. Processing returns a value of $\eta_1 = 1.34$ for Italy.

The analysis of average and marginal tax rates by income bracket in China gives instead a value of $\eta_1 = 1.14$ (source: https://taxsummaries.pwc.com/peoples-republic-of-china/individual/taxes-on-personal-income, 10 July 2021).

In summary, the estimations return the following values:

$$\eta_{1\,Italy} = 1.34;$$

$$\eta_{1\,China} = 1.14.$$

Estimates are consistent with known values from the literature, where the social values approach leads to $1 < \eta < 2$.

From the analysis of the trend of per capita GDP growth rate of the two countries, g_1 is estimated for Italy by averaging data over the last forty years, while for China the evaluation is carried out based on data over the last sixty years.

As for the estimation of the two environmental parameters, the value of η_2 is derived from (10), assuming $\eta^* = 30\%$, according to Hoel and Sterner [54], Sterner and Persson [55], and Gollier [29]. Hence, it follows that:

$$\eta_{2Italy} = 1.15;$$

$$\eta_{2China} = 1.06.$$

δ expresses the sensitivity of environmental quality q, expressed through the ETI, to changes in consumption c. The latter parameter is related to GDP per capita. For 115 countries, the index values in 2021 are related to their GDP per capita in the same year. Figure 1 gives the results of the ETI-GDP per capita regression analysis, from which δ is 0.23.

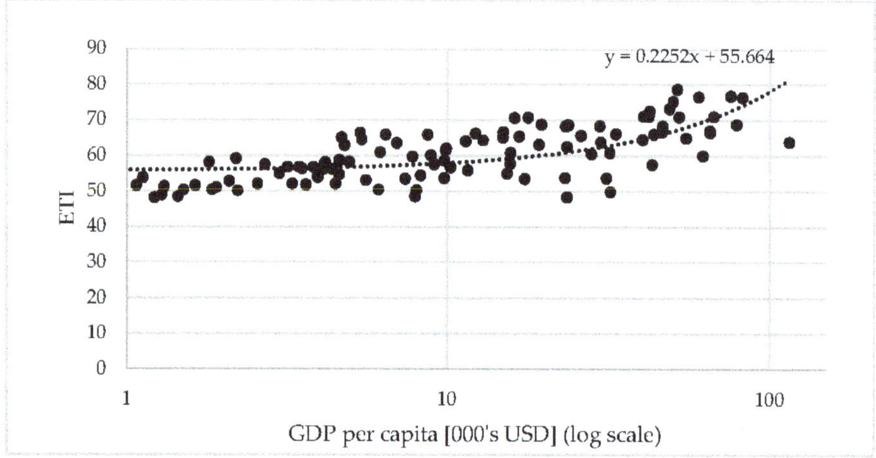

Figure 1. Regression analysis between ETI and GDP per capita.

Table 3 gives the values obtained for each parameter as well as the estimated r_c and r_q for Italy and China.

Table 3. Estimation of r_C and r_q for Italy and China.

Parameter	Value for Italy	Value for China
l	1.00%	0.68%
r	0.30%	0.30%
ρ	1.30%	0.98%
η_1	1.34	1.14
g_1	1.22%	8.17%
η_2	1.15	1.06
σ_{11}	0.03%	0.46%
δ	0.23	0.23
r_C	2.7%	9.8%
r_q	1.8%	4.02%

4.2. Estimation of Declining and Dual Discount Rates for Italy and China

To estimate time-declining r_{ct} and r_{qt} discount rates for energy projects with intergenerational effects, the reference is the approach defined in Section 3.2. Also, in this case, estimations are carried out with reference to both the Italian and Chinese economies.

g_t is estimated based on the growth rate of GDP per capita, in accordance with literature data [46]. As anticipated in Section 4.1, we consider it consistent to select data for the last forty years, i.e., from 1981 to 2020.

In fact, the data reported for the previous period reflect historical and economic contexts that can no longer be linked to either the current or foreseeable future economic, social, and cultural context of the country.

We identify the probability distribution that most closely approximates the historical series to predict the values to be associated with the growth rate of consumption, which in this context is the Weibull curve, chosen based on the Anderson–Darling test. Then, the expected values of the GDP growth rate are predicted by implementing the Monte Carlo analysis, calibrated on 10,000 random trials. The simulation was carried out using Oracle Crystal Ball software. Once the probability distribution of the consumption growth rate g_1 is defined, the probability distributions of the economic discount rate r_c and the ecological discount rate r_q are extracted by implementing (7) and (8). Table 4 shows the values of the statistical indices for the Monte Carlo simulation. The calculations indicate that: g_1 has values between -8.56% and 4.82%, and after 10,000 simulations the standard error of the mean is 0.02%; r_c and r_q have values between -10.71%–7.67% and -4.08%–3.99% respectively. In both cases, the mean standard error is acceptable as it is 0.02% and 0.01% respectively after 10,000 trials. Since negative discount rates have no economic significance, only positive values are considered in the definition of the declining structure of the two rates.

Table 4. Statistical indices on g_1, r_c, and r_q for Italy.

	Hypothesis: g_1	Forecast: r_c	Forecast: r_q
Trials number	10,000	10,000	10,000
Base case	1.17%	2.66%	1.79%
Mean	1.21%	2.71%	1.81%
Median	1.49%	3.09%	1.98%
Standard deviation	1.69%	2.33%	1.02%
Variance	0.03%	0.05%	0.01%
Kurtosis	4.75	4.75	4.75
Variation coefficient	1.40	0.86	0.56
Min	-8.56%	-10.71%	-4.08%
Max	4.82%	7.67%	3.99%
Mean standard error	0.02%	0.02%	0.01%

The analysis interval chosen for China is that between 1960–2020, in which the GDP growth trend rate tends to be steadily increasing. In this case, the Anderson–Darling test showed that the curve that best approximates the historical data is the logistic curve.

Again, the likely values of the GDP growth rate are predicted by implementing the Monte Carlo analysis, based on 10,000 random trials. Table 5 shows the values of the statistical indices for the forecast: g_1 has values between -18.96% and 40.58%, and after 10,000 simulations, the standard error of the mean is 0.06%. Furthermore, in the case of the simulations of r_c and r_q, the standard error is acceptable because it holds for the first variable at 0.07% and for the second at 0.02%. In addition, only positive values for the two discount rates are considered. This assumption is acceptable because the probability of having a positive discount rate r_c is 95.06% and the probability that the discount rate r_q is greater than 0 is 95.96%.

Table 5. Statistical indices on g_1, r_c and r_q for China.

	Hypothesis: g_1	Forecast: r_c	Forecast: r_q
Trials number	10,000	10,000	10,000
Base case	8.17%	9.86%	4.02%
Mean	8.79%	10.57%	4.262%
Median	8.81%	10.60%	4.272%
Standard deviation	5.62%	6.51%	2.18%
Variance	0.32%	0.42%	0.047%
Kurtosis	4.26	4.26	4.26
Variation coefficient	0.6399	0.6157	0.5714
Min	−18.96%	−21.55%	−6.49%
Max	40.58%	47.38%	16.58%
Mean standard error	0.06%	0.07%	0.022%

The probability distributions of r_c and r_q obtained are first discretized into 100 intervals. Then, for each of the two distributions, we estimate the probability that the average rate of each interval has of occurring. Given the set of values to be associated with the discount rates r_c and r_q and their probability, the equivalent certainty discount factors $E_c(P_t)$ and $E_q(P_t)$ are estimated using formulae (11) and (12). Finally, using (13) and (14) for each instant t leads to the estimation of the time sequence of the declining economic discount rate r_{ct} and the declining ecological discount rate r_{qt}. These are declining functions along the time horizon, assumed to be 300 years.

Figures 2 and 3 illustrate the term-structure of the economic and environmental discount rates for Italy and China respectively.

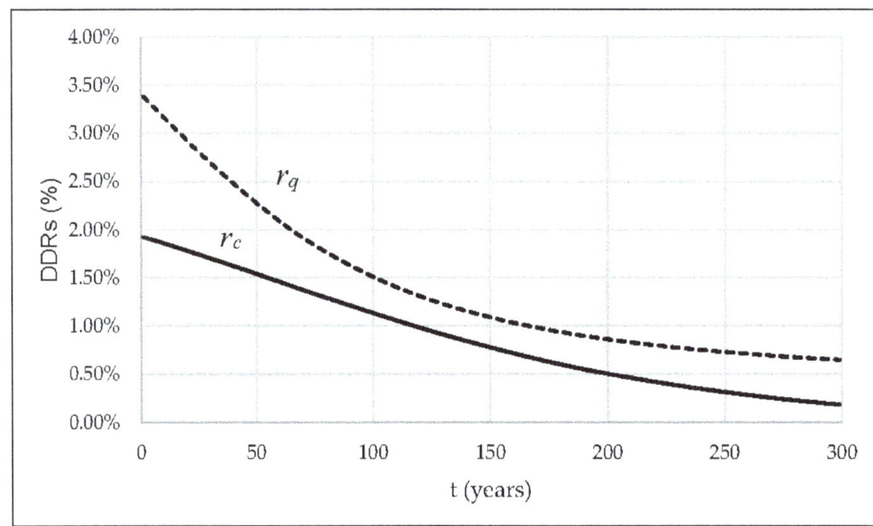

Figure 2. Term-structure of economic discount rate r_{ct} and environmental discount rate r_{qt} for Italy.

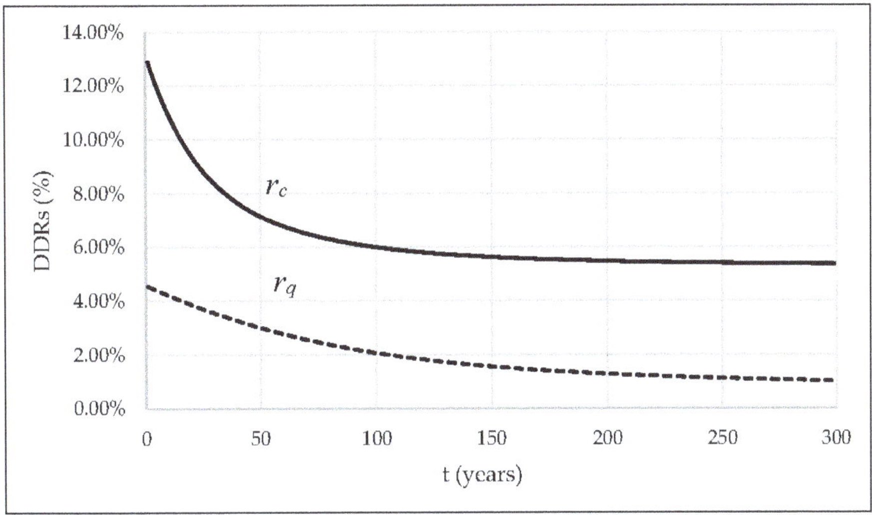

Figure 3. Term-structure of economic discount rate r_{ct} and environmental discount rate r_{qt} for China.

5. Results and Discussion

As Table 3 indicates, the values of the discount rates to be used in the analysis of intra-generational energy projects for Italy are significantly lower than those obtained for China. In fact, r_c and r_q for Italy are 2.7% and 1.8% respectively, while for China r_c is 9.8% and r_q is 4.0%. It should be noted that the difference between the environmental and economic discount rates for China is marked. On the contrary, in the Italian case, the values of the two discount rates are much closer to each other.

The implementation of the discounting model for energy inter-generational projects leads to the following results. For Italy:

- The economic discount rate function r_{ct} for Italy begins from an initial value of 3.4% to attain a value of 0.7% after 300 years, thus decreasing by about 2.6%.
- The environmental discount rate r_{qt}, on the other hand, takes on significantly smaller values of r_{ct}, starting from 1.92% and reaching 0.18% after 300 years.
- The average economic discount rate for the first 30 years is about 3.0%, which coincides with the value of the discount rate suggested by the European Commission [56].
- The average environmental discount rate for the first 20 years is 1.8%, highlighting how from the beginning of the assessment more weight is given to the damages and benefits that the investment generates on the environment.

For China:

- The economic discount rate function r_{ct} is of 12.90% and reaches a value of 5.36% after 300 years.
- The environmental discount rate r_{qt} is well below the values of r_{ct}, as it has an initial value of 4.54% and a final value at t = 300 years of 1.01%.
- The average economic discount rate for the first 30 years is about 10.2%, which is slightly higher than the value of the discount rate suggested by the Asian Development Bank [57] for economic analysis, which is 9.0%.
- The average environmental discount rate for the first 30 years is 4.0%.

Figures 4 and 5 explain the step functions (with solid lines) that approximate the functions (dashed lines) of the economic declining rate and the ecological declining rate for Italy. For practical purposes, it is useful to approximate the declining function to a step function. In other words, it may be permissible to use the same value of the discount rate for a period of thirty years in the analysis. In this time interval, the effect of present value contraction on cash flows can be considered acceptable [37–39]. Figures 6 and 7 indicate the same step functions of r_{ct} and r_{qt} for China.

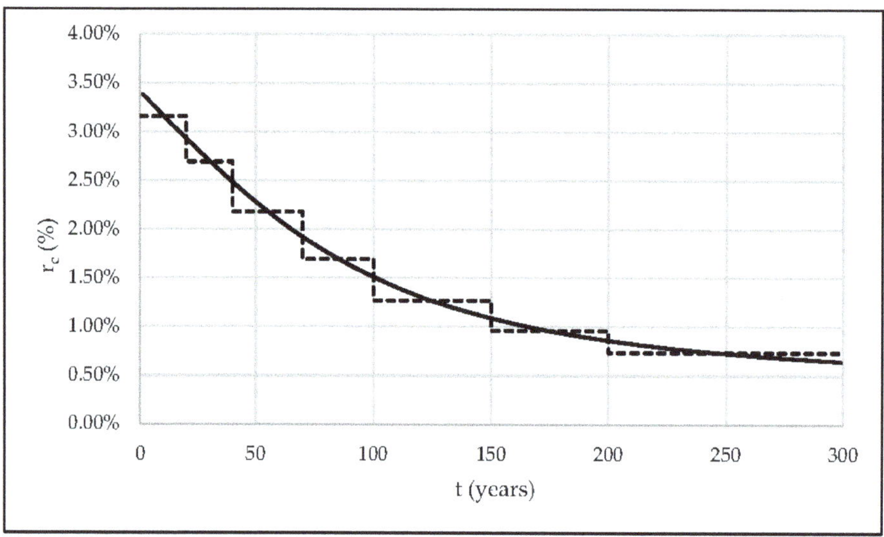

Figure 4. Step structure of the economic discount rate r_{ct} for Italy.

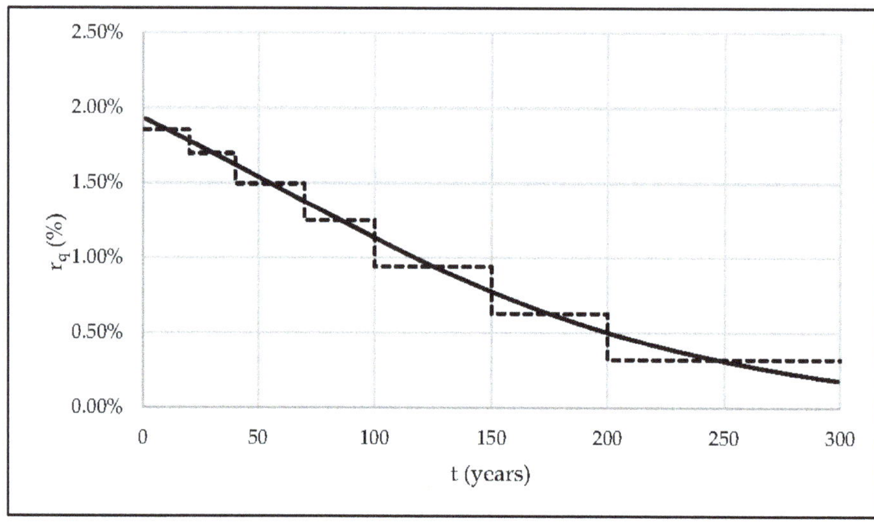

Figure 5. Step structure of the economic discount rate r_{qt} for Italy.

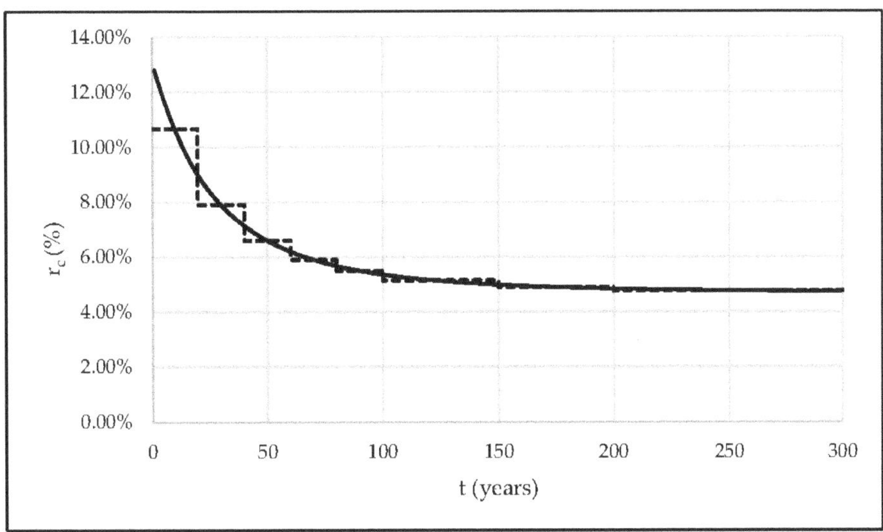

Figure 6. Step structure of the economic discount rate r_{ct} for China.

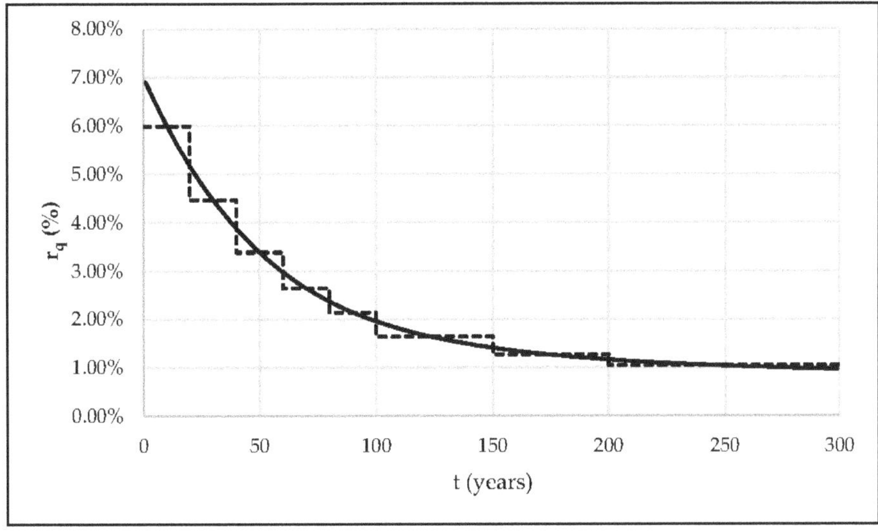

Figure 7. Step structure of the environmental discount rate r_{qt} for China.

The results indicate that the use of two different rates for discounting strictly financial and extra-financial components would allow greater weight to be given to environmental damages and benefits, thus orienting the decision-making process towards more sustainable investment choices.

It is interesting to underline that the two functions of the discount rate for China start from higher initial values than for Italy but decline much more rapidly after the early years of the period of analysis. The higher initial value is mainly due to the higher values of GDP growth rate for China compared to Italy. However, the faster decline in the term-structures of the discount rates is linked to China's 'worse' environmental condition. Indeed, as shown by the lower Energy Transition Index (ETI) value, more weight should be given to environmental impacts of energy projects in China from the early years of the assessment. This is to prioritise investment choices in line with sustainability and climate neutrality objectives to be achieved in the coming decades.

6. Conclusions

Energy transition policies aim to respond to both economic, social, and environmental challenges. Therefore, it is essential to steer the decision-making process towards policy initiatives that ensure a balance between socio-economic benefits and costs. In this context, the choice of discount rate becomes central to comparing policy strategies and investment programmes—but also to determine the speed with which an energy transition policy should be delivered to reach decarbonisation targets within the defined timeframe [2].

Thus, the discount rate affects the final judgement on the efficiency of the investment policy or project. However, there is still no unanimity in the literature as to what value of the discount rate should be used in analyses, or how it should be estimated. The question becomes even more controversial when a very long-term perspective is adopted.

With this research, we propose an innovative discounting approach for discounting energy investments, distinguishing between intra-generational and inter-generational projects.

In the first case, a constant and dual discounting approach is characterised. The discount rate used to discount the environmental components is lower than the discount rate used to weight the strictly financial contributions. However, since the effects of these projects are felt over a period of thirty years at the most, both discount rates are assumed to be time-constant.

For projects with inter-generational environmental effects, a dual and time-declining econometric model is defined to give greater weight to long-term environmental components that would be underestimated using constant rates.

For both models, the main change is that environmental quality is defined as a function of the Energy Transition Index (ETI). It is considered essential to introduce into the mathematical structure of the SDR a variable that considers the progress of countries on the path towards energy transition. In other words, a discount rate defined in this way allows decision makers to be oriented towards those projects that are in line with 2030 and 2050 climate neutrality goals. In addition, the dual and declining approach also takes macroeconomic risk into account, as the growth rate of consumption is modelled as a stochastic variable.

The defined models were implemented to estimate discount rates for both Italy and China. The results obtained show that: (i) in the case of the dual and constant approach for both Italy and China, the environmental discount rate has smaller values than the economic discount rate; (ii) in the case of the dual and declining approach, the two functions of the discount rate—economic and environmental—for China start from higher initial values than for Italy, but decline much faster from the beginning of the analysis period. The higher initial value is mainly due to the higher values of GDP growth rate for China compared to Italy. However, the application demonstrates how China's 'worse' environmental condition leads to a more rapid decline in the term-structures of the discount rates.

While the model is relatively easy to implement, for some countries it may be difficult to find the data needed to estimate each parameter of the model. In addition, estimates of discount rates need to be periodically updated. The application shows, firstly, how different discount rates can be in relation to socio-economic context. Secondly, it is clear how the use of estimated discount rates can favour more sustainable investment choices in line with UN climate neutrality objectives. The decision-making effects on energy policy investments are therefore evident and extremely important; evaluating the economic feasibility of energy projects using dual, and possibly even time-declining approaches, means attributing greater weight to extra-financial damages and benefits. On the contrary, by using the social discount rates provided by governments, which are generally unique and constant over time, policymakers would orient their choices towards investments with higher initial financial returns, without considering the short and long-term repercussions on the environment.

Finally, research perspectives may include the implementation of the model for other countries in order to provide a larger database of environmental and economic discount rates, as well as the adaptation of the model to other sectors of intervention.

Author Contributions: Conceptualization, G.M. and A.N.; Data curation, G.M. and A.N.; Formal analysis, G.M. and A.N.; Investigation, G.M. and A.N.; Methodology, G.M. and A.N.; Resources, A.N.; Software, G.M. and A.N.; Supervision, A.N.; Validation, G.M. and A.N.; Visualization, G.M. and A.N.; Writing—original draft, G.M. and A.N.; Writing—review & editing, G.M. and A.N. Both authors have read and agreed to the published version of the manuscript.

Funding: This research received no external funding.

Institutional Review Board Statement: Not applicable.

Informed Consent Statement: Not applicable.

Conflicts of Interest: The authors declare no conflict of interest.

References

1. Oxilia, V.; Blanco, G. *Energy Policy. A Practical Guidebook, Latin American Energy Organization*; OLADE: Quito, Ecuador, 2016.
2. Peñalver, D.; Poudineh, R. Social Discount Rate and the Energy Transition Policy. *Energy Insights* **2020**, *75*, 1–13.
3. Luo, L.; Chen, S.; Zou, Z. Determining the Generalized Discount Rate for Risky Projects. *Environ. Resour. Econ.* **2020**, *77*, 143–158. [CrossRef]
4. Arrow, K.J.; Cropper, M.L.; Gollier, C.; Groom, B.; Heal, G.M.; Newell, R.G.; Nordhaus, W.D.; Pindyck, R.S.; Pizer, W.A.; Portney, P.R.; et al. Determining benefits and costs for future generations. *Science* **2013**, *341*, 349–350. [CrossRef] [PubMed]
5. Arrow, K.J.; Cropper, M.L.; Gollier, C.; Groom, B.; Heal, G.M.; Newell, R.G.; Nordhaus, W.D.; Pindyck, R.S.; Pizer, W.A.; Portney, P.R.; et al. Should governments use a declining discount rate in project analysis? *Rev. Environ. Econ. Policy* **2014**, *8*, 145–163. [CrossRef]
6. Weitzman, M.L. Gamma discounting. *Am. Econ. Rev.* **2001**, *91*, 260–271. [CrossRef]
7. Evans, D.J. Uncertainty and social discounting for the very long term. *J. Econ. Stud.* **2009**, *36*, 522–540. [CrossRef]
8. Nesticò, A.; Maselli, G. A Protocol for the Estimate of the Social Rate of Time Preference: The Case Studies of Italy and the USA. *J. Econ. Stud.* **2020**, *47*, 527–545. [CrossRef]
9. Zhuang, J.; Liang, Z.; Lin, T.; De Guzman, F. *Theory and Practice in the Choice of Social Discount Rate for Cost-Benefit Analysis: A Surve*; Working Paper No. 94; RD Economics and Research Department, Asian Development Bank: Manila, Philippines, 2007.
10. Kubiak, R.J. *Decision Making in Energy Efficiency Investments—A Review of Discount Rates AND Their Implications for Policy Making*; European Council for an Energy Efficient Economy (ECEEE) Industrial Summer Study Proceedings: Brussels, Belgium, 2016.
11. Foltyn-Zarychta, M.; Buła, R.; Pera, K. Discounting for Energy Transition Policies—Estimation of the Social Discount Rate for Poland. *Energies* **2021**, *14*, 741. [CrossRef]
12. European Commission. Communication from the Commission to the European Parliament, the European Council, the Council, the European Economic and Social Committee and the Committee of the Regions. In *The European Green Deal*; European Commission: Brussels, Belgium, 2019.
13. IPCC Reports. Available online: https://www.ipcc.ch/reports/ (accessed on 8 May 2021).
14. Kula, E.; Evans, D. Dual discounting in cost-benefit analysis for environmental impacts. *Environ Impact Assess Rev.* **2011**, *31*, 180–186. [CrossRef]
15. Kossova, T.; Sheluntcova, M. Evaluating performance of public sector projects in Russia: The choice of a social discount rate. *Int. J. Proj. Manag.* **2016**, *34*, 403–411. [CrossRef]
16. World Economic Forum (WEF). Fostering Effective Energy Transition. 2021. Available online: https://www.weforum.org/reports/fostering-effective-energy-transition-2021 (accessed on 10 June 2021).
17. Lind, R. Intergenerational equity, discounting, and the role of cost–benefit analysis in evaluating global climate policy. *Energy Policy* **1995**, *23*, 379–389. [CrossRef]
18. Gollier, C. *Pricing the Future: The Economics of Discounting and Sustainable Development*; Princeton University Press: Princeton, NJ, USA, 2011.
19. Boardman, A.; Greenberg, D.; Vining, A.; Weimer, D. *Cost–Benefit Analysis: Concepts and Practice*; Cambridge University Press: Cambridge, UK, 2017.
20. Nesticò, A.; Maselli, G. Declining discount rate estimate in the long-term economic evaluation of environmental projects. *J. Environ. Account. Manag.* **2020**, *8*, 93–110. [CrossRef]
21. Moore, M.A.; Boardman, A.E.; Vining, A.R. More appropriate discounting: The rate of social time preference and the value of the social discount rate. *J. Benefit-Cost Anal.* **2013**, *4*, 1–16. [CrossRef]

22. Florio, M. *Applied Welfare Economics: Cost-Benefit Analysis of Projects and Policies*; Routledge: New York, NY, USA, 2014.
23. Emmerling, J.; Drouet, L.; van der Wijst, K.-I.; Van Vuuren, D.; Bosetti, V.; Tavoni, M. The role of the discount rate for emission pathways and negative emissions. *Environ. Res. Lett* **2019**, *14*, 104008. [CrossRef]
24. Stern, N. *The Economics of Climate Change: The Stern Review*; Cambridge University Press: Cambridge, UK, 2007. [CrossRef]
25. van den Bijgaart, I.; Gerlagh, R.; Liski, M. A simple formula for the social cost of carbon. *J. Environ. Econ. Manag.* **2016**, *77*, 75–94. [CrossRef]
26. van der Ploeg, F.; Rezai, A. Simple rules for climate policy and integrated assessment. *Environ. Resour. Econ.* **2019**, *72*, 77–108. [CrossRef] [PubMed]
27. Gollier, C. Time Horizon and the Discount Rate. *J. Econ. Theory* **2002**, *107*, 463–473. [CrossRef]
28. Weikard, H.P.; Zhu, X. Discounting and environmental quality: When should dual rates be used? *Econ. Model.* **2005**, *22*, 868–878. [CrossRef]
29. Gollier, C. Ecological discounting. *J. Econ. Theory* **2009**, *145*, 812–829. [CrossRef]
30. Almansa-Sáez, C.; Martinez-Paz, J. What weight should be assigned to future environmental impacts? A probabilistic cost benefit analysis using recent advances on discounting. *Sci. Total Environ.* **2011**, *409*, 1305–1314. [CrossRef] [PubMed]
31. Baumgärtner, S.; Klein, A.; Thiel, D.; Winkler, K.J. Ramsey Discounting of Ecosystem Services. *Environ. Resour. Econ.* **2015**, *61*, 273–296. [CrossRef]
32. Weitzman, M.L. Why the Far-Distant Future Should Be Discounted at Its Lowest Possible Rate. *J. Environ. Econ. Manag.* **1998**, *36*, 201–208. [CrossRef]
33. Newell, R.G.; Pizer, W.A. Discounting the Distant Future: How Much Do Uncertain Rates Increase Valuations? *J. Environ. Econ. Manag.* **2003**, *46*, 52–71. [CrossRef]
34. Groom, B.; Koundouri, P.; Panopoulou, E.; Pantelidis, T. Discounting the distant future: How much does model selection affect the certainty equivalent rate? *J. Appl. Econom.* **2007**, *22*, 641–656. [CrossRef]
35. Cropper, M.L.; Freeman, M.C.; Groom, B.; William, A.P. Declining Discount Rates. *Am. Econ. Rev.* **2014**, *104*, 538–543. [CrossRef]
36. Freeman, M.C.; Groom, B.; Panopoulou, E.; Pantelidis, T. Declining discount rates and the Fisher effect: Inflated past, discounted future? *J. Environ. Econ. Manag.* **2015**, *73*, 32–49. [CrossRef]
37. *HM Treasury*; The Green Book; Appraisal and Evaluation in Central Government: London, UK, 2003.
38. Rapport Lebègue. *Révision du taux D'actualisation des Investissements Publics*; Commissariat Général au Plan: Paris, France, 2005.
39. U.S. Environmental Protection Agency. *Guidelines for Preparing Economic Analysis*; U.S. Environmental Protection Agency: Washington, DC, USA, 2010.
40. Vasquez-Lavín, F.; Oliva, R.D.P.; Hernández, J.I.; Gelcich, S.; Carrasco, M.; Quiroga, M. Exploring dual discount rates for ecosystem services: Evidence from a marine protected area network. *Resour. Energy Econ.* **2018**, *55*, 63–80. [CrossRef]
41. Muñoz-Torrecillas, M.J.; Roche, J.; Cruz-Rambaud, S. Building a Social Discount Rate to be Applied in US Afforestation Project Appraisal. *Forests* **2019**, *10*, 445. [CrossRef]
42. Steinbach, J.; Staniaszek, D. *Discount. Rates Energy Syst. Analysis*; Discussion Paper commissioned by the Buildings Performance Institute Europe; BPIE: Brussels, Belgium, 2015.
43. Defrancesco, E.; Gatto, P.; Rosato, P. A 'component-based' approach to discounting for natural resource damage assessment. *Ecol. Econ.* **2014**, *99*, 1–9. [CrossRef]
44. Boardman, A.E.; Moore, M.A.; Vining, A.R.; De Civita, P. *Propos. Soc. Discount. Rate(S) Can. Based Future Growth*; Government of Canada, Policy Research Initiative: Ottawa, ON, Canada, 2009.
45. Azar, S.A. Measuring the US social discount rate. *Appl. Financ. Econ. Lett.* **2007**, *3*, 3–66. [CrossRef]
46. Percoco, M. A social discount rate for Italy. *Appl. Econ. Lett.* **2008**, *15*, 73–77. [CrossRef]
47. Ramsey, F.P. A Mathematical Theory of Saving. *Econ. J.* **1928**, *38*, 543–559. [CrossRef]
48. Florio, M.; Sirtori, E. *The Social Cost of Capital: Recent Estimates for the EU Countries*; Centre for Industrial Studies: Milano, Italy, 2013.
49. Pearce, D.; Ulph, D. A social Discount Rate for United Kingdom. In *Environmental Economics: Essays in Ecological Economics and Sustainable Development*; Pearce, D., Ed.; Edward Elgar: Cheltenham, UK, 1999; pp. 268–285.
50. Evans, D.J.; Kula, E. Social Discount Rates and Welfare Weights for Public Investment Decisions under Budgetary Restrictions: The Case of Cyprus. *Fiscal Stud.* **2011**, *32*, 73–107. [CrossRef]
51. Gollier, C. *Pricing The Planet's Future: The Economics of Discounting in An Uncertain World*; Princeton University Press: Princeton, NJ, USA, 2012.
52. Stern, N. Welfare weights and the elasticity of marginal utility of income. In *Proceedings of the Annual Conference of the Association of University Teachers of Economics in Artis M.*; Norbay, R., Ed.; Blackwell: Oxford, UK, 1977.
53. Cowell, F.; Gardiner, K. *Welfare Weights*; STICERD, London School of Economics: London, UK, 1999.
54. Hoel, M.; Sterner, T. Discounting and relative prices. *Clim. Chang.* **2007**, *84*, 265–280. [CrossRef]
55. Sterner, T.; Persson, M. An Even Sterner Report: Introducing Relative Prices into the Discounting Debate. *Rev. Environ. Econ. Policy* **2008**, *2*, 61–76. [CrossRef]
56. European Commission. *Guide to Cost-Benefit Analysis of Investment Projects. Economic Appraisal Tool for Cohesion Policy 2014–2020*; Directorate-General for Regional and Urban Policy: Brussels, Belgium, 2014.
57. Asian Development Bank. *Guidelines for the Economic Analysis of the Project*; Asian Development Bank: Manila, Philippines, 2017.

Article

Are You a Typical Energy Consumer? Socioeconomic Characteristics of Behavioural Segmentation Representatives of 8 European Countries

Sylwia Słupik [1,*], Joanna Kos-Łabędowicz [2] and Joanna Trzęsiok [3]

[1] Department of Social and Economic Policy, University of Economics in Katowice, 1 Maja 50, 40-287 Katowice, Poland

[2] Department of International Economic Relations, University of Economics in Katowice, 1 Maja 50, 40-287 Katowice, Poland; joanna.kos@ue.katowice.pl

[3] Department of Economic and Financial Analysis, University of Economics in Katowice, 1 Maja 50, 40-287 Katowice, Poland; joanna.trzesiok@ue.katowice.pl

[*] Correspondence: sylwia.slupik@ue.katowice.pl

Abstract: Scarcity of resources and their waste, as well as deteriorating quality of life and the environment, are pressing problems of modern civilisations. Rational and efficient energy consumption is one of the possibilities for preventing harmful practices and the degradation of ecosystems. Understanding the consumer's way of thinking and acting by identifying his needs and preferences are essential for effective efforts for smart, sustainable, and inclusive economic growth. Therefore, the aim of this article was a comprehensive socioeconomic analysis of particular behavioural types of energy consumers, as a continuation of the authors' previous research. The paper uses statistical methods (chi-square test and correspondence analysis) dedicated to non-metric variables for an effective analysis of the data obtained from the questionnaires. The identification of socioeconomic factors was carried out on a representative sample of $n = 4506$ respondents from eight European countries (the Czech Republic, France, Greece, Spain, Germany, Poland, Romania, and the United Kingdom). This allowed for distinguishing a typical representative of five consumer segments (EI; AE; DS; O; I), developed on the basis of motivation to save energy. The authors succeeded in combining behavioural segmentation with the socioeconomic characteristics of the created classes. The results indicated that 10 out of 12 examined factors were significantly correlated with the behavioural type. These are (in order of significance): attitude towards saving energy; age; employment status; home country; the ownership status of the premises; the number of people in a household; average monthly income per person in a household; education; gender and place of residence.

Keywords: energy consumer; behavioural model; consumer segmentation; socioeconomic characteristics; end user profile; energy awareness

Citation: Słupik, S.; Kos-Łabędowicz, J.; Trzęsiok, J. Are You a Typical Energy Consumer? Socioeconomic Characteristics of Behavioural Segmentation Representatives of 8 European Countries. *Energies* **2021**, *14*, 6109. https://doi.org/10.3390/en14196109

Academic Editor: Sergio Ulgiati

Received: 30 August 2021
Accepted: 22 September 2021
Published: 25 September 2021

Publisher's Note: MDPI stays neutral with regard to jurisdictional claims in published maps and institutional affiliations.

1. Introduction

This paper is a continuation of work on universal behavioural segmentation [1], carried out as part of a research project on energy consumer behaviour in selected European countries. Energy consumption and related behaviours, attitudes, and preferences are one of the most important social problems today. Scarcity of resources, including energy, rational use, and change of beliefs and habits of a typical consumer are becoming an issue to be solved as soon as possible, not only on a global scale (world and national) but, above all, on a regional and local scale. Local communities may become the driving force of change and a role model at a time when ecological thinking, environmental protection, and natural resources protection are becoming not just a passing fad or whim but a necessity and even a long-term strategy. Therefore, it is necessary to support local pro-ecological initiatives, build social and environmental capital, and, above all, help understand the behaviours, attitudes, and needs of ordinary people. Understanding seems to be the first step necessary to shape attitudes and change behaviours. Education and promotion of pro-environmental actions are essential to improving the public's knowledge on how to use resources more efficiently and eliminate harmful and wasteful practices.

The authors hope that the presented research results will contribute to increasing this knowledge, especially as there are still few researchers involved in a comprehensive analysis of energy consumer behaviour. In this regard, this paper fills the existing research gap and can serve as both a model and a contribution to further research for other scientists. The significant innovative contribution of the authors is both the developed behavioural segmentation and its combination with the socioeconomic characteristics of the segments. Such an approach allows for forming a full picture of a typical energy consumer, which is not a common practice in the literature. It should be emphasised that the segmentations developed so far explain a limited (geographically, size of the sample, and methodologically) research scope, hence the existing approaches characterise the energy consumer only in a fragmentary and often one-dimensional way. The authors' research presented in this paper was conducted on a wide representative research sample of $n = 4506$ respondents from eight European countries, including seven European Union member states: the Czech Republic, France, Greece, Spain, Germany, Poland, Romania, and the United Kingdom. Therefore, the obtained results can serve as a universal tool to identify the basic motivations of energy consumers as well as their socioeconomic characteristics.

Attempting to provide answers to the following research questions was the main aim of this paper:

1. Is it possible to identify a typical representative of each segment of the authors' behavioural segmentation of energy consumers, in terms of distinguishing their socioeconomic characteristics?
2. Are all the examined socioeconomic factors relevant to the characteristics of the different segments?
3. Whether the profile of the typical energy consumer obtained as a result of the analysis is convergent (similar) to other typologies existing in the literature.

Analysing the data obtained, the authors tried to identify the basic relevant socioeconomic factors that could characterise a typical representative of behavioural segmentation. This analysis can be used by local and regional decision makers as a useful tool for shaping environmental policies and campaigns, but also by consumers themselves to learn and become more aware of their own motivations and preferences, and perhaps make them reflect on the values that are worth following in their daily lives.

The study used methods enabling the analysis of non-metric variables: chi-square test and correspondence analysis. The authors chose these statistical tools because they allowed not only to indicate significant relationships between the studied variables (socioeconomic factors and respondents' behaviour towards energy saving), but also to show which categories of these variables are related to each other. Moreover, the possibility of visualising the obtained results in the correspondence analysis allows for an easy interpretation, without the need to be familiar with the method itself.

For a better understanding of the content, the article has been divided into the following six sections. An introduction of the topic and the aim of the research and the authors' original contributions are included in Section 1. Section 2 provides an overview of the literature on energy behaviour, energy saving, existing segmentations of energy users, and factors influencing their behaviour. Section 3 describes the methodology used in the paper and Section 4 presents the obtained results. The article concludes with a discussion of the results obtained with previous works (in Section 5) and a short summary in Section 6.

2. Literature Review

Energy consumption, especially by individual users and households, is an important and complicated issue, interesting not only from the point of view of the scientific community and policy and decision-makers, but also for the various market entities like energy suppliers and manufacturers of various electronical appliances. A lot of emphasis, especially in the European Union, is placed on sustainable development and the related sustainable consumption, which can take two forms: weak (increasing efficiency as a way to improve the quality of life [2,3]) and strong (seeking to change behaviours, lifestyles, and consumer decisions based on social responsibility [4–6]).

2.1. Energy Consumption and Energy Savings

Energy consumption has been, and still is, the subject of many studies attempting to discern different factors that are influencing it in order, on the one hand, to predict its future size [7], which is of key importance for the supply side of the market [8] and energy security of a given country/region [9,10], and, on the other hand, to plan and take measures aimed at balancing the load on the electricity grid and shifting part of consumption outside peak periods [11]. Research in the field of energy consumption highlights a number of factors that can be assigned to various categories, such as, the socioeconomic characteristics of the household itself (e.g., the number of people in the household and their age, economic status [12–14]), the type of dwelling (e.g., type and age of the building, floor area [15,16]), number of owned and used appliances (e.g., whether the household uses renewable energy sources RES or if the heating uses electricity or another energy source [13,17,18]), external conditions (e.g., climate [19,20]), or the level of economic development of a given area [21,22]). The influence of consumers' lifestyles on energy consumption [23,24] or childhood experiences from the family home [25] are also getting more attention from researchers.

In the case of behaviours related to energy saving, two main trends of undertaken actions, referring to the concept of strong and weak sustainable consumption, can be observed: actions aimed at increasing the efficiency of the energy used and actions aimed at persuading consumers to reduce their consumption [26]. Research concerning consumers energy-saving behaviours is trying to determine what socioeconomic features (e.g., income, education, gender [27–31]) or psychographic features (e.g., sense of duty, pro-environmental awareness [32–35]) are manifested by consumers willing to save energy and limit their consumption, and on the other hand, how to motivate other consumers to undertake such actions. In order to better understand consumers' energy-saving behaviours, attempts to link them with the consumers' lifestyles or the culture of the country of origin have also been made [36,37]. Two motivations recur most frequently in the research aimed at identifying factors that may motivate consumers to undertake energy-saving measures, one related to financial reasons (whether it is co-financing or subsidising a given type of solution or lowering the energy costs [38,39]) and the other related to pro-ecological awareness and attitude [40–42]. Additionally, the literature indicates that factors such as social pressure/influence [41,43,44] or attitudes towards technology/available technological support [45,46] may motivate consumers to save while concerns about reduced comfort and convenience resulting from reduced energy consumption [47,48] may act as demotivators.

Energy behaviours and promoting energy efficiency and energy savings are research directions that are of great interest currently due to such issues as energy scarcity, the need to change the energy mix to a more sustainable one, and climate change due to increasing CO_2 emissions. Energy consumption and energy saving behaviours are influenced by many different factors. That is a reason why when planning interventions, a certain balance between the more personalized approach to a given consumer while motivating as many consumers as possible is needed. A segmentation of energy consumers may be used as a compromise to solve that dilemma and provide interested parties with a viable tool. Attempts at energy customer segmentation is one of the approaches to understanding and managing the demand side of the energy market. Most approaches to the segmentations do not combine behavioural and socioeconomic factors, both for distinguishing the segments and then characterising them. Additionally, most of existing segmentations focus only on a very narrow set of factors. That practice was identified as a research gap that the authors attempted to fill by developing a behavioural segmentation based on consumers' motivations and beliefs and then using socioeconomic data to further characterise the different segments in order to provide more complex consumer profiles.

2.2. Segmentation of Energy Consumers Focused on Their Willingness to Save Energy

Segmentation studies aim at identifying homogeneous groups of consumers [49–51] to better understand motivations and factors influencing their decision-making process and/or in order to control and influence their future consumption behaviours. It is a complex process that requires researchers to identify and analyse the non-obvious (as they are largely ingrained in the person's mind) motivations and reasons behind manifested (and observable) consumer behaviours [52]. Any attempts at energy users' segmentation, especially those where the focus of the study is not so much on energy consumption as on consumers' energy-saving behaviours, are faced with these challenges. When describing energy consumption (and subsequently energy-saving behaviours), one should mention the developed segmentations referring to the previously mentioned sustainable consumption [53,54], as some of them will also include issues related to energy use [55–57].

There are four most common types of segmentation: demographic segmentation, psychographic segmentation, behavioural segmentation, and geographic segmentation. Each type of segmentation assigns consumers to their respective segment basing on a different category of indicators. The most common forms of energy consumer segmentations are demographic segmentation and behavioural segmentation. Psychographic segmentation, being the most difficult to perform, is not very common [58]. Nevertheless, it should be noted that certain elements of psychographic segmentation are being included in segmentations focusing mainly on consumer behaviour (behavioural segmentations) [37,59–62]. When trying to identify types of energy consumers, energy consumption (energy load profiles [63–65]), socioeconomic/infrastructural characteristics (income, occupational status, dwelling type and size [64,66]), or behavioural indicators [50,67] are most commonly used to assign consumers to particular segments.

This approach to segmentation does not always work when trying to investigate and identify types of users in terms of their willingness to take on energy-saving actions, especially as [68] has shown that factors relevant for energy consumption may not necessarily be relevant for energy-saving behaviours. For segmentations attempting to distinguish consumers in terms of their energy-saving behaviours (or potential motivation for such behaviours), behavioural segmentation is most commonly used [69]. Existing studies on energy saving behaviours have adopted, e.g., general values [59]; lifestyles [58,66,70]; general consumer behavioural patterns [62,70]; attitudes towards the environment and environmental awareness [71]; attitudes towards the use of technology [61]; contextual factors [72,73]; and rebound effect [74] as a basis for segmentation, but they usually had a rather narrow focus and only concerned particular types of action (e.g., tariff choice [67]).

It should be emphasised that, regardless of the type of segmentation and the factors differentiating the segments, in the case of most energy consumer segmentations socioeconomic factors (such as income, age, education and others) are at times used to characterise and describe individual segments, but only if they have proven to be particularly relevant and distinctive for a particular segment. This practice is so popular that it is easier to indicate studies that have dispensed with the inclusion of these factors when describing segments [19,67,75–77] than to list all those in which such an element (even though socioeconomic factors themselves were not the basis for the given segmentation study) has been included.

Table 1 presents examples of segmentation of energy consumers focusing on their tendencies towards energy-saving behaviours, with an indication of socioeconomic factors that turned out to be particularly important for the characteristics of individual segments.

Table 1. Overview of selected energy consumer segmentations with an indication of socioeconomic factors particularly relevant to segment characterisation.

Study	Focus	Relevant Socioeconomic Factors for Segments' Characteristics	Approach	Sample
Pedersen (2008) [78]	segmentation and profiling as input for preparing long-term program planning and communications strategies	• age • ethnicity • gender • income • number of cohabitants • occupational status • ownership of housing • type of housing • urban vs. rural	quantitative end-use survey, cluster analysis	4191 BC Hydro residential customers across the British Columbia province (Canada)
Accenture end–consumer observatory on electricity management (2010) [79]	identifying opinions and preferences toward electricity management programs	• age • gender • income	quantitative global survey, 17 countries conjoint analysis	9108 individuals from: Australia, Brazil, Canada, China, Denmark, France, Germany, Italy, Japan, the Netherlands, Singapore, South Africa, South Korea, Spain, Sweden, the United Kingdom, the United States
Sütterlin et. al (2011) [69]	an attempt at preparing segmentation of energy consumers using a more comprehensive way than previous attempts—advocating the need for a more behavioural based approach	• age • education • gender • income	cluster analytic approach, a mail-in survey	random sample of 1292 Swiss households (Switzerland)
Han et al. (2013) [80]	analysing preferences for interventions strategy to promote neutral urban development through energy-saving behaviour	• age • education • income • ownership of housing	latent class model analyses, an online questionnaire	1500 households of Eindhoven region (The Netherlands)
Tabi et al. (2014) [60]	searching for factors influencing adoption of green electricity	• education	latent class segmentation analysis based on choice-based conjoint data,	414 German consumers (Germany)
Yang et.al (2015) [59]	identifying household preferences for electricity products	• age • gender • income • number of cohabitants	latent class modelling, self-administered questionnaires,	Danish households, 1012 usable questionnaires (Denmark)

Table 1. *Cont.*

Study	Focus	Relevant Socioeconomic Factors for Segments' Characteristics	Approach	Sample
Albert and Maasoumy (2016) [81]	creating an intuitive segmentation and targeting process that can be used by energy utility to engage its customers	• education • ethnicity • income • number of cohabitants • occupational status • ownership of housing • religion	a-priori segmentation with use of predictive algorithm for allocation to appropriate segment, machine learning	data from enrolment and consumption from 150 consumers of energy utility (The United States)
Seidl et. al. (2017) [82]	analysing the potential for behavioural change through the cities' intervention through links between current behaviours and potential for future change	• age • gender • income • number of cohabitants • ownership of housing	cluster analysis, survey	706 respondents from cities of Baden, Winterthur, and Zug (Switzerland)
Tumbaza and Moğulkoç (2018) [83]	investigating attitudes and behaviours concerning energy efficiency	• age • education • income	two-step cluster analysis based on online survey	526 Turkish households (Turkey)
Smart Energy Consumer Collaborative (2019) [61]	analysing the relationship between technology and energy-saving decisions	• age • education • gender • income • occupational status • ownership of housing • number of cohabitants • type of housing	online survey, update of SECC's consumer segmentation from 2015	2451 residential energy consumers (The United States)
Słupik et. al. (2021) [62]	identifying underlaying motivation for energy saving behaviours	• income • number of cohabitants	cluster analysis, survey	1237 Silesia, Poland

Two differences can be observed in the case of segmentations presented in Table 1. Firstly, they differ by the number (between 1 and 9) of socioeconomic factors that were found to be important for characterisation, and secondly by the factors that were found to be important themselves. Among the most commonly included socioeconomic factors were income (10 out of 11), age (8 out of 11), education (6 out of 11), number of people in the household (6 out of 11), ownership of the housing (5 out of 11), and gender (5 out of 11). The selected and presented examples of segmentation do not include all of the previous scientific outputs related to the attempts at dividing the energy consumers into more homogeneous groups. Nevertheless, it can be assumed that they are a representative sample that allows for distinguishing three main trends, dividing segmentations (regardless of their type) into those that:

- refer to a specific type of solution (RES, green energy), action, or potential intervention [59,60,77–81,83–87],
- attempt to indicate how the characteristics of individual householders and their mutual dynamics will influence the behaviours related to energy consumption and saving [88,89],

- focus on the reasons, conditions, and motivations for taking or not taking actions aimed at saving energy [61,62,69,82,90,91].

Focusing on motivations rather than specific types of interventions provides greater insight into how and why consumers make decisions. That knowledge may be utilized to develop a universal strategy for engaging and motivating consumers to take different types of action tailored to their abilities and preferences. This was the goal of the authors for preparing their own behavioural segmentation for the eco-bot project (more details about the project itself [1,92,93]). The inclusion of socioeconomic factors in the characteristics of the segments (although they do not constitute the basis for distinguishing individual types of consumers per se) may allow for a better understanding and prediction of their behaviours. A more detailed description of the segments may allow stakeholders to better design and prepare their interventions.

In the following sections, socioeconomic factors relevant to the behavioural segmentation of energy consumers proposed by the authors will be indicated (in Section 4) and confronted with examples of factors relevant for previous segmentations (Section 5).

3. Materials and Methods

The analysis presented in this article is based on empirical data obtained during a quantitative survey carried out by the CAWI method using a structured questionnaire survey. The gathering of data funded by the eco-bot project and conducted between March and June 2021 was commissioned to two research agencies: DRB Polonia and SW Research. A total of 4506 interviews were conducted with a representative sample of energy consumers in 8 European countries: the Czech Republic ($n = 500$), France ($n = 500$), Germany ($n = 572$), Greece ($n = 500$), the United Kingdom ($n = 512$), Poland ($n = 900$), Romania ($n = 500$), Spain ($n = 522$). The sample was selected using the quota method, taking into account the selection of individual participants in terms of gender, age, and region of residence (urban, rural). The assumptions of the conducted survey were based on a sample size of a minimum of 500 respondents in each country. The survey questionnaire was collected over a two-week period, and due to a high interest in the survey, in some countries more questionnaires were received than initially specified. The additional data received were analysed and selected to meet the conditions of quota representativeness. The authors therefore decided that the obtained material would enrich the conducted analysis, hence the extra questionnaires were not rejected. It should be mentioned here that we worked with relative and not absolute values, therefore the analysis performed and its results were not affected by the disproportionate sample. The average time to complete the questionnaire was 15 min. The countries of the consortium partners were primarily selected for the study due to the nature of the eco-bot project and the participation of some project partners in the pilot phase (Spain, Germany, the United Kingdom). In addition, the sample was expanded geographically, taking the following criteria into account:

- inclusion of the widest possible range of countries in the study in terms of geographical location, hence representatives of Central and Eastern Europe were also selected;
- inclusion of countries with a diversified energy mix;
- inclusion of countries due to sociocultural, income, lifestyle, climate, and energy price differences.

The survey mainly focused on identifying the main motivations, opinions, and declared behaviours regarding energy saving, as well as consumers' attitudes towards various IT tools supporting energy management at home. The authors received a raw database of aggregated and anonymised data in the SPSS/MS Excel format for analysis. The survey contractors have all the remaining data, i.e., the completed online survey questionnaires of individual respondents, which they are obliged to archive for a period of one year.

3.1. Behavioural Segmentation—Novel Approach and Assumptions

The authors' behavioural segmentation of consumers was performed on the basis of respondents' answers to selected questions (prepared in terms of the hypotheses set—a priori approach to segmentation). All the assumptions of this segmentation, along with the methodology and the individual stages of the procedure used, are described in detail in our paper [1]. On this basis, five segments were distinguished, differentiating respondents in terms of their motivation to save energy:

Ecological Idealist (EI)—characterised by the highest environmental awareness, which is the main motivation for their actions. They manifest leadership qualities—they can be leaders and initiators as well as ambassadors inspiring others to take pro-environmental actions. They are very often financially involved in ecological activities.

Aspiring Ecologist (AE)—characterised by high environmental awareness, however, they are also very prone to follow trends, fashion, and the example of other social groups. Consumers assigned to this segment are willing to pay more for ecological products but are less motivated to behave and look for pro-environmental solutions on their own.

Dedicated Saver (DS)—representatives of this segment show average environmental awareness but are mainly motivated to act by potential financial benefits. They are often very well informed and have a wide knowledge of ecology which they are able to use (devoting their time) if it provides a chance at cost savings, even achievable in the long term. Potentially, over time, they could become representatives of EI or AE segments.

Opportunist (O)—consumers with very low ecological awareness, relatively uninvolved in pro-ecological behaviours and activities, are assigned to this segment. They may act occasionally, irregularly, and show pro-environmental behaviours under the influence of financial or ecological incentives, but only under the condition that these actions are easy to perform and convenient. Consumers in this segment are satisfied with their attitude and very often do not want to change it.

Indifferent (I)—this segment is characterised by a complete lack of environmental awareness, showing a complete lack of interest in and concern for energy consumption levels. Representatives assigned to this segment do not show any signs of motivation to change their behaviours to more sustainable ones. Hence, this segment is extremely resistant to any financial or worldview incentives that could influence the attitudes and behaviours of this type of people.

According to the behavioural segmentation procedure, the following distribution of the respondents was obtained (see Figure 1): EI—28.9%, AE—15.3%, DS—43.5%, O—4.2% and I—4.1%. As can be seen, 4332 people were unambiguously assigned to the individual segments. Only 3.8% of the total number of respondents (174 persons) were not unambiguously classified into the indicated groups, which is a very good result and validates the applied assumptions of segmentation procedure.

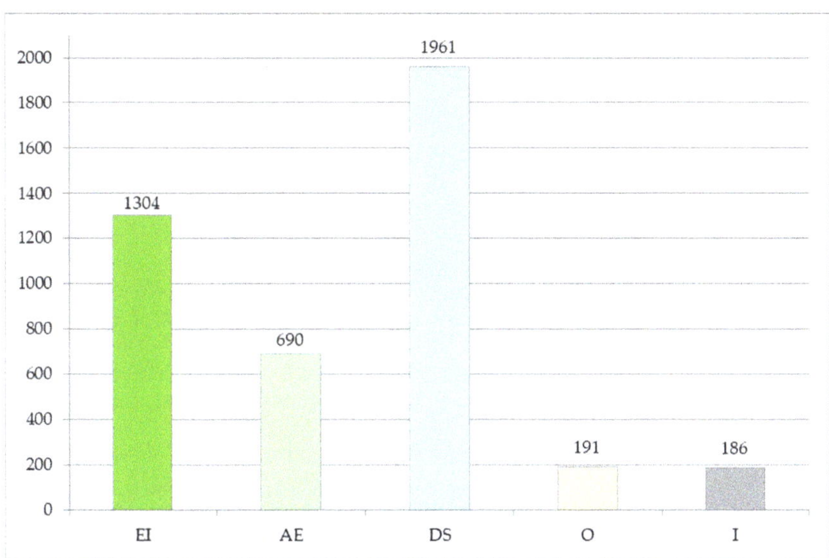

Figure 1. Distribution of respondents assigned to particular behavioural segments.

3.2. Identifying the Relationship between a Behavioral Type and Socioeconomic Factors—Statistical Approach

The aim of this paper is to characterise the created behavioural types of energy consumers by taking into account socioeconomic factors that were not applied during the segmentation procedure. The actual analysis was therefore performed for pairs of variables, where each time one of them was segment membership (representing the respondent's behavioural type) and the other was a selected socioeconomic factor describing the respondents. The intention of the study was both to identify the relationship between these variables and to measure similarities between their categories. Therefore, the study proceeded in the following steps:

Step 1. The hypothesis about the independence of the studied variables was verified using the χ^2 test, assuming a significance level of $\alpha = 0.05$. Moreover, in the case of dependencies, the strength of the relationship between the studied variables was determined using the Cramer's V coefficient.

Step 2. Correspondence analysis was performed for pairs of variables that were found to be dependent to indicate which categories of these variables are related.

Step 3. An auxiliary dendrogram to facilitate the interpretation of the results of the correspondence analysis was made using Ward's method.

Achievement of the paper's stated goal was possible mainly through the use of correspondence analysis. Benzécri (1973) [94] was the precursor of this method, but it was popularised mainly by Greenacre (1984) [95]. Hoffman and Franke (1986) [96], Carrol, and Green and Schaffer (1986) [97] proposed the first economic applications concerning marketing research.

Correspondence analysis is an exploratory technique for examining contingency tables, which aims to transform the points representing the rows and columns of this table into a space with a lower dimension, in which it is easier for the researcher to observe certain regularities. The purpose of this analysis is to graphically present the relationships between the categories of variables under study, which permits forming conclusions about the relationships occurring between these categories.

It is worth emphasising that the statistical methods the authors used for the analysis are dedicated to non-metric variables, as they often allow interesting conclusions to be drawn from empirical surveys. The possibility to visualise the results, which leads to easy interpretation, is another advantage and reason for using these methods.

The general scheme of operation in correspondence analysis can be presented by the following points:

1. A correspondence table, or relative frequencies matrix, is created from the contingency table.
2. The columns and rows of the correspondence table are transformed separately to obtain points (called row and column profiles) representing the categories of non-metric variables under study.
3. A space with a smaller dimension is designated, to which the points (profiles) obtained in a previous phase are projected (with possible rotation). The choice of space and its rotation is made in such a way as to minimise the loss of information contained in the original data.
4. The so-called correspondence map, which is a graphical presentation of the relationship between the categories of the studied variables is created.
5. Appropriate conclusions about the dependencies involved are drawn on the basis of the map, which constitutes the interpretation of the obtained results.

The stage of inferring dependencies between categories of examined nominal variables takes place on the basis of the arrangement of points, representing these categories, presented on the correspondence map. Unfortunately, it is not always possible to draw clear conclusions from the obtained maps. In such cases, cluster analysis methods are generally used as an auxiliary. In this work, Ward's hierarchical method was used [98].

As this article focuses on the applicability of the proposed segmentation procedure, the authors do not describe the methodology of the statistical tools used: the chi-square test and Cramer's V coefficient, nor do they present in detail the subsequent steps of correspondence analysis. These methods are well known, and their detailed description can be found in many references. The chi-square test and measures of dependence for nominal variables are described, among others, in the works of Cramér (1946) [99] and Brzezińska (2011) [100], while correspondence analysis are described in papers by Greenacre (2021) [101], Rozmus (2004) [102], and Stanimir (2005) [103].

4. Results

The respondents divided, according to the authors' behavioural segmentation [1], into five segments (EI, AE, DS, O, I) which differed in terms of their motivation to save energy, since this was the basis for assigning them to the right groups. However, the question is whether they are also significantly differentiated by other characteristics, such as age, employment status, or country of residence. The results of the analysis, carried out according to the procedure described in Section 3, show that most of the studied socioeconomic factors are significantly related to the behavioural type of energy users, represented by the segment to which they have been classified. As the results of the chi-square test (presented in Table 2) indicate, 10 out of 12 socioeconomic characteristics (empirically extracted factors) show a significant correlation with assigning the consumer to a specific segment (in each case the p–Value was lower than the adopted significance level, equal to 0.05). In addition, the calculated Cramer's V coefficient allows for measuring the strength of this relationship, which is, however, weak in each of the studied cases ($V \in (0.04; 0.3)$).

Only the type of the respondent's dwelling (house, apartment, etc.), as well as the average monthly electricity costs in a household, showed no significant correlation (factor 11 and 12 in Table 2). The first factor (type of housing) can be explained by the fact that there are now more opportunities to invest in RES or take other energy-saving measures in apartments and other types of collective housing and not only in detached or terraced houses, as was the case only a few years ago. An example of such energy-saving measures is the investment by the French energy company Électricité de France (EDF) in the town of Alès, where a photovoltaic installation was located on the roof of a residential block. This photovoltaic installation will directly supply energy to 100 households on a self-consumption model and the estimated annual savings could be around €100 for each family [104,105]. Another solution is the introduction of the so-called virtual prosumer option, where a person who does not have sufficient space to instal their own RES may join with another prosumer or a person who is considering such an investment and has the possibility to install required utilities. Such solutions available in the USA, Lithuania, Greece, Italy, Cyprus, and France allow more flexibility for energy consumers in terms of energy-saving measures regardless of the type of property they own. Hence, this factor may not have an impact on consumer profiling, as shown by our study.

Table 2. Results of the chi-square test between the respondent's type/segment and individual socioeconomic factors, along with the values of the Cramer's *V* coefficient.

Factor	χ^2	p-Value	V-Cramer
(1) attitude towards saving energy	372.87	0	0.293
(2) age	248.81	0	0.138
(3) employment status	209.04	0	0.110
(4) country	181.82	0	0.102
(5) the ownership status of the premises	26.42	< 0.001	0.078
(6) number of people in a household	98.14	< 0.001	0.075
(7) average monthly income per person in a household	62.85	< 0.001	0.060
(8) education	53.49	< 0.001	0.056
(9) gender	25.25	0.001	0.054
(10) place of residence	16.24	0.04	0.043
(11) type of a dwelling (a flat/a detached house, etc.)	6.15	0.63	–
(12) average monthly electricity costs/bills in a household	23.16	0.51	–

The cost of electricity for households depends on its price and level of consumption. However, within the EU, both very large price differences and variations in consumption between countries can be observed [106]. In both cases, this is mainly due to the impact on price and consumption of a number of supply and demand conditions, such as the geopolitical situation of a given country, the characteristics of the national energy mix, diversification of energy imports, varying weather conditions, and all kinds of end-user taxes, network charges, or environmental protection costs [107]. Furthermore, the price of electricity relative to purchasing power parity can significantly alter the perceived cost of electricity for the individual consumer. At the same time, high energy consumption when it is the so-called green energy may not be perceived by consumers as something negative (the rebound effect). All these conditions may result in the lack of a demonstrated significant relationship between electricity costs and segment-specific consumer motivation to save electricity.

The results presented in Table 2, showing a significant relationship of the behavioural type of energy user with the 10 studied factors (1–10 in Table 2), indicate that conducting a correspondence analysis for these variables would be valid. This will allow to combine categories of the studied variables. As already mentioned, one of them is the segment into which the respondent is classified, and the other is one of the ten socioeconomic factors. The interpretation of the obtained results allows for characterisation of each of these segments, which in turn forms the basis for creating profiles of typical representatives of these classes.

The strongest dependence determining the classification of an energy consumer into a segment, as shown in Table 2, is whether the surveyed respondent declares taking any energy-saving measures (factor 1 from Table 2). Due to methodological limitations of correspondence analysis, it was not possible to create a correspondence map in this case (this is due to the fact that the variable representing the answers to the question about energy saving has only two categories: "yes" and "no"; the reduction of multidimensionality which is necessary here leads to the creation of only one dimension, which precludes the drawing of a two-dimensional correspondence map).

From the distribution of answers, it can be seen that energy saving is most often declared by respondents assigned to the DS segment, as well as by ecological idealists (EI segment). The group of consumers who declared not taking any steps towards energy saving included those assigned to groups: EI (1.3% of the total sample), AE (1.4%), and DS (2.3%). However, their percentage share in each segment does not differ significantly. When analysing the reasons given by respondents for not taking any steps to save energy, a certain consistency can be observed between the segments of consumers characterised by ecological motivation. Namely, both idealists (EI) and aspiring ecologists (AE) gave similar reasons for not saving energy. Lack of time was indicated as by far the main reason for not taking action, and to a large extent, representatives from both segments declared that both lack of relevant information on how to save energy and lack of capacity to make changes are important obstacles for taking energy-saving actions. Aspiring ecologists (AE) also strongly indicated that other household members were not interested in energy saving, which influenced their attitude. This should not come as a surprise since, as shown by the characteristics of the aspirants (AEs),consumers classified in this segment are very often guided in their actions by the opinions of other people, especially their family and friends. Therefore, the negative example of household members could be considered as a deterrent to making any effort. The last reason indicated by the segment of idealists (EI) was, to a large extent, the lack of technical or practical support. The reasons for not saving energy provided by consumers classified in the financially motivated segment (DS) seem to be interesting. Here, the main reason was the lack of technical and practical support and, as in the case of the AEs, the lack of interest in energy saving by other family members. It can be assumed that the representatives of this segment decided that a single effort, without the support of other household members, to reduce consumption would not significantly affect the reduction of bills, and therefore they did not take any actions in this direction.

The obtained results can also be analysed by extracting the individual countries participating in the survey from the sample. Interestingly, from this angle, in almost all countries the greatest number of people declaring energy savings were in the DS segment. Countries such as Poland, the Czech Republic (over 50%), France, Greece, and Romania (over 40%) recorded a clear advantage of DSs over other segments. Only in Spain, among energy-saving consumers, were ecological idealists (EI) the dominant segment (above 40%). When looking at non-savers (excluding the Indifferent segment, which is generally characterised by a passive attitude), Germany (14.1% of the total) and the United Kingdom (8.1%) recorded the highest number of such cases, and Poland (1.9%) and the Czech Republic (2.7%) the lowest. In all surveyed countries, the largest number of people declaring to be non-saving energy were also classified in the DS sector. The high percentage of non-savers classified as segments other than (I) in Germany can be explained, for example, by the high share of renewable and green energy in the overall energy mix, which may translate into a tendency to consume more energy and not to think about saving it in part of the population. On the other hand, energy costs in Germany are the highest in the entire EU (€0.3 per kWh), but at the same time, Germany is characterised by a dynamic increase in investment in small-scale photovoltaic installations (5 GW capacity additions in 2020 compared to 3.8 GW in 2019), which translates into a reduction in the costs associated with the purchase of electricity by households [108].

Another factor that is significantly related to the behavioural type of the energy user is the age of the respondent (factor 2 of Table 2). On the basis of the correspondence map (Figure 2), it is easy to notice that the ecological attitude to energy saving (EI and AE segments) is most often characteristic of people aged 26–40 years. In contrast, reducing energy consumption for financial reasons is most notable for people over the age of 40. The youngest people (aged 18–25) most often display opportunistic behaviour or are not interested in specific activities that could reduce their energy consumption.

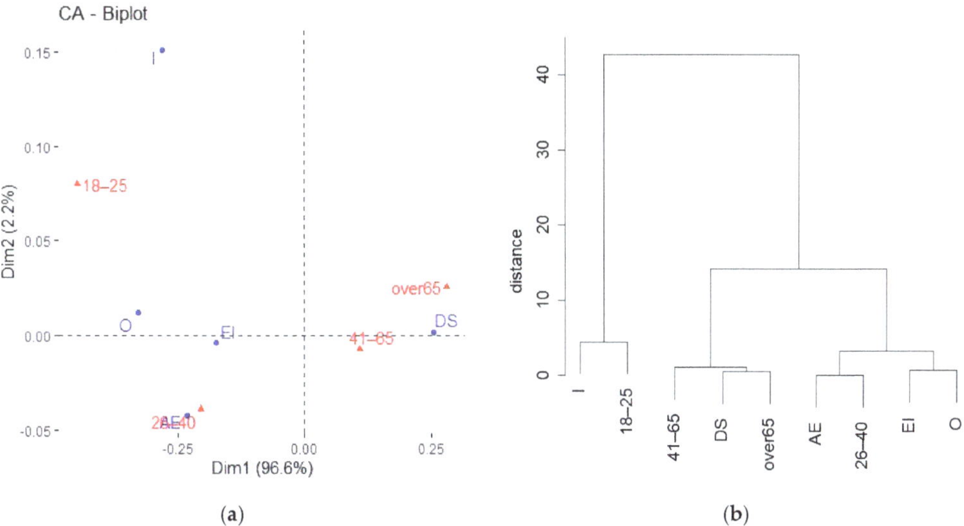

(a)   (b)

Figure 2. Correspondence map (**a**) and an auxiliary dendrogram (**b**) showing the dependence of each behavioural type of energy consumer on their age groups.

Looking again at the results of individual countries, it can be noticed that in almost all countries the characteristics of the segments in terms of age groups coincide with the typical representative of the classes (see Table 3). Interesting differences can be observed in the case of Poland and Spain. Poland is dominated by the DS sector in each age bracket, which should not come as a surprise, since the analysis of correspondence of individual segments with the respondent's country shows that the Polish are the closest to consumers motivated to save energy by financial factors. This can be explained by the fact that Poland is among the EU countries with the highest energy price calculated according to the purchasing power parity. In Spain, on the other hand, the dominant segment in almost every age group is the ecological idealist (EI). Only those over 65 are characterised by a financial approach and were mostly classified in the DS sector. However, this may change as Spain has been experiencing significant energy price rises recently, with a new time-of-day billing system being introduced across the country in June 2021. Differences in rates during the day can reach up to 50%. It is therefore likely that the Spanish will probably start to pay more attention to their electricity bills [109].

Analysing another factor significantly influencing the behavioural type of energy consumer, employment status (factor 3 from Table 2), the following relationships can be identified (Figure 3):

- people who save energy for ecological reasons (EI and AE) most often work full-time and part-time; respondents who are self-employed and even unemployed also share an ecological idealist attitude (EI);
- financial motivation (specifically for the DS segment) is typical for retired people, which partially confirms the results of the survey taking into account the respondents' age. This is also easily explained by the fact that in this occupational group, in most cases, there is a decrease in income. As a result, consumers are becoming more financially sensitive, also in relation to environmental and especially energy issues. Those two aspects are usually closely linked to energy prices, which are often the biggest burden on household budgets;
- housekeepers are very often indifferent (I) to the motion of reducing energy consumption;
- students are the most opportunistic (O) consumers.

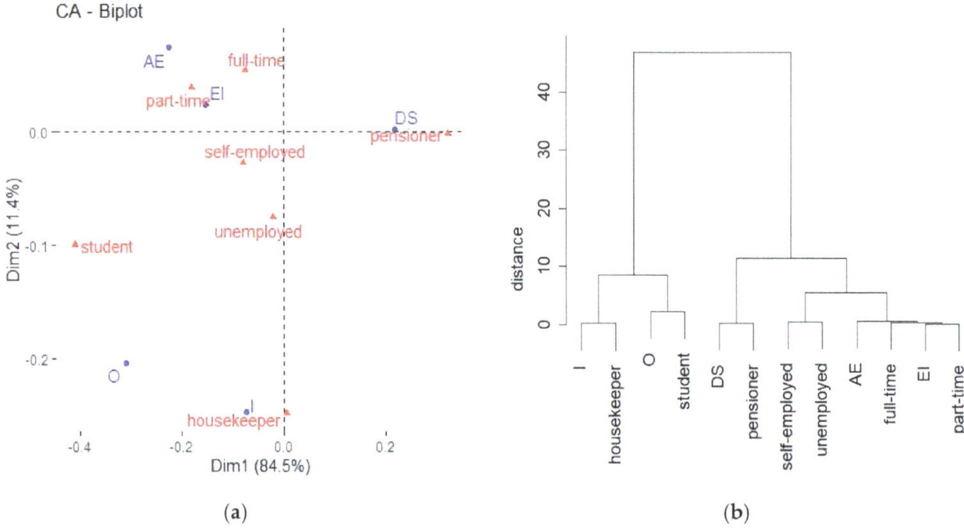

(**a**)   (**b**)

Figure 3. Correspondence map (**a**) and an auxiliary dendrogram (**b**) showing the dependence of each behavioural type of energy consumer on their professional status.

Studying individual countries in detail, the results point to Germany and Spain as the countries that have the closest segment distribution to the typical representative obtained by correspondence analysis. In Central European countries (Poland, the Czech Republic) a slightly different distribution of respondents according to segment affiliation is observed. In these countries, the dominant segment, in almost every occupational group, is the DS. It should be emphasized that in these countries ecological, environmental, and climate protection issues are becoming relatively recent themes in public debates. Hence, it can be assumed that the level of environmental awareness and the ideological motivations of consumers to save energy will only become more apparent in the future.

It is worth emphasising again that, due to surveys being carried out in eight European countries, it was possible to find out what motivates respondents from different European countries to save energy (factor 4 of Table 2). As the results of the correspondence analysis illustrated in the chart (Figure 4) show, financial motivation (DS) is most often the key motivation for the Poles, French, and Czechs. The distribution of answers indicates that the characteristics of aspiring ecologists are relevant for the Greek population, but also for Romania. Respondents from the UK and Spain, on the other hand, generally reduce their energy consumption out of concern for the environment and ecology (EI). People from Germany were found to be the most opportunistic, however, it should be noted that the behaviours characteristic of both ecological idealists and opportunists are not as clear-cut as those of aspiring environmentalists and dedicated savers. This is indicated by the greater distance between the points representing the different categories of examined characteristics. Interestingly, the behaviour of indifferent consumers cannot be linked to any of the countries under study. Such people are certainly present in the analysed group, but it is difficult to unambiguously assign them to any nationality.

The assignment of energy consumers to the relevant behavioural segment is also significantly related to whether the respondent owns their flat or house or lives in a flat/building that they rent (factor 5 in Table 2). This is quite an understandable relation, as it is logical that non-owners are much less inclined to invest in solutions that can significantly reduce their energy consumption. These people are more likely to carry out simpler and less costly activities, which, however, are not as efficient or are perceived as natural everyday habits/routines and are not associated with significant energy savings.

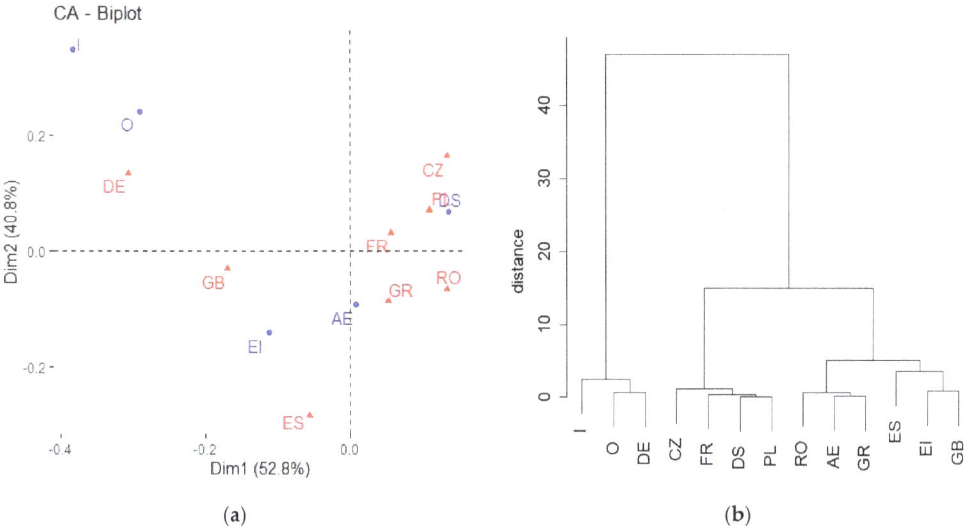

(**a**) (**b**)

Figure 4. Correspondence map (**a**) and an auxiliary dendrogram (**b**) showing the dependence of each behavioural type of energy consumer on their country of residence.

For this factor, it was also impossible to create a correspondence map (factor 5 has only two categories, which precludes making a two-dimensional map). However, the analysis of the distribution of answers shows that, among those who declare to be flat and house owners, most show the characteristics of a dedicated saver, although this group also includes respondents with a pro-environmental attitude towards energy saving (both EI and AE). Tenants are more often opportunists and indifferent people. Due to the relatively weak influence of this factor on segment profiling, the links between the different categories of variables studied are not very obvious.

Considering individual countries, the case of Germany is interesting, where 63.8% of respondents declare that they rent their flats or houses. The dominant tenant sector in this case turned out to be DS, unlike the owners classified as EI. In the other surveyed countries, the situation is quite the opposite, as most respondents own their property. Romania and Poland are the leaders with 90.7% and 81.9% of declarations claiming ownership, respectively. Moreover, in these countries, both the owner and the tenant have mostly been assigned to the DS sector. This distribution is not surprising, looking at the current housing and cultural situation occurring, e.g., in Poland, where there is still a deficit of housing stock, especially for renting. Poles prefer taking out loans to buy their own property rather than rent it, which is still perceived as a temporary solution in extraordinary situations. Moreover, Poles have little knowledge of long-term investing. Combining those facts with a lack of ability to assess the real and full costs of buying and maintaining a flat, most often bought on credit, compared to the cost of renting, as well as high rental prices, translate into this type of preference [110]. On the other hand, highly developed countries are characterised by a flexible approach to rental housing that can adapt to changing housing needs and living situations, including work situation or the family's financial capabilities.

The number of persons in the household (factor 6 in Table 2) is another one significantly differentiating respondents assigned to particular segments. The results of the correspondence analyses (Figure 5) show that:

- two-person households are characteristic to the dedicated saver segment; however, those living alone, as well as those forming multi-person households (seven persons), are most financially motivated to save energy;

- aspiring environmentalists (AE) are most often representatives of five-person households and environmental idealists (EI) of four-person households; however, members of three- or six-person households are also characterised by pro-environmental motivation to reduce energy consumption and can be classified as both AE and EI;
- for this analysis, it is not easy to indicate how many people the opportunist households consist of; moreover, it is not possible to assign to any category those in the indifferent segment, nor those in households consisting of eight or more people.

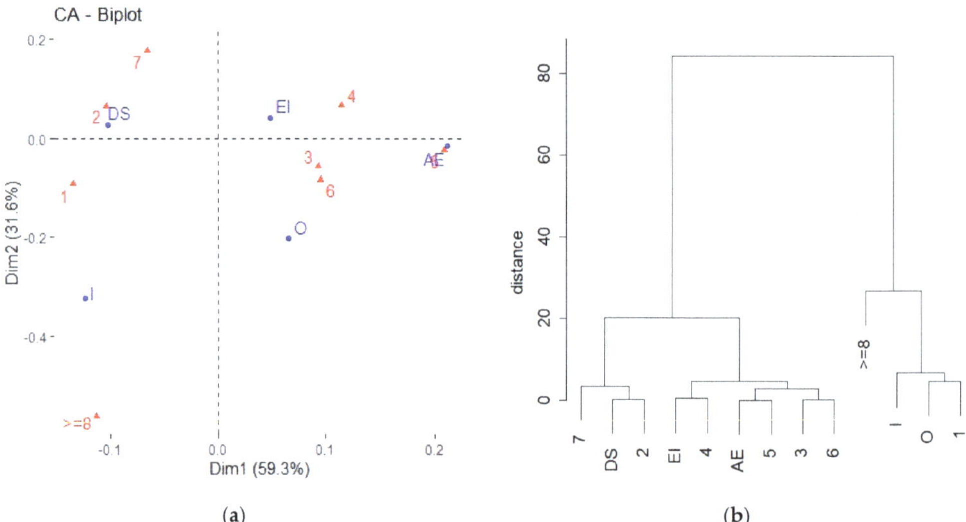

Figure 5. Correspondence map (**a**) and an auxiliary dendrogram (**b**) showing the dependence of each behavioural type of energy consumer on the number of persons in the household.

The results of the chi-square test showed that the income of the respondents was also related to the segments into which they were classified (factor 7 from Table 2). Income ranges shown in the graph have been calculated against the average earnings in each country, as reported by Eurostat [111]. In the course of the study, four income brackets were assumed (0–0.4 of the average earnings, 0.41–0.7 of the average, 0.71–1.4 of the average, and above 1.4 of the average), taking into account the analysis of minimum and average earnings per person in a household in each country. The authors attempted to set the brackets in such a way that the distribution of respondents by income would reflect as closely as possible a cross-section of the population in a given country in terms of earnings.

The arrangement of points representing categories of these two variables (Figure 6), indicates the occurrence of the following relationships:

- monthly per capita income in a household below 0.7 of the average earnings in a given country is characteristic of people in the DS segment;
- people who save energy for pro-ecological reasons (EI and AE) generally have a higher disposable income than DS; they range from 0.71 to 1.4 of the average earnings;
- Opportunists and indifferent respondents most often refused to provide answers about their income.

When analysing individual countries, the highest percentage of respondents who did not answer the question about income were from Poland (11.3% of all Poles participating in the survey) and Greece (10.5% of all Greeks). In both cases, most were classified to the DS sector. Taking the income factor into account, it can be noted that in all surveyed countries the distribution of respondents with respect to sectors was very close to the obtained typical representative (Table 3).

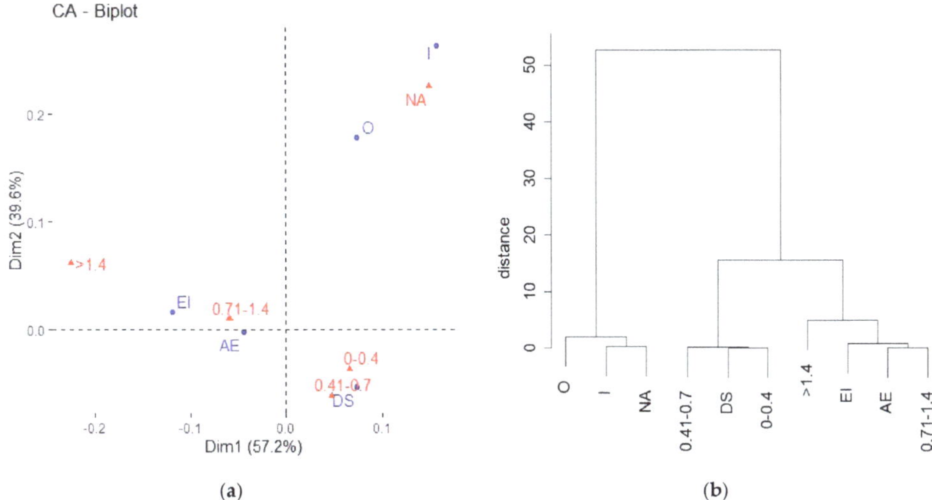

Figure 6. Correspondence map (**a**) and an auxiliary dendrogram (**b**) showing the dependence of each behavioural type of energy consumer on their average monthly income.

Examination of the relationship between the behavioural type of energy consumer and their education (factor 8 of Table 2, Figure 7) shows that people whose motivation to save energy comes from pro-environmental considerations tend to be better educated than dedicated savers or opportunists. Respondents assigned to the EI and AE segments most often have a bachelor's or master's degree, while most people in the DS segment have secondary education. Indicating the level of education for the typical opportunist is not so clear-cut, although the correspondence analysis, as well as the auxiliary dendrogram, most closely associate these individuals with primary education. Indifferent respondents most often declared other education. It can also be added that a doctoral degree was not typical for any of the segments.

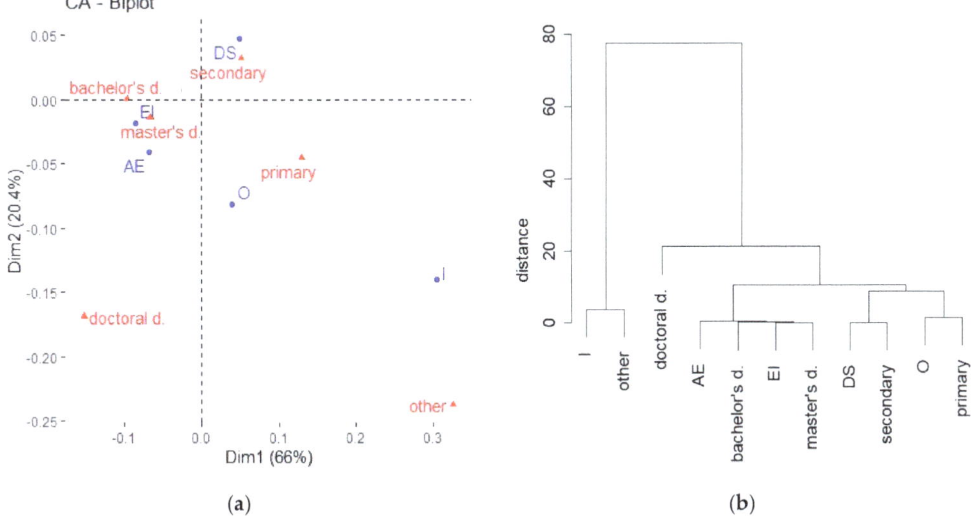

Figure 7. Correspondence map (**a**) and an auxiliary dendrogram (**b**) showing the dependence of each behavioural type of energy consumer on their education.

The obtained distribution coincides to a large extent with the results of studies by other authors [61,69,78,80,81], where an increase in environmental awareness and attitudes among consumers is observed as they attain higher degrees of education. Environmental and ecological education helps people to make informed choices. Activities designed in the educational policies of European countries are aimed at shaping people's behaviours to be more sustainable and environmentally friendly. Better educated consumers more often show pro-ecological attitudes, intentions, and behaviours. On the other hand, the obtained results confirm the necessity to reach those less educated and excluded with the appropriate message and reliable knowledge. The authors see great possibilities for action and positive influence of regional administration together with other local stakeholders, who have the potential to become focus and facilitators for communities of residents and family circles. Applications dedicated to energy management, such as the eco-bot [93], may, thanks to their simplicity, be used as an effective educational tool that will change the user's habits.

When analysing another factor showing a significant relationship (factor 9 from Table 2), it was noticed that due to the presence of people declaring a different gender among the surveyed energy consumers, arrangement of points presented on the correspondence map (Figure 8) is not very clear. For this reason, the auxiliary dendrogram was used to interpret the results: It shows that when it comes to saving energy, mainly women are driven by financial motivation (DS), while menare most often motivated by pro-ecological considerations (EI and AE).

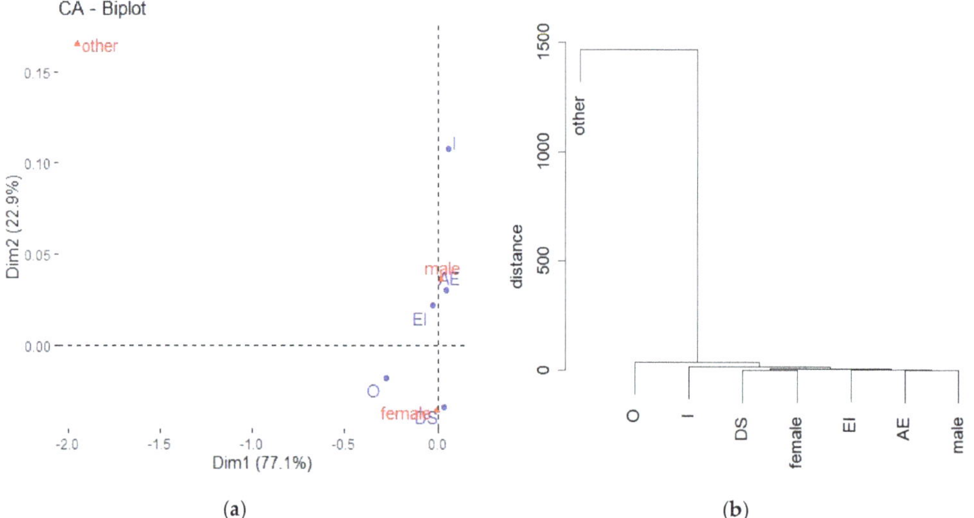

Figure 8. Correspondence map (**a**) and an auxiliary dendrogram (**b**) showing the dependence of each behavioural type of energy consumer on their gender.

It can be assumed that such classification results from the fact that in households it is mostly women who take care of household matters, including finances. They control family expenditures and initiate pro-saving measures. Hence, they are very familiar with energy-saving opportunities and benefits, so their actions, behaviours, and even opinions may be financially motivated. Men, on the other hand, may look at environmental issues from a broader, long-term perspective, concerned about family security, and their pro-environmental motivations may stem from their desire to ensure healthy and better living conditions for their family.

The last factor that shows a significant dependence with the analysed behavioural segment is the place of residence (factor 10 from Table 2). The results of the correspondence analysis (Figure 9) show that:

- people from the ecological idealist segment usually live in the suburbs; it turns out, however, that this location is typical for the opportunists as well;
- dedicated savers are more likely to live in rural and urban areas than in the suburbs;
- the place of residence of a typical aspiring ecologist is not as obvious as in the case of respondents from the already mentioned segments, although people from this group most often indicated that they live in the rural areas;
- once again, concerning their place of residence, the indifferent respondents are ambiguously assigned.

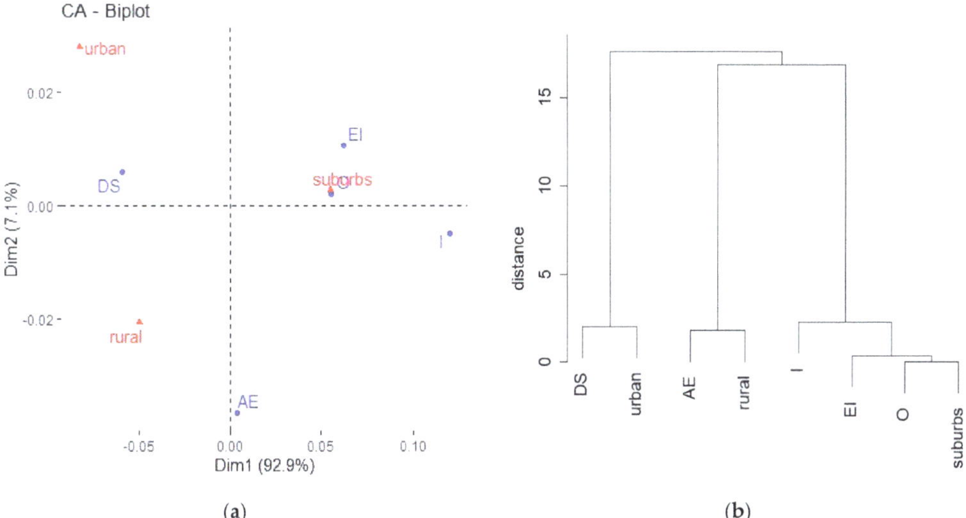

Figure 9. Correspondence map (**a**) and an auxiliary dendrogram (**b**) showing the dependence of each behavioural type of energy consumer on their place of residence.

The location of residence of people characterised by pro-ecological motivations (suburbs, rural areas) results mainly from greater ecological awareness and following the currently fashionable trends, like the so-called "escape to the countryside", the desire for greater contact with nature, seeking silence and peace, as well as breaking away from the hustle and bustle of the city and switching to the increasingly popular slow lifestyle. The dedicated saver is a city dweller. The financial motivation is not surprising here either—the rising prices of real estate and maintenance costs, shrinking areas suitable for habitation, and scarcity of natural resources (often in cities there are no or only limited possibilities to introduce ecological solutions, e.g., water recycling or RES installations), result in increased expenses for satisfying basic living needs. As [112] noted, despite the fact that city dwellers adopt and declare, to a large extent, pro-environmental values, their attitudes, behaviours, and actions are less pro-ecological than those of rural residents. The actions of the latter are more consistent with their expressed worldviews. This is also confirmed by the presented results.

5. Discussion

Using the results obtained from the analysis of socioeconomic characteristics of individual behavioural segments, an attempt was made to identify the profile of a typical representative for each segment. At the same time, the obtained profiles, presented in Table 3 (in the first row), were compared with previous works on behavioural segmentation of energy consumers. Unfortunately, it quickly turned out that such a comparison is only possible to a limited extent. In order to ensure the highest possible comparability of the obtained profiles, only segmentations with a certain level of similarity between the designated segments and the novel segmentation approach presented by the authors were selected. Despite the earlier indication (Section 2 and [1]) of such similarities, it turned out that for some segmentations a comparison of typical segment representatives in terms of socioeconomic characteristics is not possible due to insufficient representation of this aspect of the consumers studied by other researchers. For example, the Accenture end-consumer observatory on electricity management (2010) [79] identifies the following characteristics of six proposed segments: Eco-rationals—more often women, Proactives—no significant socioeconomic characteristics, Cost conscious—more often women, Pragmatics—more often men (two segments with financial motivation), Skepticals (equivalent to Opportunist)—higher income, and Indifferents—more often men with lower income under 24 years old. However, these descriptions mainly contain a reference to one socioeconomic characteristic, which was considered insufficient to include this segmentation in the comparisons. In the case of another segmentation that parallels the currently proposed one in terms of motives, the results of the study described by Tabi et al. (2014) [60] indicated that most socioeconomic characteristics (age, gender, income, household size) were evenly distributed across all segments, and only education was a distinguishing factor for the segment motivated by pro-ecological beliefs. In other segmentations, despite the occurrence of partial similarity between segments and motivation, the number and nature of factors motivating consumers to save energy were so different from those taken into account by the authors that this supported the removal of these works from the compilation [59,61]. Finally, for the comparison included in Table 3, three of the behavioural segmentations indicated in Section 2 were selected, which included, in addition to the factors used to prepare the segmentation, socioeconomic characteristics [69,78,80]. Segmentations of energy consumers that did not allow a comparison of segments due to differences with the approach to motivation were not taken into account.

Table 3. Characteristics and comparison of a typical representative in selected existing energy consumer segmentations.

	A Typical Representative of the Behavioural Segments				
Study	**Ecological by Conviction**	**Eco-Friendly But with Other Focus**	**Focused on Costs and Money Saving**	**Comfort and Convenience Focused**	**Indifferent**
Proposed segmentation	(EI) most often declared energy-saving male	(AE) male	(DS) most often declare energy-saving female	(O)	(I)
	age 26–40	age 26–40	over 40 years of age, especially over 65 years of age	age 18–25	
	resident of the suburbs	rural resident	urban and rural dweller	resident of the suburbs	
	employed full-time or part-time	employed full-time and part-time	pensioner	student	housekeeper
	flat or house owner	flat or house owner	flat or house owner	tenant	tenant
	bachelor or master's degree	bachelor or master's degree	secondary education	primary education	other education
	4 people in household	5 people in household	1 or 2 people in household		
	0.71–1.4 of earnings average	0.71–1.4 of earnings average	0–0.7 of earnings average	refuses to answer questions about earnings	refuses to answer about earnings
	The United Kingdom, Spain	Greece, Romania	France, the Czech Republic, Poland	Germany	

Table 3. *Cont.*

Study	A Typical Representative of the Behavioural Segments				
	Ecological by Conviction	**Eco-Friendly But with Other Focus**	**Focused on Costs and Money Saving**	**Comfort and Convenience Focused**	**Indifferent**
Han et al. (2013) [80]	Environmentally minded residents (EMR)	Conscious residents (CR)	Cost focused residents (CFR)	Ease-driven residents (EDR)	
	age does not differentiate	35 years and more	age 27–35	age 35–59	
	highest income	income average and higher	lower income on average	income slightly above average	
	highest education		average education		
	more often owners	more often tenants	more often tenants	more often owners	
Sütterlin et al. (2011) [69]	Idealistic energy savers (IES)	Selfless inconsequent energy savers (SIES) and problem-aware wellbeing-oriented energy consumers (PAWOEC)	Thrifty energy savers (TES)	Convenience-oriented indifferent energy consumers (COIEC)	
	well educated (higher technical education)	two types of semi-environmental consumers but only one (PAWOEC) was distinguished by higher general education	more often vocational or secondary education		
	more often female			more often male	
			the oldest lowest income		
Pedersen (2008) [78]	Devoted conservationists (DC)	Stumbling proponents (SP)	Cost-conscious practitioners (CCP)	Comfort seekers (CS)	Tuned-out and carefree (TOaC)
	the oldest segment—6 out of 10 consumers are over 65 years old		"the average segment"	age 35–44	younger than 45 years
	lowest income	high income			highest income
	retired				
			slightly more females		mainly male
		highest education	least people with higher education		
		rather house dwellers		more often flat dwellers more often urban resident more often live with children and more (3+) people in the household other nationalities (languages)	rather tenant rather from urban areas
	rather from out of town				

Table 3. *Cont.*

	A Typical Representative of the Behavioural Segments				
Study	Ecological by Conviction	Eco-Friendly But with Other Focus	Focused on Costs and Money Saving	Comfort and Convenience Focused	Indifferent
Similarities between segment characteristics	mostly higher education	more often higher education	on average lower education	no similarities—the segments here may actually differ in their motivation/approach to comfort and convenience	problems with distinguishing that segment or lack of distinctive features in this segment
	highest income	average and higher income	on average lower income		
	more often owner		more often older		
			more often female		

The last row of the Table 3 identifies, on the basis of the prepared comparison, the socioeconomic characteristics common to the different behavioural segments distinguished in terms of the motivation behind energy-saving behaviours. It should also be noted that certain characteristics indicated as important in the segmentations which are not included in Table 3 overlap to some extent with the authors' findings (e.g., women with low incomes are more likely to represent mainly financially motivated segments [59,79], or higher education is characteristic of the pro-environmental segment [59,60]). The prepared analysis of socioeconomic factors, which complements the basic characteristics of behavioural segments determined on the basis of the distinguished motivation of the consumer, allows for a better definition of individual segments. It can be considered an attempt to extend the behavioural type of segmentation to include socioeconomic factors. It should be emphasised that the aim was not to determine segments on this basis alone (it is not a demographic type of segmentation) but to extend behavioural segmentation with an additional aspect that may make it easier for potential stakeholders to use the tool prepared by the authors.

By analysing and comparing the existing research, it can be noticed that the image of energy consumers is constantly changing. This is related not only to changing fashion or culture but also to changing socioeconomic and environmental conditions. Perception and sensitivity to the problems of modern civilisation, as well as knowledge and willingness to prevent them, play an important role. Consumers increase their environmental awareness through education and the influence of local communities and associations of residents.

However, there is still little research into identifying motivations and attitudes, as well as comprehensive characterisation, of energy consumers. This seems crucial at a time when behavioural change and greater care for the natural environment are becoming a necessity, not only for the present but also for future generations. In order to achieve this, it is necessary to understand the needs, aspirations, and motives of consumers' behaviours as well as prioritise the preservation of natural heritage and encourage the building of the social capital. The authors hope that the results of the presented research will contribute significantly to promoting behavioural changes and raising environmental awareness of consumers as well as serving as a starting point for other researchers' further considerations.

6. Conclusions

The article aimed to identify the main socioeconomic factors specific to the distinct behavioural segments of the energy consumer. The identified factors were used to characterise a typical representative of the created segments. It should be emphasised that the segmentation presented in the paper is an original and innovative concept of the authors, as well as a result of continuing research on the motives of energy consumers' behaviours. The combination of behavioural segmentation with socioeconomic characteristics of the created classes is an approach rarely seen in the literature, hence the results provide a comprehensive picture of the energy consumer. Moreover, the statistical tools used in this paper (chi-square test and correspondence analysis), which allow us to find out significant relationships between behavioural type and the indicated factors, have not been used so far in studies on energy consumption.

In the course of the analysis, the authors managed to positively verify the first, and partially the second, posed research question. Namely, as the research results showed, it is possible to indicate a typical representative for each of the obtained segments, as well as its socioeconomic characteristics. However, it turned out that not all the socioeconomic factors assumed in the study are relevant for describing the typical representative. Two of the twelve characteristics (type of housing and electricity costs) are irrelevant for the differentiation of the segments. In the case of the third research question, it turned out that the picture of the typical energy consumer has changed over time and a comparison of typologies is not entirely possible. The authors were able to identify three other existing behavioural segmentations whose class characteristics are relatively close to those developed and presented in this paper. However, it should be emphasised that in each of the identified segments, only a small number of factors coincided with previous proposals by other authors and that the similarity of factors was not the same for all three compared segmentations.

The approach to a comprehensive analysis of the energy consumer presented in the paper is not free from certain limitations. Some of them have been described in detail in [1] and concern the constructed research tool, which was used both to validate the segmentation proposed by the authors and to characterise the segments. First of all, it concerns the impossibility of including in the questionnaire all contextual factors influencing the attitudes and motivations of consumers and the fact that the answers of the respondents reflect their self-assessment of behaviour and not actually the observed behaviour. Moreover, it should be noted that in the case of some demographic questions, the respondents had the option not to provide the answer, which they sometimes took. Although it was not a large percentage of the respondents, it could have influenced the final results. Of course, the countries chosen for the study can also be seen as a kind of limitation, but the authors tried to select representatives of Europe, taking into account existing economic, social, and cultural differences.

According to the authors, the most important limitation of the presented approach is the necessity of repeating the research and verifying the segmentation obtained. Along with the changing political-economic and sociocultural conditions, it may turn out that the main motivations of energy consumers can change dramatically—which will have an impact on their segmentation.

It is also important to note that the segmentation presented is based on the results of extensive empirical research, including several European countries, which is not a common practice. This comprehensive analysis of individual behavioural types of energy consumers can serve many different stakeholders as a baseline tool for the construction of policies, instruments, and plans that take into account the problems of energy saving, climate change, and dwindling natural resources. Consumers from different European countries themselves, for whom the results of the study can provide interesting insights into their own internal motivations for saving energy and can also be a starting point for considering changes in their behaviour and attitudes, should also be mentioned as stakeholders. In addition, consumer organisations and energy suppliers can use the results to better reach a diverse customer base. This will increase their efficiency in operating, providing tailored products and services, and helping to effectively promote sustainable consumption and attitudes that reduce energy consumption. We also hope that our publication will be of interest to other researchers, and by referring to and discussing our results they will further expand the common knowledge and understanding of the consumers' energy behaviours.

The authors see the possibility of extending the comprehensive analysis of the energy consumer further, for example, by examining the relationship of individual behavioural types with the various types of instruments, measures, and incentives used to support energy management at home, as well as enabling energy savings and behavioural change in a pro-environmental direction. Furthermore, it might be interesting to investigate the influence of social groups on particular segments and their approaches to modern technologies. The authors plan to extend the research in those directions and publish the results in future articles.

Author Contributions: Conceptualization, S.S., J.K.-Ł. and J.T.; methodology, S.S., J.K.-Ł. and J.T.; formal analysis, S.S., J.K.-Ł. and J.T.; investigation, S.S., J.K.-Ł. and J.T.; data curation, J.T.; writing—original draft preparation, S.S., J.K.-Ł. and J.T.; writing—review and editing, S.S., J.K.-Ł. and J.T.; funding acquisition, S.S. All authors have read and agreed to the published version of the manuscript.

Funding: This research was funded by European Union's Horizon 2020 research and innovation program: "Reducing energy consumption and carbon footprint by smart and sustainable use", as a part of the currently implemented project "Personalised ICT-tools for the Active Engagement of Consumers Towards Sustainable Energy. Eco-bot" under grant agreement No. 767625.

Institutional Review Board Statement: Not applicable.

Informed Consent Statement: Not applicable.

Data Availability Statement: Not applicable.

Conflicts of Interest: The authors declare no conflict of interest.

References

1. Słupik, S.; Kos-Łabędowicz, J.; Trzęsiok, J. An Innovative Approach to Energy Consumer Segmentation—A Behavioural Perspective. The Case of the Eco-Bot Project. *Energies* **2021**, *14*, 3556. [CrossRef]
2. Neale, A. Zrównoważona konsumpcja. Źródła koncepcji i jej zastosowania. *Pr. Geogr.* **2015**, *141*, 141–158. [CrossRef]
3. Seyfang, G. The New Economics of Sustainable Consumption. In *Seeds of Change*; Palgrave Macmillan: New York, NY, USA, 2011.
4. Matel, A. Przesłanki ekologizacji konsumpcji z perspektywy zachowań konsumenckich. *Zarządzanie. Teor. Prakt.* **2016**, *16*, 55–61.
5. Kiełczewski, D. Wpływ ekologizacji konsumpcji na zmiany w zarządzaniu organizacjami. *Han. Wew.* **2016**, *6*, 55–63.
6. Lorek, S.; Fuchs, D. Strong sustainable consumption governance—Precondition for a degrowth path? *J. Clean. Prod.* **2013**, *38*, 36–43. [CrossRef]
7. IEAE. *Energy, Electricity and Nuclear Power Estimates for the Period up to 2050, No. 1*; International Atomic Energy Agency: Vienna, Austria, 2019.
8. Steinbuks, J.; de Wit, J.; Kochnakyan, A.; Foster, V. *Forecasting Electricity Demand: An Aid for Practitioners*; World Bank: Washington, DC, USA, 2017.
9. Leal-Arcas, R.; Ríos, J.A.; Grasso, C. The European Union and its energy security challenges: Engagement through and with networks. *Contemp. Politics* **2015**, *21*, 273–293. [CrossRef]
10. Kovacic, Z.; Di Felice, L.J. Complexity, uncertainty and ambiguity: Implications for European Union energy governance. *Energy Res. Soc. Sci.* **2019**, *53*, 159–169. [CrossRef]
11. Yang, Y.; Wang, M.; Liu, Y.; Zhang, L. Peak-off-peak load shifting: Are public willing to accept the peak and off-peak time of use electricity price? *J. Clean. Prod.* **2018**, *199*, 1066–1071. [CrossRef]
12. McLoughlin, F.; Duffy, A.; Conlon, M. Characterising domestic electricity consumption patterns by dwelling and occupant socio-economic variables: An Irish case study. *Energy Build.* **2012**, *48*, 240–248. [CrossRef]
13. Huebner, G.; Shipworth, D.; Hamilton, I.; Chalabi, Z.; Oreszczyn, T. Understanding electricity consumption: A comparative contribution of building factors, socio-demographics, appliances, behaviours and attitudes. *Appl. Energy* **2016**, *177*, 692–702. [CrossRef]
14. Cayla, J.-M.; Maizi, N.; Marchand, C. The role of income in energy consumption behaviour: Evidence from French households data. *Energy Policy* **2011**, *39*, 7874–7883. [CrossRef]
15. Bartusch, C.; Odlare, M.; Wallin, F.; Wester, L. Exploring variance in residential electricity consumption: Household features and building properties. *Appl. Energy* **2012**, *92*, 637–643. [CrossRef]
16. Estiri, H. Building and household X-Factors and energy consumption at the residential sector. *Energy Econ.* **2014**, *43*, 178–184. [CrossRef]
17. Karatasou, S.; Laskari, M.; Santamouris, M. Determinants of high electricity use and high energy consumption for space and water heating in European social housing: Socio-Demographic and building characteristics. *Energy Build.* **2018**, *170*, 107–114. [CrossRef]

18. Sakah, M.; Can, S.D.L.R.D.; Diawuo, F.A.; Sedzro, M.D.; Kuhn, C. A study of appliance ownership and electricity consumption determinants in urban Ghanaian households. *Sustain. Cities Soc.* **2019**, *44*, 559–581. [CrossRef]

19. Mohamed, A.M.A.; Al-Habaibeh, A.; Abdo, H.; Elabar, S. Towards exporting renewable energy from MENA region to Europe: An investigation into domestic energy use and householders' energy behaviour in Libya. *Appl. Energy* **2015**, *146*, 247–262. [CrossRef]

20. Laetitia, M.; Jiyun, S.; Alan, S.C.; Shuqin, C.; Jindong, W.; Wei, Y.; Jie, X.; Qiulei, Z.; Jian, G.; Meng, L.; et al. The hot summer-cold winter region in China: Challenges in the low carbon adaptation of residential slab buildings to enhance comfort. *Energy Build.* **2020**, *223*, 110181. [CrossRef]

21. Thapar, S. Energy consumption behavior: A Data-Based analysis of urban Indian households. *Energy Policy* **2020**, *143*, 111571. [CrossRef]

22. Fouquet, R. Lessons from energy history for climate policy: Technological change, demand and economic development. *Energy Res. Soc. Sci.* **2016**, *22*, 79–93. [CrossRef]

23. Sanquist, T.F.; Orr, H.; Shui, B.; Bittner, A.C. Lifestyle factors in U.S. residential electricity consumption. *Energy Policy* **2012**, *42*, 354–364. [CrossRef]

24. De Meester, T.; Marique, A.-F.; De Herde, A.; Reiter, S. Impacts of occupant behaviours on residential heating consumption for detached houses in a temperate climate in the northern part of Europe. *Energy Build.* **2013**, *57*, 313–323. [CrossRef]

25. Hansen, A.R. 'Sticky' energy practices: The impact of childhood and early adulthood experience on later energy consumption practices. *Energy Res. Soc. Sci.* **2018**, *46*, 125–139. [CrossRef]

26. Umit, R.; Poortinga, W.; Jokinen, P.; Pohjolainen, P. The role of income in energy efficiency and curtailment behaviours: Findings from 22 European countries. *Energy Res. Soc. Sci.* **2019**, *53*, 206–214. [CrossRef]

27. Poruschi, L.; Ambrey, C.L. On the confluence of city living, energy saving behaviours and direct residential energy consumption. *Environ. Sci. Policy* **2016**, *66*, 334–343. [CrossRef]

28. Belaïd, F.; Garcia, T. Understanding the spectrum of residential energy-Saving behaviours: French evidence using disaggregated data. *Energy Econ.* **2016**, *57*, 204–214. [CrossRef]

29. Schleich, J. Energy efficient technology adoption in low-income households in the European Union—What is the evidence? *Energy Policy* **2019**, *125*, 196–206. [CrossRef]

30. Mills, B.; Schleich, J. Residential energy-Efficient technology adoption, energy conservation, knowledge, and attitudes: An analysis of European countries. *Energy Policy* **2012**, *49*, 616–628. [CrossRef]

31. Liu, X.; Wang, Q.; Wei, H.H.; Chi, H.L.; Ma, Y.; Jian, I.Y. Psychological and demographic factors affecting household energy-saving intentions: A TPB-based study in northwest China. *Sustainability* **2020**, *12*, 836. [CrossRef]

32. Urban, J.; Ščasný, M. Exploring domestic energy-Saving: The role of environmental concern and background variables. *Energy Policy* **2012**, *47*, 69–80. [CrossRef]

33. Diaz-Rainey, I.; Ashton, J.K. Investment inefficiency and the adoption of eco-Innovations: The case of household energy efficiency technologies. *Energy Policy* **2015**, *82*, 105–117. [CrossRef]

34. Gupta, S.; Ogden, D.T. To buy or not to buy? A social dilemma perspective on green buying. *J. Consum. Mark.* **2009**, *26*, 378–393. [CrossRef]

35. Pothitou, M.; Hanna, R.F.; Chalvatzis, K.J. Environmental knowledge, pro-Environmental behaviour and energy savings in households: An empirical study. *Appl. Energy* **2016**, *184*, 1217–1229. [CrossRef]

36. Hori, S.; Kondo, K.; Nogata, D.; Ben, H. The determinants of household energy-Saving behavior: Survey and comparison in five major Asian cities. *Energy Policy* **2013**, *52*, 354–362. [CrossRef]

37. Botetzagias, I.; Malesios, C.; Poulou, D. Electricity curtailment behaviors in Greek households: Different behaviors, different predictors. *Energy Policy* **2014**, *69*, 415–424. [CrossRef]

38. Vassileva, I.; Campillo, J. Increasing energy efficiency in low-Income households through targeting awareness and behavioral change. *Renew. Energy* **2014**, *67*, 59–63. [CrossRef]

39. Mi, L.; Qiao, L.; Du, S.; Xu, T.; Gan, X.; Wang, W.; Yu, X. Evaluating the effect of eight customized information strategies on urban households' electricity saving: A field experiment in China. *Sustain. Cities Soc.* **2020**, *62*, 102344. [CrossRef]

40. van der Werff, E.; Steg, L.; Keizer, K. It is a moral issue: The relationship between environmental self-Identity, obligation-Based intrinsic motivation and pro-Environmental behaviour. *Glob. Environ. Chang.* **2013**, *23*, 1258–1265. [CrossRef]

41. Sloot, D.; Jans, L.; Steg, L. Can community energy initiatives motivate sustainable energy behaviours? The role of initiative involvement and personal pro-Environmental motivation. *J. Environ. Psychol.* **2018**, *57*, 99–106. [CrossRef]

42. Martinsson, J.; Lundqvist, L.J.; Sundström, A. Energy saving in Swedish households. The (relative) importance of environmental attitudes. *Energy Policy* **2011**, *39*, 5182–5191. [CrossRef]

43. Nakamura, H. Effects of social participation and the emergence of voluntary social interactions on household Power-Saving practices in Post-Disaster Kanagawa, Japan. *Energy Policy* **2013**, *54*, 397–403. [CrossRef]

44. Fornara, F.; Pattitoni, P.; Mura, M.; Strazzera, E. Predicting intention to improve household energy efficiency: The role of Value-Belief-Norm theory, normative and informational influence, and specific attitude. *J. Environ. Psychol.* **2016**, *45*, 1–10. [CrossRef]

45. Wittenberg, I.; Blöbaum, A.; Matthies, E. Environmental motivations for energy use in PV households: Proposal of a modified norm activation model for the specific context of PV households. *J. Environ. Psychol.* **2018**, *55*, 110–120. [CrossRef]

46. Corradi, N.; Priftis, K.; Jacucci, G.; Gamberini, L. Oops, I forgot the light on! The cognitive mechanisms supporting the execution of energy saving behaviors. *J. Econ. Psychol.* **2013**, *34*, 88–96. [CrossRef]
47. Sweeney, J.C.; Kresling, J.; Webb, D.; Soutar, G.N.; Mazzarol, T. Energy saving behaviours: Development of a Practice-Based model. *Energy Policy* **2013**, *61*, 371–381. [CrossRef]
48. Pukšec, T.; Mathiesen, B.V.; Duić, N. Potentials for energy savings and long term energy demand of Croatian households sector. *Appl. Energy* **2013**, *101*, 15–25. [CrossRef]
49. Choi, S.; Feinberg, R.A. The LOHAS lifestyle and marketplace behavior: Establishing valid and reliable measurements. In *Handbook of Engaged Sustainability*; Springer International Publishing: Cham, Switzerland, 2018; Volume 2, pp. 1069–1086. [CrossRef]
50. Zhang, T.; Siebers, P.O.; Aickelin, U. A three-dimensional model of residential energy consumer archetypes for local energy policy design in the UK. *Energy Policy* **2012**, *47*, 102–110. [CrossRef]
51. Witek, L. Typologia konsumentów na rynku produktów ekologicznych. *Probl. Zarządzania, Finans. Mark.* **2014**, *35*, 209–218.
52. Kotler, P.; Wong, V.; Saunders, J.; Armstrong, G. *Principles of Marketing*, 4th ed.; Pearson Education Limited: Essex, UK, 2005.
53. Shukla, R.K.; Upadhyaya, A. Perceptual Mapping of Green Consumers: An Innovative Market Segmentation Strategy. *J. Int. Mark. Strategy* **2014**, *2*, 28–38.
54. Finisterra do Paço, A.M.; Raposo, M.L.B. Green consumer market segmentation: Empirical findings from Portugal. *Int. J. Consum. Stud.* **2010**, *34*, 429–436. [CrossRef]
55. Poortinga, W.; Darnton, A. Segmenting for sustainability: The development of a sustainability segmentation model from a Welsh sample. *J. Environ. Psychol.* **2016**, *45*, 221–232. [CrossRef]
56. Newton, P.; Meyer, D. Exploring the attitudes-action gap in household resource consumption: Does 'Environmental Lifestyle' segmentation align with consumer behaviour? *Sustainability* **2013**, *5*, 1211–1233. [CrossRef]
57. Graczyk, A.M. Households behaviour towards sustainable energy management in poland—The homo energeticus concept as a new behaviour pattern in sustainable economics. *Energies* **2021**, *14*, 3142. [CrossRef]
58. Axsen, J.; TyreeHageman, J.; Lentz, A. Lifestyle practices and pro-environmental technology. *Ecol. Econ.* **2012**, *82*, 64–74. [CrossRef]
59. Yang, Y.; Solgaard, H.S.; Haider, W. Value seeking, price sensitive, or green? Analyzing preference heterogeneity among residential energy consumers in Denmark. *Energy Res. Soc. Sci.* **2015**, *6*, 15–28. [CrossRef]
60. Tabi, A.; Hille, S.L.; Wüstenhagen, R. What makes people seal the green power deal?—Customer segmentation based on choice experiment in Germany. *Ecol. Econ.* **2014**, *107*, 206–215. [CrossRef]
61. Smart Energy Consumer Collaborative. Consumer Pulse and Marketing Segmentation Wave. Available online: https:// smartenergycc.org/consumer-pulse-and-market-segmentation-wave-7-report/ (accessed on 28 August 2021).
62. Słupik, S.; Kos-Łabędowicz, J.; Trzęsiok, J. Energy-related behaviour of consumers from the silesia province (Poland)—Towards a low-carbon economy. *Energies* **2021**, *14*, 2218. [CrossRef]
63. Albert, A.; Rajagopal, R. Smart meter driven segmentation: What your consumption says about you. *IEEE Trans. Power Syst.* **2013**, *28*, 4019–4030. [CrossRef]
64. Gouveia, J.P.; Seixas, J. Unraveling electricity consumption profiles in households through clusters: Combining smart meters and door-to-door surveys. *Energy Build.* **2016**, *116*, 666–676. [CrossRef]
65. Pan, S.; Wang, X.; Wei, Y.; Zhang, X.; Gál, C.; Ren, G.; Yan, D.; Shi, Y.; Wu, J.; Xia, L.; et al. Cluster analysis for occupant-behavior based electricity load patterns in buildings: A case study in Shanghai residences. *Build. Simul.* **2017**, *10*, 889–898. [CrossRef]
66. Hayn, M.; Bertsch, V.; Fichtner, W. Electricity load profiles in Europe: The importance of household segmentation. *Energy Res. Soc. Sci.* **2014**, *3*, 30–45. [CrossRef]
67. Maréchal, K.; Holzemer, L. Unravelling the 'ingredients' of energy consumption: Exploring home-related practices in Belgium. *Energy Res. Soc. Sci.* **2018**, *39*, 19–28. [CrossRef]
68. Abrahamse, W.; Steg, L. How do socio-demographic and psychological factors relate to households' direct and indirect energy use and savings? *J. Econ. Psychol.* **2009**, *30*, 711–720. [CrossRef]
69. Sütterlin, B.; Brunner, T.A.; Siegrist, M. Who puts the most energy into energy conservation? A segmentation of energy consumers based on energy-related behavioral characteristics. *Energy Policy* **2011**, *39*, 8137–8152. [CrossRef]
70. Santin, O.G. Behavioural patterns and user profiles related to energy consumption for heating. *Energy Build.* **2011**, *43*, 2662–2672. [CrossRef]
71. Vicente, P.; Reis, E. Segmenting households according to recycling attitudes in a Portuguese urban area. *Resour. Conserv. Recycl.* **2007**, *52*, 1–12. [CrossRef]
72. Lutzenhiser, L. Social and Behavioral Aspects of energy use. *Annu. Rev. Energy Environ.* **1993**, *28*, 247–289. [CrossRef]
73. Steg, L. Promoting household energy conservation. *Energy Policy* **2008**, *36*, 4449–4453. [CrossRef]
74. Sorrell, S.; Dimitropoulos, J.; Sommerville, M. Empirical estimates of the direct rebound effect: A review. *Energy Policy* **2009**, *37*, 1356–1371. [CrossRef]
75. Sapci, O.; Considine, T. The link between environmental attitudes and energy consumption behavior. *J. Behav. Exp. Econ.* **2014**, *52*, 29–34. [CrossRef]
76. van Der Werff, E.; Steg, L. One model to predict them all: Predicting energy behaviours with the norm activation model. *Energy Res. Soc. Sci.* **2015**, *6*, 8–14. [CrossRef]

77. Frankel, D.; Heck, S.; Tai, H. *Using a Consumer Segmentation Approach to Make Energy Efficiency Gains in the Residential Market*; McKinsey and Company: Chicago, IL, USA, 2013; pp. 1–9.

78. Pedersen, M. Segmenting residential customers: Energy and conservation behaviors. In Proceedings of the ACEEE Summer Study on Energy Efficiency in Buildings, Pacific Grove, CA, USA, 7–22 August 2008; pp. 229–241.

79. Guthridge, G.S. *Understanding Consumer Preferences in Energy Efficiency: Accenture end-CONSUMER Observatoryon Electricity Management*; Accenture: Dublin, Ireland, 2010.

80. Han, Q.; Nieuwenhijsen, I.; de Vries, B.; Blokhuis, E.; Schaefer, W. Intervention strategy to stimulate energy-saving behavior of local residents. *Energy Policy* **2013**, *52*, 706–715. [CrossRef]

81. Albert, A.; Maasoumy, M. Predictive segmentation of energy consumers. *Appl. Energy* **2016**, *177*, 435–448. [CrossRef]

82. Seidl, R.; Moser, C.; Blumer, Y. Navigating behavioral energy sufficiency. Results from a survey in Swiss cities on potential behavior change. *PLoS ONE* **2017**, *12*, e0185963. [CrossRef] [PubMed]

83. Tumbaz, M.N.M.; Moğulkoç, H.T. Profiling energy efficiency tendency: A case for Turkish households. *Energy Policy* **2018**, *119*, 441–448. [CrossRef]

84. Michelsen, C.C.; Madlener, R. Motivational factors influencing the homeowners' decisions between residential heating systems: An empirical analysis for Germany. *Energy Policy* **2013**, *57*, 221–233. [CrossRef]

85. Sanguinetti, A.; Karlin, B.; Ford, R. Understanding the path to smart home adoption: Segmenting and describing consumers across the innovation-decision process. *Energy Res. Soc. Sci.* **2018**, *46*, 274–283. [CrossRef]

86. Ropuszynska-Surma, E.; Weglarz, M. Profiling end user of renewable energy sources among residential consumers in Poland. *Sustainability* **2018**, *10*, 4452. [CrossRef]

87. Abrahamse, W.; Steg, L.; Vlek, C.; Rothengatter, T. The effect of tailored information, goal setting, and tailored feedback on household energy use, energy-related behaviors, and behavioral antecedents. *J. Environ. Psychol.* **2007**, *27*, 265–276. [CrossRef]

88. Bell, S.; Judson, E.; Bulkeley, H.; Powells, G.; Capova, K.A.; Lynch, D. Sociality and electricity in the United Kingdom: The influence of household dynamics on everyday consumption. *Energy Res. Soc. Sci.* **2015**, *9*, 98–106. [CrossRef]

89. Boudet, H.S.; Flora, J.A.; Armel, K.C. Clustering household energy-saving behaviours by behavioural attribute. *Energy Policy* **2016**, *92*, 444–454. [CrossRef]

90. Sardianou, E. Estimating energy conservation patterns of Greek households. *Energy Policy* **2007**, *35*, 3778–3791. [CrossRef]

91. Wang, Z.; Zhang, B.; Yin, J.; Zhang, Y. Determinants and policy implications for household electricity-saving behaviour: Evidence from Beijing, China. *Energy Policy* **2011**, *39*, 3550–3557. [CrossRef]

92. Słupik, S.; Kos-Łabędowicz, J.; Trzęsiok, J. Report on Findings from Consultations and Online-Survey. Personalised ICT-Tools for the Active Engagement of Consumers Towards Sustainable Energy; Report on Findings from Consultations and Online-Survey. Eco-Bot. Project. Available online: http://eco-bot.eu/publications/ (accessed on 28 August 2021).

93. Eco-Bot. 2021. Available online: http://eco-bot.eu/ (accessed on 28 August 2021).

94. Benzécri, J.P. *L'Analyse des Données. Volume II. L'Analyse des Correspondances*; Dunod: Paris, France, 1973.

95. Greenacre, M.J. *Theory and Applications of Correspondence Analysis*; Academic Press: London, UK, 1984.

96. Hoffman, D.L.; Franke, G.R. Correspondence Analysis: Graphical Representation of Categorical Data in Marketing Research. *JMR* **1986**, *23*, 213–227. [CrossRef]

97. Carrol, D.J.; Greek, P.E.; Schaffer, M.C. Interpoint distance comparisons in correspondence analysis. *JMR* **1986**, *23*, 271–280. [CrossRef]

98. Ward, J.H. Hierarchical Grouping to Optimize an Objective Function. *J. Am. Stat. Assoc.* **1963**, *58*, 236–244. [CrossRef]

99. Cramér, H. *Mathematical Methods of Statistics*; Princeton University Press: Princeton, NJ, USA, 1946. [CrossRef]

100. Brzezińska, J. Analiza logarytmiczno-liniowa. In *Teoria i Zastosowania z Wykorzystaniem Programu R*; Wydawnictwo C.H. Beck: Warszawa, Poland, 2015.

101. Greenacre, M. *Correspondence Analysis in Practice*, 3rd ed.; Chapman and Hall/CRC Press: London, UK, 2021. [CrossRef]

102. Gatnar, E.; Walesiak, M.; Rozmus, D. Analiza korespondencji. In *Metody Statystycznej Analizy Wielowymiarowej w Badaniach Marketingowych*; Gatnar, E., Walesiak, M., Eds.; Wydawnictwo AE we Wrocławiu: Wrocław, Poland, 2004; pp. 283–315.

103. Stanimir, A. *Analiza Korespondencji Jako Narzędzie Badania Zjawisk Ekonomicznych (Correspondence Analysis as a Tool for the Study Economic Factors)*; Wydawnictwo AE we Wrocławiu: Wrocław, Poland, 2005.

104. Électricité de France (EDF). A Rochebelle, Une Résidence Pionnière en Autoconsommation Collective. 2019. Available online: https://www.youtube.com/watch?v=LYsIQb22ZVM&t=29s (accessed on 24 August 2021).

105. Gramwzielone.pl. "Energia Ze Wspólnego PV Dla Mieszkańców Bloku. W Polsce to Na Razie Nierealne". 2019. Available online: https://www.gramwzielone.pl/energia-sloneczna/100687/energia-ze-wspolnego-pv-dla-mieszkancow-bloku-w-polsce-to-na-razie-nierealne (accessed on 22 August 2021).

106. Eurostat. Electricity Prices for Household Consumers—Biannual Data (from 2007 Onwards). 2020. Available online: https://appsso.eurostat.ec.europa.eu/nui/submitViewTableAction.do (accessed on 18 August 2021).

107. Eurostat. Electricity Price Statistics. 2021. Available online: https://ec.europa.eu/eurostat/statistics-explained/index.php?title=Electricity_price_statistics (accessed on 25 August 2021).

108. gramwzielone.pl. W Niemczech Przybyło 100 tys. Prosumentów z Magazynami Energii. 2021. Available online: https://www.gramwzielone.pl/magazynowanie-energii/105012/w-niemczech-przybylo-100-tys-prosumentow-z-magazynami-energii (accessed on 22 August 2021).

109. Opińska, G. Hiszpanie Protestują Przeciwko Wzrostowi Opłat za Prąd i ich Zróżnicowaniu w Ciągu Doby. 2021. Available online: https://www.bankier.pl/wiadomosc/Hiszpanie-protestuja-przeciwko-wzrostowi-oplat-za-prad-i-ich-zroznicowaniu-w-ciagu-doby-8138776.html (accessed on 22 August 2021).
110. Cipiur, J. Dlaczego Niemcy Wolą Wynajmować Mieszkania a Polacy Kupować? 2015. Available online: https://forsal.pl/artykuly/872555,dlaczego-niemcy-wola-wynajmowac-mieszkania-a-polacy-kupowac.html (accessed on 27 August 2021).
111. Eurostat, Wages and Labour Costs. 2021. Available online: https://ec.europa.eu/eurostat/statistics-explained/index.php?title=Wages_and_labour_costs (accessed on 27 August 2021).
112. Berenguer, J.; Corraliza, J.A.; Martin, R. Rural-Urban differences in environmental concern, attitudes, and actions. *Eur. J. Psychol. Assess.* **2005**, *21*, 128–138. [CrossRef]

Article

The Transition to Clean Energy: Are People Living in Island Communities Ready for Smart Grids and Demand Response?

Dana Abi Ghanem * and Tracey Crosbie

Centre for Sustainable Engineering, School of Computing, Engineering and Digital Technologies, Teesside University, Middlesbrough TS1 3BX, UK; t.crosbie@tees.ac.uk
* Correspondence: d.abighanem@tees.ac.uk

Abstract: Islands are widely recognised as ideal pilot sites that can spearhead the transition to clean energy and development towards a sustainable and healthy society. One of the assumptions underpinning this notion is that island communities are more ready to engage with smart grids (SGs) than people on the mainland. This is believed to be due to the high costs of energy on islands and the idea that the sense of community and collective action is stronger on islands than on the mainland. This paper presents findings from a survey conducted to assess people's perception of, and readiness to engage with, SG and demand response (DR) in the communities of three islands taking part in a H2020 project called REACT. The main objective of the survey, conducted in 2020, was to inform the recruitment of participants in the project, which is piloting different technologies required for SGs and DR with communities on the three islands. The results show that many island residents are motivated to take part in SG, to engage with energy saving, and are willing to change some energy-related behaviours in their homes. However, the results also indicate that levels of ownership of, and knowledge and familiarity with, the SG and DR related technologies are extremely low, suggesting that the expected uptake of DR in islands might not be as high as anticipated. This brings into question the readiness of island dwellers for the SG, their role in the deployment of such schemes more widely and the validity of the assumptions often made about island communities. This has significant implications for the design of SGs and DR solutions for islands, including devoting sufficient efforts to build knowledge and awareness of the SG, investing in demonstration projects for that purpose and tailoring interventions based on island communities' motivations.

Keywords: smart grids; demand response; island communities; social acceptance; technology readiness; sustainability; Spain; Italy; Ireland

Citation: Abi Ghanem, D.; Crosbie, T. The Transition to Clean Energy: Are People Living in Island Communities Ready for Smart Grids and Demand Response?. *Energies* **2021**, *14*, 6218. https://doi.org/10.3390/en14196218

Academic Editor: Sergio Ulgiati

Received: 20 August 2021
Accepted: 24 September 2021
Published: 29 September 2021

Publisher's Note: MDPI stays neutral with regard to jurisdictional claims in published maps and institutional affiliations.

1. Introduction

In response to the global challenge of climate change, a green energy transition is needed, based on replacing fossil fuels with renewable energy sources [1,2]. These alternatives, which include wind power and solar photovoltaics (PV), generate intermittent energy. To be able to power electricity provision using renewables, smart energy networks known as smart grids (SG) coupled with demand response (DR) principles are key [3–5]. Achieving a SG entails the engagement of users in managing peak loads, whereby they have to change their electricity consumption patterns and shift their daily activities to when demand on the grid is lower [6]. It is envisaged that homes will become spaces where smart home energy management systems (HEMS) and other ICT-enabled technologies are deployed so that householders either manage their electricity consumption or give permission to third parties to do so on their behalf, and consequently plan their everyday activities accordingly.

Islands are widely recognised as ideal pilot sites which can spearhead the transition to clean energy through SGs and DR technologies that enable localised and renewable energy production and the optimised management of load on the network [7–9], to function as "laboratories for technological, social, environmental, economic and political innovation" [8].

One of the assumptions underpinning this notion is that people living in island communities are more ready to engage with SGs and DR than people on the mainland, mainly due to the higher costs of energy on islands [10,11]. In this regard, the SG can have financial benefits. The geographic characteristics of islands and the challenges related to the provision of electricity services from the mainland [12] make them ideal candidates for deployment from a technical point of view. Another reason is the perceived stronger sense of community among smaller populations on islands compared to mainland communities and neighbourhoods [11,13], suggesting that the typical social barriers attributed to reluctance of DR uptake will be weaker in the case of islands. This posits the question of whether island communities are more likely to adopt these technologies and adapt their lifestyles to provide flexibility through DR and SG technologies. Energy research that has pointed out the role of islands in testing the SG with various DR technologies [14–18] has not adequately considered societal challenges [19]. Meanwhile, policy assessments emphasise the importance of autonomy and energy sustainability for Europe's outer regions including islands, suggesting that public engagement and acceptance remain important challenges to overcome [20], thus maintaining the need for promotional and educational efforts.

The interest in developing SG in islands [8] and the increasing efforts in financing research and development projects for islands in Europe [21] raise an important question of whether island communities are ideally suited for such interventions. Studies have generally found that island residents' knowledge and attitudes towards renewable energy technologies or DR to be positive. Insights from a study on the uptake of solar photovoltaics in French island territories suggest that decisions to install such technologies may be driven by energy insularity (power cuts and comparatively expensive energy bills), though also complicated by other life-course events and a desire to maintain high quality energy services [22]. Where projects on energy management are successful, this also depends on island energy utility companies' efforts to maintain trust with the community, and building on this relationship to encourage participation in DR schemes [23]. Another element is knowledge and understanding of DR and flexibility, which are necessary for people to be motivated. In relation to islanders' knowledge of how DR technologies work and how intermittency can be managed, a distinction is made between formal technical knowledge and other forms of sensory or practical knowledge that can be effective in helping residents adapt to systems in order to offer flexibility to the grid [24].

These findings indicate similarities to mainland communities, where knowledge of the technologies and attitudes to RETs and energy consumption are important determinants for DR adoption. However, exploring the notion of islands as test beds for energy technologies, Skjølsvold et al. [13] noted that islands are often exoticized on a social level, perceived as being "distinct sociomaterial places with a particular form of topography, location in global value chains, and being associated with a form of social exoticism" (p. 7). As such, the presence of a sense of community and kinship is emphasised and presented as an advantage to the implementation of new energy technologies and innovations. However, in some cases, the history of economic and political dependence and peripherality of islands can be a hurdle for economic and sustainable development [25]. Other studies [26] have pointed out that island communities' close-knit nature offers a potential for community energy but also a challenge for gaining the trust and acceptance of sustainable-focused development interventions. These findings focus on what distinguishes island communities from mainland populations. This itself raises the question of what experiments in SG can tell us about their potential for wider deployment. However, existing and future investments in island communities for energy using smart solutions [8], as well as concerns for energy provision for islands [20,21], indicate that the readiness of island dwellers for SG and DR programmes remains an important question requiring further research.

As indicated above, a large number of research and innovation projects have and are focusing on piloting renewable energy technologies and smart energy grids on islands, including the demonstration of several SG technologies, storage and the introduction of renewable energy sources [21]. The research presented in this paper was conducted as part of one of these projects called REACT [27]. The REACT project is seeking to demonstrate the potential of renewable energy systems (RES) on geographical islands to bring economic benefits, decarbonise islands' energy systems, reduce greenhouse gas (GHG) emissions and improve environmental air quality. This is to be achieved by integrating existing and emerging technologies to create a cloud-based solution enabling a SG with the potential to support the energy autonomy of geographical islands, demonstrating the solution on three different islands with differing climate and market contexts, and developing plans for large-scale replication of the REACT solution on five follower islands. The REACT project pilot islands are La Graciosa (Spain), San Pietro (Italy) and the Aran Islands (Ireland). Within these islands, the REACT solution is being piloted in the following towns or population centres: Caleta del Sebo in La Graciosa, Carloforte in San Pietro and Inis Mór, one of the Aran Islands. This paper presents the findings of a survey conducted with a sample of the people living in these towns. An analysis of the survey responses from the survey participants is presented to show whether people living on these islands are motivated to engage in SGs and DR and how ready and willing they are to engage with DR principles and technologies in their homes.

Following this introduction, the remainder of this paper is split into five sections. To contextualise the research, Section 2 presents a discussion of what motivates people to take part in SG and DR programmes, and how the technologies people have in their homes impact on their readiness to do so. Section 3 discusses the methodology applied in the survey research presented. Section 4 presents the findings of the survey. Section 5 discusses whether the findings suggest that people living in island communities are ready to engage with SGs and DR. It then goes on to discuss what the findings suggest are the best strategies to encourage island communities to engage in SGs and DR, including considerations in relation to the design of SG and DR solutions intended for use by island communities. The final section of the paper draws some conclusions as to whether island communities are ready for the SG and the implications of this on broader plans and projects focusing on smart solutions intended for geographical islands.

2. Elements of Smart Grid Readiness: Technical, Economic and Social Considerations

2.1. Technical Requirements for Householder Readiness to Take Part in Smart Grids and Demand Response

When it comes to users and consumption, SGs entail the introduction of demand side management [28]. Traditionally, demand side interventions focused on behavioural change and improving the energy efficiency of appliances in the home, e.g., energy efficient light bulbs [29,30]. In the context of the SG, consumers are expected to engage through DR, which offers them a significant role in the delivery of flexibility by reducing or shifting their electricity usage during periods of stress or constraint [31].

Participation in SG-related DR programmes by household consumers has two key technical requirements. Firstly, households must have a smart meter installed to enable some form of dynamic pricing based on real-time or near real-time consumption patterns [32,33]. However, whilst these enable the implementation of dynamic pricing schemes, on their own they do not ensure significant improvements in demand-side activity [33]. As such, HEMS are a second key requirement if household electricity consumers are to participate actively in DR programmes [34]. Essentially, HEMS are a technology platform comprised of both hardware and software that allow the user to monitor energy usage and production and to manually control and/or automate the use of energy within a household [33]. In this regard, the integration of new ICT capabilities is instrumental in facilitating wider engagement with the SG to achieve various benefits [35].

This is particularly the case in the domestic sector, where automation and direct load control were found to be not only acceptable but also key in broadening engagement with DR programmes [36]. To enable automated DR across all electric loads, heating and cooling systems as well as wet appliances will have to be smart and electric vehicles will require smart charging and discharging [28]. Research has highlighted numerous technical barriers to DR in the domestic sector in relation to ICT [37]. These include the lack of the necessary SG infrastructure in many regions, the diversity of ICT devices on the consumer side from different market providers that follow different communication protocols, creating interoperability issues, as well as security risks and scalability challenges [38].

Research on blocks of buildings and their readiness for demand response has shown similar challenges and potential barriers, which would indicate the propensity to engage with implicit or explicit DR depends on the level of integration of various electricity loads via energy management systems [39]. Accordingly, the heating and cooling services in homes can have different impacts, such as whether they include a thermostat that can be programmed externally, or whether the householders themselves are accustomed to controlling or programming the heating and cooling systems in their homes [40–42]. Given that 64% of the residential sector's energy consumption is used for space heating (Eurostat available at: https://ec.europa.eu/eurostat/web/products-eurostat-news/-/DDN-201906 20-1 (accessed on 20 February 2020)), the types of heating and cooling systems that people have installed in their homes are also key to the level of flexibility they can offer the grid in terms of DR. In this context, the penetration of electric, as opposed to gas or wood powered heating, is an important consideration.

2.2. Motivation to Take Part in the Smart Grid and Demand Response

2.2.1. Economic Motivations

Lower energy costs are often cited as the main reason why people might engage with demand side interventions such as DR [43–46]. The expectation is that consumers in their homes will be incentivised through the implementation of time-of-use (TOU) tariffs that vary the cost of electricity throughout the day [40]. Earlier work on the SG has argued that the two main reasons why consumers adopt SG-related technologies are financial benefits and environmental motivations, with the majority expecting economic pay back, suggesting that environmental concerns are not sufficient on their own to affect adoption [47]. A survey of student motivations found similar trends, where the main motivation was to save money, followed by positive environmental benefits, such as reducing energy demand and supporting the integration of renewables into energy networks [48]. However, many factors influence the extent to which price signals can be effective, including the type of pricing mechanism used, climate zone and season, ownership of air-conditioning equipment, and household characteristics such as income [49]. For example, studies have shown that higher income households do not respond well to price signals alone [45]. The way an economic incentive is framed is also important, e.g., is the incentive to use the electricity from the battery communicated as payment or saving [50]. Whilst further research and analysis is needed to assess how well price signals work [51,52], it remains an important factor to consider when assessing the lowering of bills as a motivation for households when choosing to take up TOU tariffs as a way of managing their electricity demand.

2.2.2. Attitudes and Social Norms

Studies based on social norms have examined energy consumption and participation in DR as driven by pro-social motives [53]. Attitudes that pertain to the uptake of DR relate to environmental motivations for individuals and communities to reduce their carbon emissions and contribute to climate change mitigation. Studies have shown that environmental motivations have prompted people to change their energy demand [46,54] and develop new energy-related habits [54]. Studies focusing on personal attributes have identified motivations related to an interest in cutting edge technology [55], awareness of energy consumption and better quality feedback [46], as well as the desire to better control energy consumption [56,57]. Other pro-social motivations that have been highlighted include the ability of households to contribute to the reliability of the grid [58] and to improve their local community [59], as well as feeling empowered to manage and take responsibility as a citizen over their electricity consumption [60]. For example, in a study on the acceptance of TOU tariffs, the interest in national energy independence was a stronger factor than the expected economic benefits for individual households [61]. More broadly, how climate change is framed within national policy has been shown to impact on the likelihood for support and uptake of energy technologies including DR solutions [62].

Feedback technologies play a crucial role in developing awareness with regards to energy consumption, and the energy savings that could be achieved by using it [63]. Motivation to reduce energy consumption by allowing households to compare their consumption to other households in their neighbourhood or community, called comparative feedback, has been developed [64] based on research that works on activating social norms in individuals, i.e., how does an individual behaviour compare to others [65]. Several interventions in this field have utilized social media platforms to enable a comparison of ecological footprint and energy behaviour [66]. This work developed into dedicated online communities to promote energy-saving behaviour through public pledges and competition [67], whilst other research—focused on social media based applications—found that energy consumption reduced significantly through socially mediated incentives and competition [68]. In a similar vein, community-based projects that focused on comparative feedback found longer-term engagement with in home displays when it was coupled with weekly email-based newsletters and sustained communication [69]. In summary, research on energy consumption suggests economic, environmental and norm-activated motivations can all play a part in incentivising behavioural changes at the individual and household levels in relation to energy consumption. These have informed our survey questions pertaining to people's motivation in taking part in DR programmes and participating in the SG, to better understand the readiness of island households and communities for technologies such as the REACT solution, the results of which are presented in Section 4.

2.2.3. Knowledge and Familiarity with SG

Familiarity with the principles and technologies of DR and the SG is a frequently identified barrier to the uptake of TOU tariffs or active DR management across European electricity markets [70]. Familiarity is a key factor in increasing or decreasing social acceptance of technology [71]. In a study on the acceptance of smart meters in the USA, familiarity along with climate change risk perception were found to be the strongest predictors of smart meter acceptance [72]. Arguably, a lack of familiarity with DR is likely to lead to a lack of trust in the interventions of energy utility companies and further reluctance to take up other DR services [73]. Crucially, familiarity has a determining effect on uptake as well, which is very important when considering contexts where DR programmes and innovations are being introduced [74,75]. Li et al. [76] found that lower levels of reported familiarity with DR and SG concepts was correlated with less reported willingness to postpone the use of different appliances in the homes. As such, in the research reported in this paper, we consider familiarity with the concept of the SG and familiarity with the SG and DR related technologies to be an important factor when assessing the readiness of households for the SG.

2.3. Flexible Energy Demand

Considering the household level, the potential for flexible energy demand depends on the level of interest in smart appliances and devices [77,78], attitudes in relation to lifestyle [79] and anxiety over smart technology installation and use difficulty [80]. Earlier research found that in some cases, flexibility was related to the design of the heating system. For example, in homes fitted with insulated underfloor heating (that keeps the warmth for a longer time period), householders are more flexible in the timing of their energy demand [81]. In another example it was found that the inclusion of 'buffer' heating, such as fireplaces, made households happier to be flexible in their energy use, because they could save on their bills whilst enjoying the warmth and comfort in their homes [82]. Overall, earlier research found that householders were able to offer flexibility in relation to their use of several appliances. However, the level of flexibility depends on the appliance and the type of activity associated with it. Mostly, people are more willing to shift the use of their washing machines, dryers and dishwashers more than freezers, fridges and bathroom heaters [82]. Inflexibility is reported in the literature in relation to dining and cooking, where mealtimes and the hours of the day allocated to food preparation are the least flexible, and therefore the use of appliances for those purposes is not amenable to shifting [83]. To summarise, whilst heating flexibility is related to the type of heating equipment and level of insulation, in relation to the different everyday activities, those that pertain to cleaning, household care and laundry are generally more flexible than those pertaining to family life, such as mealtimes and social gatherings.

3. Methods

3.1. Survey Design and Distribution

Taking the literature discussed in the previous section into consideration, it was decided to conduct a survey as part of the REACT project to inform the recruitment of participants in the project, which is piloting different technologies required for SGs and DR with the communities on the three islands that are testing the REACT solution. The paper-based survey was adapted to the specific needs of the REACT project, from a survey conducted by Li et al. (2017). The surveys were distributed to the communities in the towns or population centres where the technology will be deployed as part of the REACT project. These are Caleta del Sebo in La Graciosa, Carloforte in San Pietro and Inis Mór, on one of the Aran Islands. In total, the survey questionnaire included 31 questions, including questions on demographic and household characteristics, home heating and cooling systems, knowledge of and familiarity with SG concepts and DR technologies and motivators for taking part in the SG and DR programmes (see Appendix A, Table A1). The data collected enabled the researchers to assess the communities' perception of, and readiness to engage with, SG-related technologies and DR in order to inform the recruitment strategy for the later stages of the REACT project. The surveys were distributed via schools, community meetings related to the REACT project and by going door to door in the case of Inis Mór.

3.1.1. Sample Size

In Inis Mór in the Aran Isles roughly 230 households are permanently occupied throughout the year. In total, 81 surveys were collected from households residing there during the winter months. Therefore, our sample is roughly 35% of the permanent residents on the island. The average number of households occupied in Caleta de Sebo on the island of La Graciosa can be conservatively estimated at 150 [84]. In total, 21 surveys were collected in Caleta de Sebo. Therefore, the sample is roughly 13% of the total number of households. In San Pietro on the island of Carloforte there are 2800 households. A total of 77 surveys were collected in San Pietro. Therefore, the sample size is approximately 3% of the total number of households (anecdotal evidence suggests that less people live in Carloforte than is officially indicated in the population census).

3.1.2. Sample Representativeness

To check the representativeness of our samples where possible, we compared the socio-demographic data from the survey to the available population demographics. If there are no data available at the local or regional level, the data from our sample were compered to data at the national level. As shown in Table 1, this comparison indicates that, on the whole, our samples do not differ too greatly from the wider populations to which they were compared.

Table 1. Comparison of survey demographic data with wider community regional and national demographic data in the three islands.

	Socio-Demographic Variable	Survey Results	Wider Community (Regional Level)
Inis Mór, Aran	Age	65% or older: 30% 55 to 64: 20% 45 to 54: 21% 35 to 44: 26% 25 to 34: 2% 18 to 24: 1%	65% or older: 25% 55 to 64: 15% 45 to 54: 18% 35 to 44: 20% 25 to 34: 13% 18 to 24: 8%
	Gender	64% Female; 36% Male	51% Female; 49% Male
	Education	Primary: 7% Lower secondary: 13% Leaving certificate: 23% Post-leaving cert.: 16% Third level: 41%	Primary: 7% Lower secondary:15% Leaving certificate: 24% Post-leaving cert.: 14% Third level: 40%
Carloforte, San Pietro	Age	65% or older: 22% 55 to 64: 19% 45 to 54: 30% 35 to 44: 16% 25 to 34: 13% 18 to 24: 0%	65% or older: 34% 55 to 64: 17% 45 to 54: 19% 35 to 44: 14% 25 to 34: 11% 18 to 24: 5%
	Gender	44% Female; 56% Male	50.9% Female; 49.1% Male
	Education	Primary: 0% Middle school: 5% Diploma: 56% University level: 39%	Primary: 0% Middle school: 28% Diploma: 29% University level: 43%
La Graciosa	Age	65% or older: 14% 55 to 64: 19% 45 to 54: 24% 35 to 44: 24% 25 to 34: 5% 18 to 24: 10%	65% or older: 21% 55 to 64: 11% 45 to 54: 24% 35 to 44: 20% 25 to 34: 15% 18 to 24: 9%
	Wider community (National level)		
	Gender	Female: 52%; Male: 48%	Female: 50%; Male: 50%
	Education	Secondary: 5% Diploma/vocational: 24% University level: 58%	Secondary: 28% Diploma: 27% University level: 45%

3.2. Data Analysis

Data from the collected surveys were entered into Microsoft® Excel. This spreadsheet is used for two purposes. Firstly, to check and revise the responses ensuring missing data and invalid responses were corrected, and secondly to conduct a comparative frequency analysis of the responses from each of the island communities. By conducting the comparative analysis of the responses from each of the islands, we were able to see if there were any noteworthy differences in the survey responses of the islands residents on each of the REACT pilot islands. This was key to enabling us to inform the recruitment strategy for the later stages of the REACT project on each of the pilot islands. It also enabled a consideration of whether our findings might be typical of island communities in general. Given their geographical location, the main differences in the island pertained to the heating and cooling systems installed in respondents' homes, with respondents on Inis Mór mostly using heating and Carloforte largely using cooling appliances. For most variables relating to the SG and DR perceptions and attitudes, the findings show that respondents from the three islands are not significantly different from each other. This enables us to consider the results as indicative of island communities in Europe more generally.

4. Results

In this paper, we present the results from survey responses on the three islands pertaining to respondents' motivations for taking part in DR and the SG, and their readiness to take part in that. To identify respondents' motivations, we asked questions about the impact of energy bills on household expenditure, the importance of saving energy and RET's to respondents and the different factors that might motivate them to take part in the SG. To address their readiness to take part in the SG, we asked questions about the technologies they have in their homes, the SG and DR technologies they would like to adopt, their familiarity with these technologies and how flexible they are willing to be across the use of different household appliances.

4.1. Motivations to Take Part in Demand Response and the Smart Grid

As discussed in Section 2.2, different factors can motivate people to take part in the SG and adopt DR technologies, including financial motivations (e.g., lower bills) and environmental concerns. To gauge how important energy costs are to the respondents, our survey asked what impact their energy bill has on their household budget. The answer categories were provided on a scale from "Very High" to "Very Low". As Figure 1 illustrates, in the case of all three islands, by far most respondents (71% in La Graciosa, 84% in Carloforte and 88% in Inis Mór) indicated that the impact of their energy bill on their monthly expenditure is very high, high or medium. Respondents were also asked to indicate the importance they gave to energy saving and for having RET technologies on their island, using a five-point scale (from "Very Important" to "Not important at all"). As can be seen in Figure 2, the majority of respondents on all three islands indicated a high level of interest in saving energy, with 85% in La Graciosa, 74% in Carloforte and 77% in Inis Mór believing that saving energy is very important. The percentages of respondents reporting that the use of RETs is very important is also high across the three islands, with 80% of respondents in La Graciosa, 71% of respondents in Carloforte and 63% of respondents in Inis Mór indicating that the use of RET is very important on their island (see Figure 2).

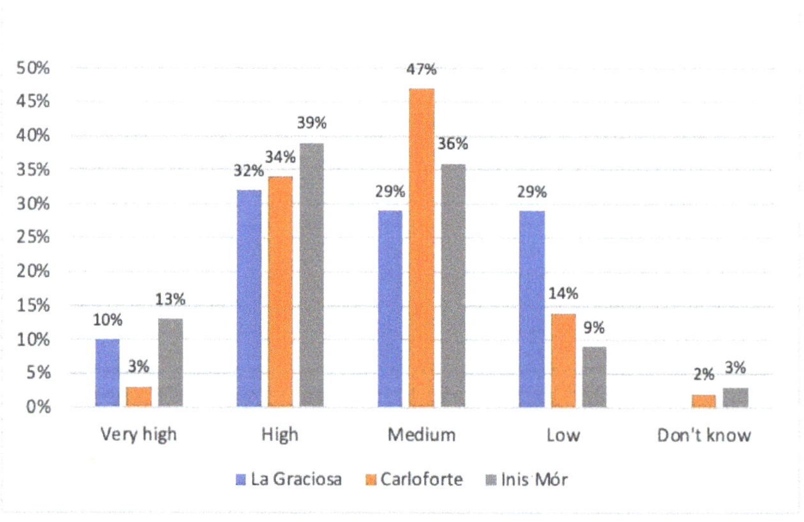

Figure 1. Impact of energy expenditure of households in the three islands.

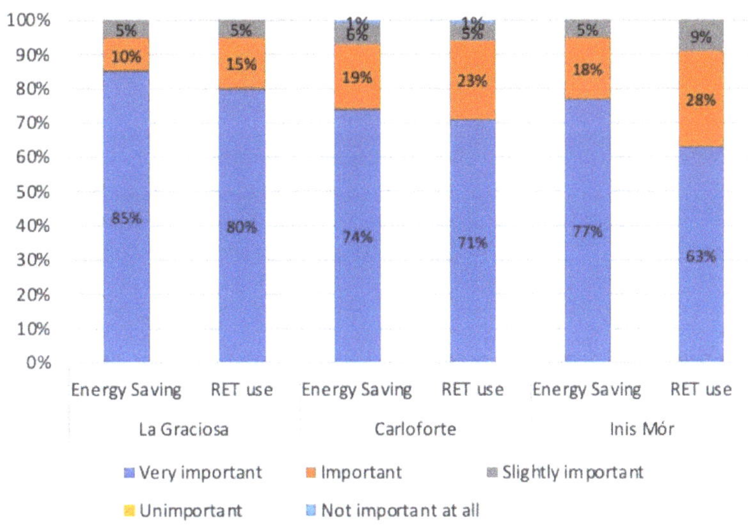

Figure 2. Importance of energy saving and RET use for households in the three islands.

In our survey, we asked respondents if the factors listed in Figure 3 would motivate them to accept SGs and use smart appliances. The answer categories provided were "Strongly motivating", "Motivating", "Slightly motivating", "Not motivating" and "I don't care." In terms of respondents' motivations to take part in the SG and DR programmes, our findings echo those of the earlier studies discussed in Section 2, which have found that people are motivated by economic and environmental concerns. By far, the majority of respondents from all three islands (88% in La Graciosa, 78% in Carloforte and 83% in Inis Mór) would be strongly motivated or motivated by a reduction in their energy bills to accept SGs and the use of smart appliances (see Figure 3). In relation to environmental and altruistic motivations, the results show that the majority of respondents (75% in La Graciosa, 65 % in Carloforte and 83% in Inis Mór) would be strongly motivated to accept SGs and the use of smart appliances to reduce CO_2 emissions (see Figure 3). As illustrated in Figure 3, giving your house a more sustainable character was also found to be strongly motivating or motivating by a majority of respondents (75% in La Graciosa, 70% in Carloforte and 52% in Inis Mór).

Interestingly, respondents were not as strongly motivated to accept SGs and the use of smart appliances by the prospect of sharing results on social media or comparing their household's energy consumption to other households as the literature in this field often assumes (see Figure 3). Only 15% of respondents in Carloforte and 17% of respondents in Inis Mór indicated that they would be strongly motivated by sharing results on social media to accept SGs and the use of smart appliances. The number of people motivated by this is slightly more encouraging in the case of La Graciosa, with 41% of respondents indicating that they would be strongly motivated by the prospect of sharing results on social media. However, this approach to motivating people to accept SGs and the use of smart appliances is unlikely to be successful in the other two islands, where the majority of respondents (59% in Carloforte and 64% in Inis Mór) did not care about sharing results on social media or are not motivated by this to adopt SGs and the use of smart appliances. The findings relating to the potential of motivating respondents by enabling them to compare their household's energy consumption to that of other households is also less encouraging than might be expected, with only 18% of respondents in Carloforte and 19% of respondents in Inis Mór indicating that they would be motivated by this to accept SGs and the use of smart appliances. Again, the number of respondents that would be motivated by this is slightly more encouraging in the case of La Graciosa, with 38% of respondents indicating that they would be strongly motivated to accept SGs and the use of smart appliances if this enabled them to compare their household's energy consumption to that of other households.

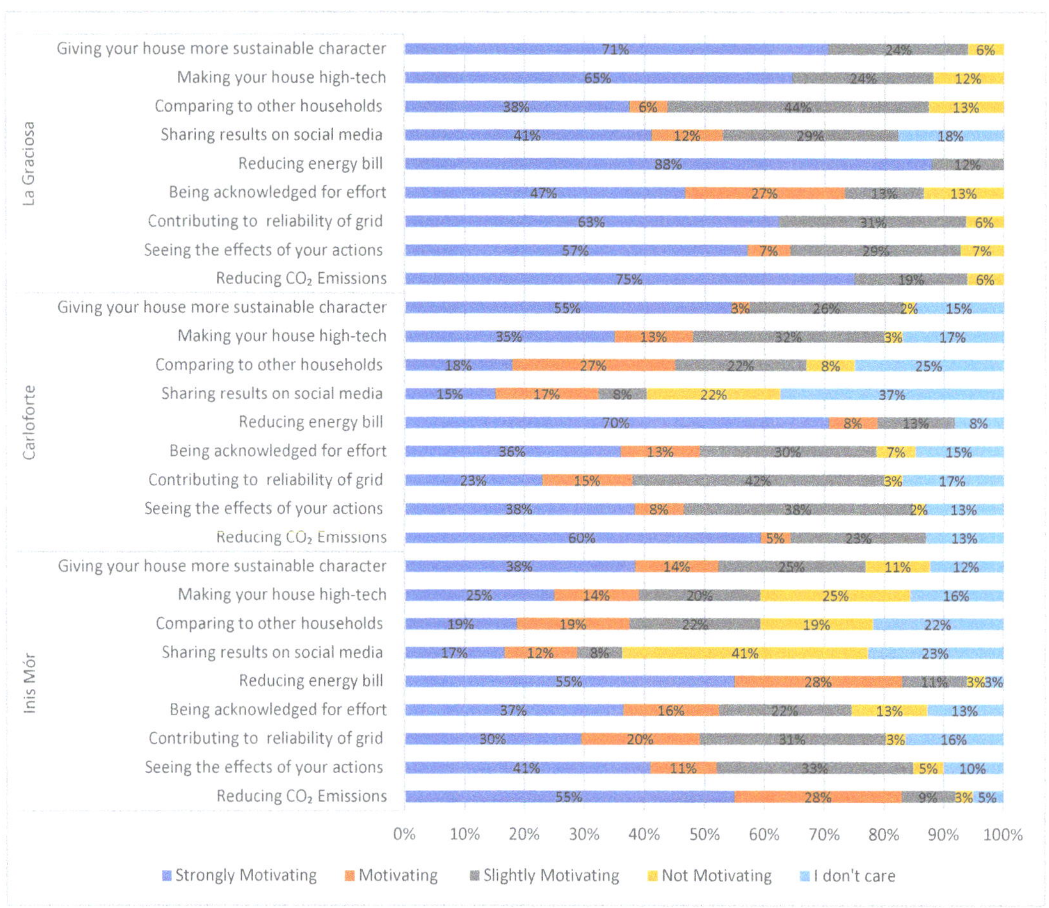

Figure 3. Motivating factors for the uptake of DR technologies.

The high percentages of respondents that perceive the impact of energy costs on household expenditure to be significant and the use of RETs to be important, combined with participants' high levels of interest in saving energy, are encouraging in terms of island residents' motivations to take part in the SG and DR programmes. Similarly encouraging are the high numbers of respondents reporting that they would be motivated by the potential of reducing their energy bills and lowering their CO_2 emissions to use the technologies required for DR. However, respondents' 'readiness' to take part in the SG and DR depends on the technologies they currently have in their homes, how they use them and whether they are prepared to adopt enabling technologies such as smart meters, HEMs and smart appliances. These issues were explored in our survey and the findings are presented in the following section.

4.2. Readiness to Take Part in Demand Response and the Smart Grid

This section presents the responses from the survey questions related to the respondent's existing heating and cooling systems, their current ownership of SG technologies and their willingness to adopt them, their familiarity with the SG concept and their knowledge and familiarity with the SG and DR technologies.

4.2.1. Household Heating and Cooling Systems

The existence of heating and cooling systems and their type (central or individual), as well as whether they use a thermostat, are good indicators of the likelihood for people to shift their heating/cooling or turn it off to offer flexibility. As discussed in Section 2, how people heat or cool their homes is key to the level of flexibility they can offer the grid in terms of DR. None of the respondents in Inis Mór use air-conditioning whilst homes in Carloforte use both heating and cooling to achieve thermal comfort. Unexpectedly, only four households in La Graciosa reported having cooling systems whilst only one household has a mobile heating unit. The results are summarised in Table 2.

Table 2. Heating and cooling systems of the respondents in the three islands.

	Heating			**Cooling**		
La Graciosa	*With Heating*	5%		*With Cooling*	20%	
	Central		–	Central		50%
	Individual		100%	Individual		25%
	Both types		–	Both types		25%
	Other		–	Other		–
Carloforte	*With Heating*	81%		*With Cooling*	67%	
	Central		11%	Central		85%
	Individual		74%	Individual		9%
	Both types		10%	Both types		2%
	Other		5%	Other		4%
Inis Mór	*With Heating*	100%		*With Cooling*	5%	
	Central		78%	Central		75%
	Individual		4%	Individual		25%
	Both types		6%	Both types		–
	Other		12%	Other		–

In La Graciosa, only 5% of households have individual heaters and only 20% of those surveyed had air-conditioning installed in their homes. Of those, half (50%) have a centralized cooling system, 25% have individual units and another 25% have both systems. Therefore, the findings of our survey in relation to the potential for DR control for thermal load is not encouraging in the case of La Graciosa.

For Carloforte, 81% of those surveyed have heating in their homes, split across central, individual or other types (biomass burners). The majority have individual heating units, with only 11% having a centralized heating system, indicating that a minority would have the option to include automated heating controls as part of the REACT solution in their homes. For cooling, 67% of respondents have air-conditioning, and of those, 85% have a central cooling system in their homes, whilst 9% have individual air-conditioning for separate rooms, 2% have both and 4% have other cooling devices such as fans.

In Inis Mór, all respondents have heating equipment installed in their home and none of them have air-conditioning equipment. The majority (78%) there have central heating, while 4% have only individual heating units and 6% from Inis Mór have both central and individual heating equipment. The individual heating units used by the respondents from Inis Mór are typically storage heaters, stoves or solid fuel heaters.

With respect to thermostats for heating, none of the households surveyed in La Graciosa had any thermostats since they did not have central heating. In Carloforte, only 41% of respondents reported having a thermostat for heating. Of those, 32% have the heating on when at home set at different temperatures and 21% have it set at a constant temperature. In Carloforte, 16% of respondents have the heating on all the time at a constant temperature whilst 16% have their heating at different temperatures for each room. The remaining 16% of respondents reported that they did not use the thermostat often.

In Inis Mór, 63% of respondents relied on thermostats to control their heating systems. Of those, the majority have the heating turned on when they are at home at a constant temperature (43%) or at different temperatures across the house (24%). The remainder have the heating always on, including 18% at different temperatures and only 6% at a constant temperature, whilst 9% reported not using heating often. Figure 4a compares heating control behaviours for Carloforte and Inis Mór. Overall, this shows that the majority do use the thermostats albeit in different ways. In this context, there is potential to consider having their heating controlled remotely if needed, due to their familiarity with controlling the temperature in their homes centrally. This is an encouraging indicator in terms of the familiarity with temperature or heating control remotely. However, the majority of the heating systems in Inis Mór are not electric and moving them to electric heat pumps would require significant investment, alongside related barriers to the electrification of heating [85].

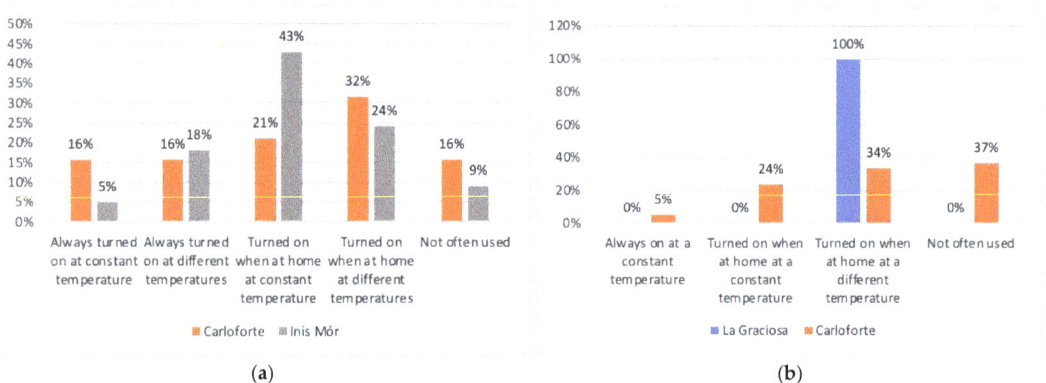

Figure 4. (**a**) Heating system use among respondents in Carloforte and Inis Mór islands; (**b**) Cooling system use among respondents in La Graciosa and Carloforte islands.

For the control of cooling systems, in La Graciosa, only four respondents reported having an air-conditioning system in their homes, and of those, three homes have a centralised cooling system. Further, 10% of the respondents have a thermostat that they use to turn on the cooling when at home, allowing for different temperatures for each room (see Figure 4). Therefore, in relation to heating and cooling, the potential of DR control is not encouraging. In Carloforte, only 5% of those surveyed have the cooling always on, 24% have the cooling on when they are at home at a constant temperature and 34% have it set at a different temperature. None of the respondents in Inis Mór have air-conditioning in their homes. Figure 4b compares cooling control behaviours of La Graciosa and Carloforte.

4.2.2. Familiarity with Smart Grids and Demand Response and Their Enabling Technologies

The survey asked respondents how familiar they are with the concept of the SG prior to being contacted by members of the REACT project team. The responses offered were 'Never heard of it', 'Heard a little but don't understand the concept', 'Heard a lot but don't understand the concept', 'Know a little about the concept' or 'Know a lot about the concept.' A large proportion of respondents from all three islands said that they had never heard of the concept of the SG prior to being contacted by the REACT project team or they had heard of it but didn't understand the concept (58% in La Graciosa, 80% in Carloforte and 74% in Inis Mór). Very few of the respondents (5% in La Graciosa, 3% in Carloforte and 8%

in Inis Mór) said they knew a lot about the concept of the SG prior to being contacted by the project team (see Figure 5).

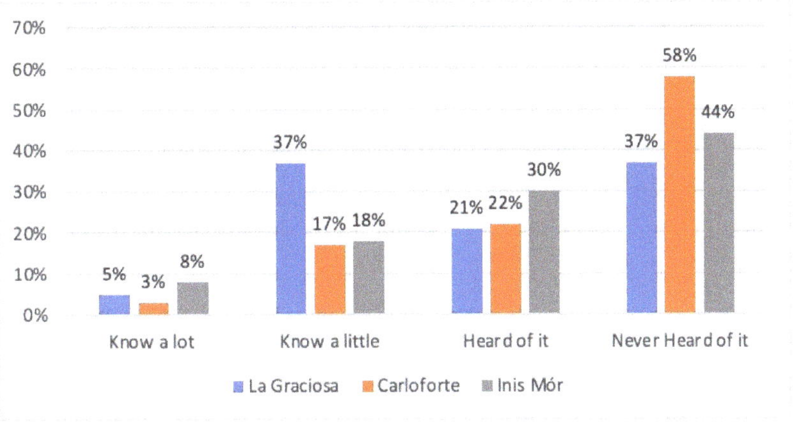

Figure 5. Familiarity of the three island residents with the Smart Grid concept.

Respondents were also asked how familiar they are with the different SG and DR enabling technologies listed in Figure 6. With regards to each of the technologies, they were asked to state whether they had 'Never heard of it', 'Heard of it but do not understand it', 'Know a little about it', 'Know a lot about it' or whether they own such technology. In La Graciosa, only 5% of research participants own a smart meter and only 5% know a lot about smart meters. In the case of Carloforte, only 1% have a smart meter and only 1% know a lot about them. In Inis Mór only 1 % have a smart meter and only 4% know a lot about them. In addition, many of the respondents had never heard of a smart meter (30% in La Graciosa, 65% in Carloforte and 49% in Inis Mór). Given that smart meters are a key enabling technology for engaging with the SG and DR, these figures are not encouraging. In the case of HEMs, the figures are no more encouraging, with no respondents in Carloforte owing a HEMs and only 1% of respondents from Inis Mór and 5% of respondents in La Graciosa owning a HEMs. Again, for in-home displays, the situation is similar, with no installations amongst the sample in La Graciosa and Carloforte, and only 10% in La Graciosa and 3% in Carloforte knowing a lot about them. In Inis Mór, only 1% have an in-home display and only 4% reported confident levels of knowledge and understanding of these technologies. This suggests that the existing level of knowledge and understanding of the concepts of SG or DR and the technologies and devices that would have to be installed or deployed in homes is limited. Therefore, concentrated efforts are needed to increase the awareness and familiarity of island residents with these technologies to ensure the success of any deployment plans for islands communities to be realised.

Overall, as illustrated in Figure 6, very few of the respondents from all three islands owned or had a significant knowledge of any of the SG and DR enabling technologies. The majority of the respondents reported that they have not heard of most of the SG and DR technologies included in the survey (see Figure 6). In Inis Mór and Carloforte, around half of the respondents reported never to have heard of the technologies with the exception of electric vehicles, solar photovoltaic panels and smart washing machines. Levels of familiarity are slightly higher in La Graciosa. Of note (see Figure 6), is the very small number of respondents that own one of the DR technologies or report to know a lot about them, especially in Carloforte and Inis Mór. In all three islands, the very low levels of knowledge and familiarity with smart meters, in-home displays and HEMs is concerning, given their importance as enabling technologies for DR and the SG [33].

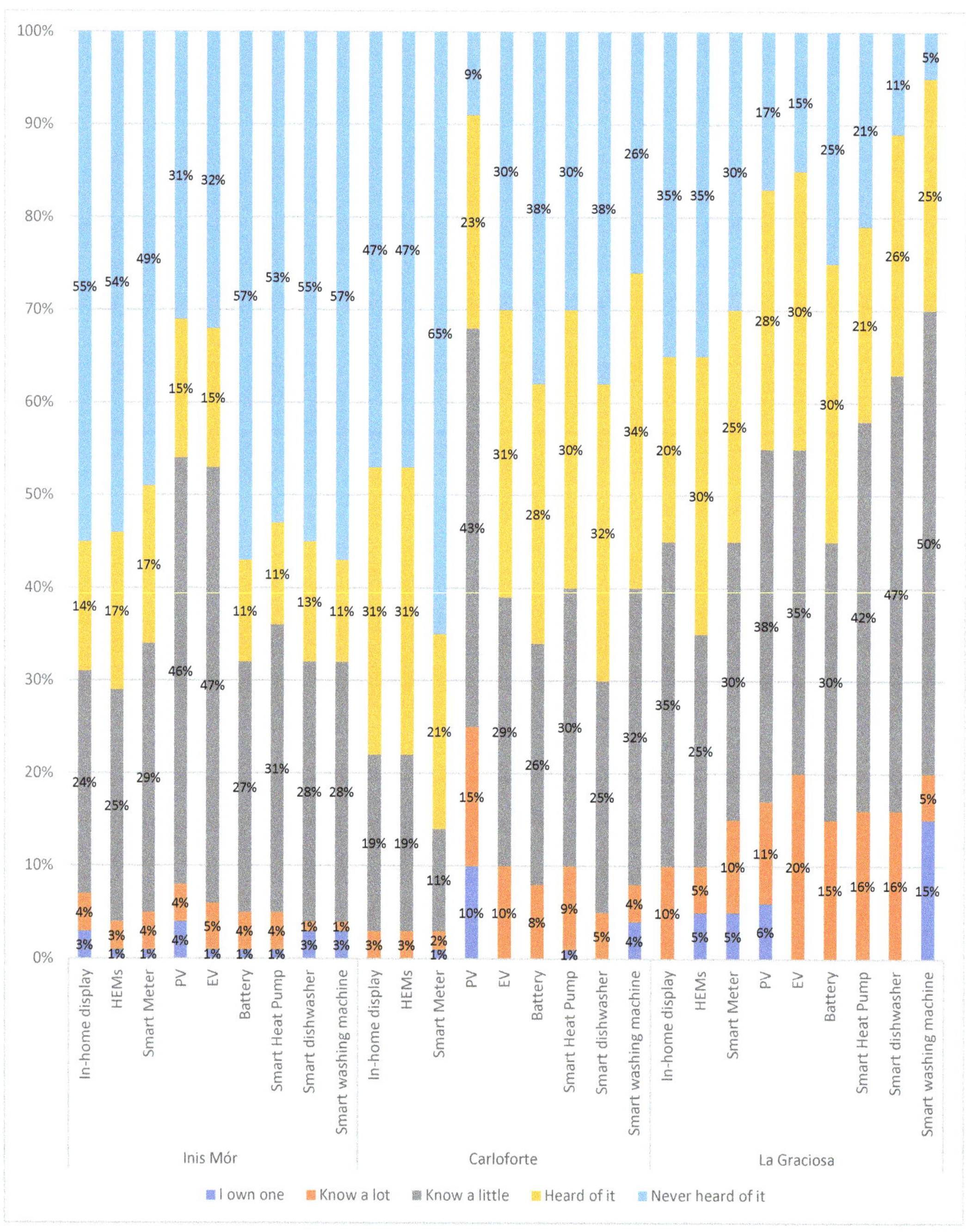

Figure 6. Familiarity of the three island residents with various Smart Grid technologies.

4.2.3. Respondents' Willingness to Adopt SG Enabling Technologies

Given the very low levels of ownership of SG and DR technologies reported by respondents, their willingness to adopt these technologies in the future is key to their engagement in the SG and DR in the future. Respondents were asked which of the different SG enabling technologies listed in Figure 6 they would like to use in their home. The results (see Figure 7) show that respondents in La Graciosa were the most positive about these technologies. However, even in La Graciosa, only 52% of respondents indicated that they would like to use a smart meter or HEMs in their homes, which as discussed in Section 2 are key enabling technologies for the SG. The lowest support for SG and DR enabling technologies comes from Carloforte, where respondents are the least willing of the respondents on the three islands surveyed to adopt such technologies. In Carloforte, only 14% of respondents said they would like to use a smart meter in their home, although a more encouraging percentage of respondents (39%) said they would like to use a HEMS. In Inis Mór, the percentages of respondents that indicated they would like to use SG enabling technologies is more encouraging overall than is the case for Carloforte, with 41% indicating that they would like to use a smart meter and 40% indicating they would like to use a HEMs. Overall, the results suggest a reluctance amongst a significant number of island residents when it comes to adopting or installing various SG and DR related technologies, such as heat pumps, EVs, smart metering or technologies that would allow the management of the energy loads in their homes, such as HEMs or in-home displays.

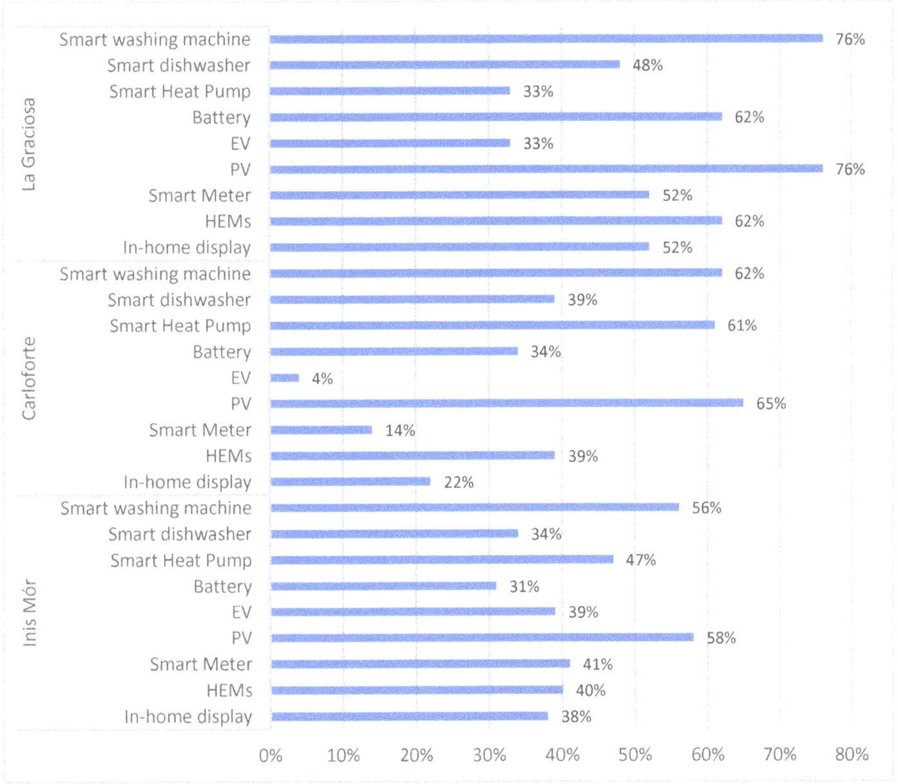

Figure 7. Willingness of island respondents to use SG enabling technologies.

4.2.4. Willingness to Invest in Smart and Renewable Energy Technologies

Given the low levels of ownership of SG and DR enabling technologies, the level of investment that respondents are willing to make to install RETs or SG-related technologies is important to consider. Respondents were asked how much they are willing to invest (as one-off investment) for installing RETs or smart energy control systems in their homes. The responses offered were "less than €99", "between €100 and €499", "between €500 and €999", "between €1000 and €4999", "€5000 or more", "I don't know" and "Not willing to invest". Figure 8a below shows that in Inis Mór, almost half (52%) of the respondents were willing to invest, with the majority (39%) preferring to pay between €1000 and €4999, followed by 25% who prefer to pay between €500 and €999 (Figure 8b).

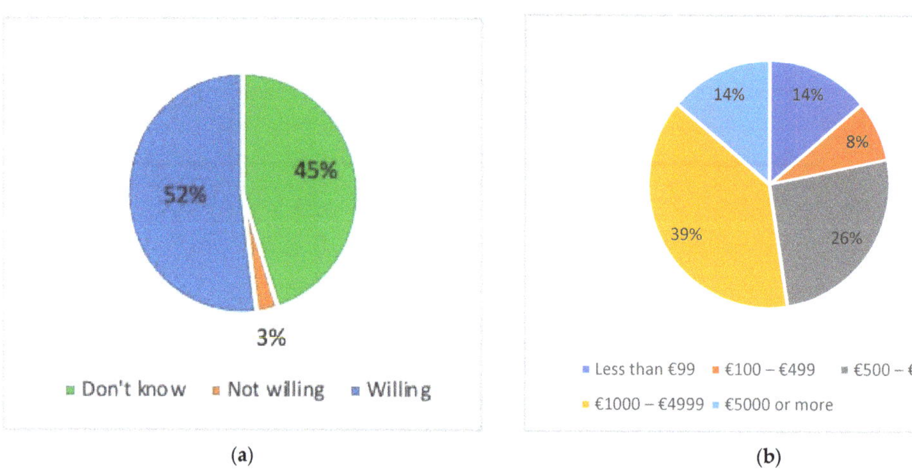

Figure 8. (**a**) Willingness to pay for SG technologies in Inis Mór, Aran Isles; (**b**) Range of payment respondents are willing to make in Inis Mór, Aran Isles.

Figure 9a, below, shows that in Carloforte the majority of the respondents were willing to invest (69%), with 33% preferring to pay between €1000 and €4999, followed by 24% who prefer to pay between €500 and €999 (Figure 9b).

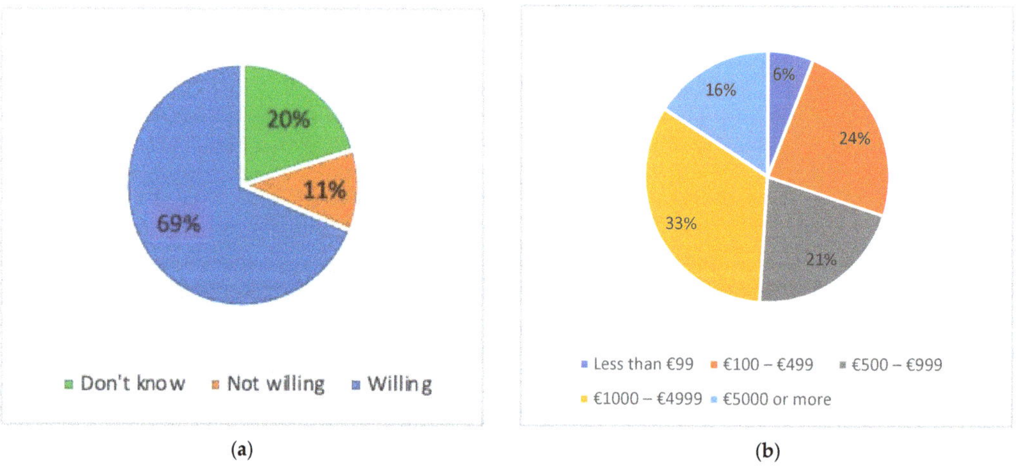

Figure 9. (**a**) Willingness to pay for SG technologies in Carloforte; (**b**) Range of payments respondents are willing to make in Carloforte.

In La Graciosa, as shown in Figure 10a, 59% will consider investing in SG technologies compared to 41% who do not know whether they are willing or not. Of these, 30% are willing to pay less than €99, 10% are willing to pay between €100 and €499, 20% are willing to pay between €500 and €999 and 40% are willing to pay between €1000 and €4999 (as shown in Figure 10b). The results suggest that more transparent communication is needed with respect to how much these technologies cost and what levels of investments and subsidies should be considered if more RET assets are desired. Therefore, for the majority of respondents, it can be seen that people are willing to invest in DR and RET technologies. However, a large proportion of the respondents in the three surveys were not willing to pay more than €500. This is important to consider when designing residential solutions, which should have realistic and feasible costs.

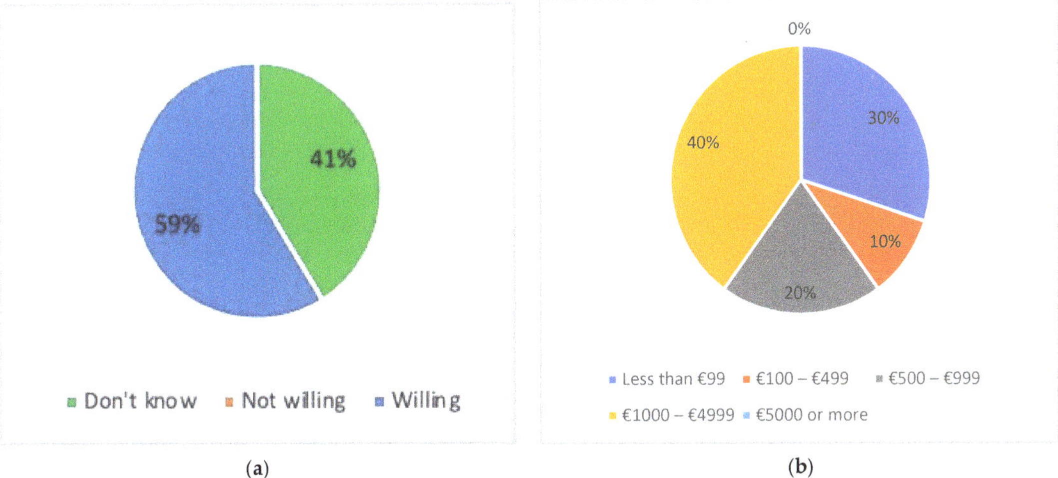

Figure 10. (**a**) Willingness to pay for SG technologies in La Graciosa; (**b**) Range of payment respondents are willing to make in La Graciosa.

5. Discussion

The survey results from the three islands are important for considering how ready and likely residents are in islands in Europe to participate in the SG and to adopt the various enabling technologies in their homes. Given the results discussed in Section 4, there is potential insofar as the respondents' motivations are concerned. The high importance of energy bills for households on islands and the high levels of reported willingness to save energy and adopt RETs are encouraging and correspond with the literature on islands as communities generally motivated to take part in sustainable energy transitions [9]. In this section, we discuss the results in comparison to findings from the literature exploring acceptance of TOU, DR and SG-related technologies, and knowledge and perceptions of the SG and smart appliances. Although the majority of these studies were conducted on mainland communities and households, we consider their findings relevant to our results, highlighting the salient issues that should be considered in assessing the readiness of island communities for the SG and when implementing DR and SG-related interventions in islands.

The cost of energy is argued to be a key motivator for the uptake of new and innovative energy technologies for the home and many argue that financial incentives are the most important motivators for people to change their behaviour, and respond to DR programming and flexibility requirements [40,86]. This is also evident in our results, showing that environmental motivations are also important for respondents, corresponding with findings on residents in general in relation to non-economic factors increasingly reported in the literature as a determinant in people's motivation for DR uptake [46,87]. However, previous studies on TOU have also highlighted the limitations of economic motivations alone [44,45], especially for higher income groups or those that do not use electricity for space heating and cooling [88]. Equity and fairness concerns were also raised in earlier research [89], particularly in relation to low-income groups and price rises where households do not have the means to respond to mechanisms such as TOU, e.g., not having a thermostat to control temperature settings [89]. Whilst wealthier households are able to invest in energy saving solutions for their homes [90], the transition to greener and smarter homes risks driving poorer households into further poverty and exclusion. Accordingly, DR implementation should only follow sufficient investment in energy efficiency for lower-income homes and maintaining the voluntary contracting based of TOU. Importantly, the context within which DR interventions are made, i.e., the household appliances and equipment and the extent to which homes on the island require investment, should also be considered. In this regard, the readiness of the three islands with respect to the SG should carefully consider how far can the economic motivations of the islanders be effective in incentivising them to respond positively to DR requests. Importantly, equity and justice considerations should be taken into account to ensure energy transitions for island communities are just.

As mentioned earlier, familiarity and knowledge with the principles and technologies of SG and DR is imperative for participation in and adoption of such systems [70,73]. However, our survey found that many people living on the REACT pilot islands have little or no understanding of the smart grid and the technologies required to interact with it, with very small proportions of respondents reporting any familiarity or experience of using it. This is reflected in their lack of willingness to install most of the DR technologies in their homes, particularly the enabling technologies, as they do not know enough about how these technologies work or how they might impact their everyday life. We note that smart meters, in-home displays and HEMs for households that wish to participate in DR are the necessary infrastructure that is required, and therefore the reluctance of individual households to install these technologies in their homes suggests that island communities may not be as ready to lead in sustainable energy transitions as has been argued in the literature. This low level of readiness is also reflected in the limited penetration of controllable heating and cooling equipment in the homes of the respondents in our survey and the challenges associated with transforming the domestic sector towards electric heating such as heat pumps [85]. Even though the respondents' reported behaviours for saving and thermostat use are encouraging, the limited proportion of households with programmable thermostats suggests a low level of technological readiness.

A further barrier to readiness to engage in the SG and DR that emerges from our survey results is that respondents are generally not motivated by mechanisms that activate social norms related to energy consumption [65], such as comparing their energy performance to others or using social media platforms to motivate and compete, in order to reduce or change their electricity consumption. In this regard, whilst the community aspect of islands relative to social cohesion [25] should not be ignored, relying on competition and comparison as a driver for engagement will have limited traction with islanders. Therefore, although islands possess characteristics that can make them ideal testbeds for new the SG, efforts should focus on how DR and SG interventions are designed, and caution should be taken when making assumptions of what can motivate householders to adopt new technologies in their homes and to adapt their lifestyles to accommodate the requirements of the SG and localised renewable energy generation.

Finally, a further barrier to consider regarding readiness for the SG and DR would be people's everyday lifestyles and practices, and how far these interventions can succeed in changing people's energy consumption patterns and behaviours. Studies on energy consumption informed by perspectives from the social sciences have shown that, in general, people do not often think about or calculate how much electricity they use until they see their energy bill [91], as people use their electrical appliances to conduct normal everyday activities, including cooking and eating, leisure, as well as convenience and cleanliness [92,93]. As such, electricity consumption is largely inconspicuousness, and different energy consuming behaviours become habitual [94]. In contrast, energy related practices are expected to change—in their timing as well as in their constituent elements (the different technologies and artefacts that shape them)—if the SG and the uptake of associated DR technologies is to be achieved. Further challenges related to the uptake of DR include the nature of some everyday practices. An examination of user experiences with DR has shown where many of those activities can be inflexible [95,96]. In this way, participating in the smart grid can be perceived as an inconvenience, particularly when the economic incentives are unable to overcome these issues [97]. A family's convenience is dependent on schedules and the simultaneous timing of different events and activities that include using different appliances. Family structures influence uptake, where single-person households and childless couples are more flexible than families with children or pets [82,98,99]. For families with children, the morning routine would be too busy a time to insert those activities into [82]. Therefore, designer expectations on how much flexibility can be obtained from users in DR interventions such as REACT should take into consideration barriers that pertain to normal everyday practices, and the extent of shifting action and flexibility is therefore bounded by household variables. Hence the type of households that are most suited to interventions like REACT could be further limited. Therefore, providing householders with the option of opting out of certain DR requests during certain hours of the day or over different times of the year will help circumvent those difficulties and encourage householders to participate in DR programmes.

To build on the economic and environmental motivations emerging from the analysis, efforts are needed to improve knowledge and understanding of the technologies of the SG and the implications these have on people's homes, their everyday lives and communities. The current low levels of adoption of DR and SG technologies forces us to think about the different barriers for the diffusion of innovation in societies, from becoming aware of an innovation, deciding whether to accept or reject it, to adopting and using it [100,101]. From this perspective, users who adopt an innovation before others are considered "early adopters", who as individuals might be more open to change and "earlier in adopting new ideas than other members of a system" [100] (p. 267). These individuals are typically younger in age, enjoy a higher social status, and are more educated. As such, they can see the benefits of adopting new technologies earlier than others. Similarly, in a study on the likelihood of accepting sustainable energy technologies, Stephanides et al. [9] found that the group more supportive of change were male, knowledgeable and concerned about the environment. This in turn suggests that to instigate the diffusion of DR technologies and the SG, efforts should aim at increasing knowledge about these technologies in island communities. This is especially important given the overall low familiarity reported from the survey results.

The implications for DR technologies such as the REACT solution are discussed on the basis of five factors, as developed by Garling and Thogersen [102], namely relative advantage, compatibility with existing context, complexity, trialability and observability. Relative advantage pertains to how much is an innovation perceived to be better than existing ideas or solutions. Compatibility concerns the innovations is suitable and fitting with the lifestyle, needs, values and experiences of potential users. Complexity refers to how relatively difficult or challenging to understand or use the innovation is, whereas trialability refers to how easy it is to demonstrate or experiment with the innovation. Finally, observability pertains to how visible the outcome of the innovation would be to potential users. The relevance of these implication to DR technologies and the case of the REACT solution is outlined in Table 3.

Table 3. Dimensions of technology adoption for the REACT solution.

Dimension for Adoption	Dimension of Adoption for the REACT Solution
Relative advantage	The advantages of DR are well known; however, knowledge of the principles and technologies of SG is low overall.
Compatibility	Previous research indicates that the flexibility that SG requires might not be easy to implement. The reluctance of respondents for adopting SG technologies suggests that perceived compatibility is low.
Complexity	The low levels of understanding of SG concepts and technologies is likely to be a result of the complexity of the concept and its application.
Trialability	The possibility to trial the REACT solution is an advantage here, particularly given the high costs if the REACT solution was to be available in an open market.
Observability	Financial benefits to households can be observed via lower electricity bills. The wider outcomes of SG technologies are not directly visible as these pertain to efficiencies to the grid, lower CO_2 emissions and lower costs for the islands.

The findings from our survey show the limited knowledge that residents in the three islands have of DR concepts and the SG. As such, demonstrations and trials are necessary for engaging communities with energy demand management, including DR and other technologies of the SG. In this context, the ability to trial DR projects like REACT can be an advantage. However, the complexity involved in the operation of various technological components and in being able to understand the principles underpinning the SG is a challenge. Direct lower electricity bills might be an observable outcome of the REACT project, which is advantageous. However, the inability to observe wider effects on the island will mean that other non-economic and community benefits can lose their significance. Therefore, to increase the observability of REACT and make it more visible in the communities, the demonstration of its solutions in community and public buildings should be an opportunity to showcase the benefits to the island residents. In turn, these will have to be communicated effectively to members of the community and households, using local networks that are trusted and that are embedded in the respective communities. As examples from other energy projects have shown, partnering with local community organisations and institutions [103], as well as engaging with the islanders with transparency in order to build trust [23] are crucial for the success of interventions for DR. Earlier research has also shown that if the designers of technologies intended to reduce or shift energy consumption are not sensitive to how people live and work in buildings, a gap occurs between the expected and actual performance of those technologies [104,105]. Therefore, it is essential to the success of SG interventions on islands that they are designed in ways which are sensitive to island residents' expectations.

6. Conclusions

In this paper, we introduced the H2020 REACT project that is being piloted in the three European islands: La Graciosa, Carloforte and Inis Mór. We outlined the main objectives of the project and presented the findings from a survey we conducted with a sample of households on the three islands. Conducting the survey presented us with a unique opportunity to assess the readiness levels for the SG among island dwellers and to measure familiarity and knowledge of the SG concept and DR technologies among island dwellers by employing a quantitative approach. Our results show that high levels of economic and environmental motivation and a willingness to change electricity consuming behaviours indicate households in those communities are ready to participate in the SG and adopt DR technologies. However, other variables suggest that more efforts are needed to succeed in engaging with communities with technological solutions such as REACT. Insofar as people's experience and familiarity with DR technologies is concerned and their lack of familiarity with the concept of the SG, coupled with the very low levels of current adoption of such technologies, these island communities might not be as receptive as expected.

This has implications for the design and development of interventions in DR and the future of smart grids. Firstly, more effort should be devoted to building awareness and knowledge of the SG by effectively utilising local networks and partnering with community-based organisations and, where possible, to demonstrate these technologies in community buildings. The design of the user interface should also take into consideration the different factors that motivate people to take part in DR programmes. For example, tailoring interventions based on people's stated preferences and motivations based on conducting consumer surveys will improve the roll-out of these solutions. Finally, the ability to opt-out of DR requests should be a built-in feature of any DR intervention to overcome the reticence of households who are concerned about how shifting their consumption patterns might affect their routines and the smooth running of their homes. Considering the processes required for the successful diffusion of innovation [102], the demonstration of technological interventions such as REACT represents a crucial phase in their development. Sufficient attention to make these projects effective models of innovation and engagement is therefore imperative to success in transitioning islands to greener and more sustainable futures. Our study indicates that the readiness of island dwellers for the SG should not be assumed, while careful social indicators in relation to knowledge, familiarity and motivations should be assessed when planning smart solutions for islands.

Author Contributions: Conceptualization, T.C. and D.A.G.; methodology, D.A.G.; formal analysis, D.A.G. and T.C.; investigation, D.A.G. and T.C.; data curation, D.A.G.; writing—original draft preparation, D.A.G.; writing—review and editing, T.C.; visualization, D.A.G.; supervision, T.C.; project administration, T.C. and D.A.G.; funding acquisition, T.C. All authors have read and agreed to the published version of the manuscript.

Funding: The research presented in this paper is partly financed by the European Union (H2020 REACT project, Grant Agreement No.: 824395).

Institutional Review Board Statement: Ethics approval was granted to D.A. and T.C. by the School of Computing, Engineering and Digital Technologies Research Ethics Sub-Committee [SRESC] held on 14 November 2019, subject to minor amendments. These amendments were submitted on 3 September 2019 and final approval was granted on 6 December 2019.

Informed Consent Statement: Informed consent was obtained from all subjects involved in the questionnaire study.

Data Availability Statement: Not applicable.

Acknowledgments: The authors would like to express their gratitude towards the partners of the REACT project show supported data collection efforts on the three islands, reviewed earlier versions of the questionnaire and translated the questionnaire to the local languages. Therefore, special thanks go to Feníe Energía in La Graciosa, R2M Solutions especially Giulia Carbonari in Carloforte and to Údaras na Gaeltachta and Comharchumann Forbartha Árann Teo (CFAT) in Inis Mór, especially Cathy Ní Ghóill for their tremendous efforts that made this research possible.

Conflicts of Interest: The authors declare no conflict of interest. The funders had no role in the design of the study; in the collection, analyses, or interpretation of data; in the writing of the manuscript, or in the decision to publish the results.

Appendix A

The survey questions are described in Table A1 below. The questions on energy attitudes, familiarity with SG and DR technologies and motivating factors were adapted from the work of Li et al. [76]. Questions on demographics, households, building characteristics, heating and cooling technologies and behaviours, as well as willingness to invest in RET and SG enabling technologies were developed by the authors.

Table A1. Questions in the survey.

Aspects	Survey Questions
Demographic, household and building characteristics	Age; gender; household size; age of household members; education level (primary, secondary, technical training/education, university, postgraduate); employment (student, part-time work, full-time work, self-employed, unemployed, retired); type of dwelling (detached, semi-detached, row house, apartment in building, shared house); number of bedrooms.
Energy consumption and impact on household	impact of energy bill on household budget (very high impact, high impact, medium impact, low impact, no impact, Don't know), average household energy bill (€50 or less, €50–€100, €100–€150, €150–€200, €200 or more, I don't know), heating and air-conditioning systems: type (central, individual units, both, Don't know); rooms heated; method of use (always on at constant temperature, always on at varied temperatures, turned on when someone is at home at constant temperature, turned on only when someone is at home in varied temperatures, Not often used); thermostat availability).
Energy attitudes	Importance of energy saving (very important, important, slightly important, not important, not important at all); having RETs (very important, important, slightly important, not important, not important at all)
Familiarity with SG and DR	How familiar are you with the SG concept before contact by REACT? (never heard of it, heard a little of it but don't understand the concept, heard a lot of it but don't understand the concept, know a little about the concept, know a lot about the concept); How familiar are you with the following SG technologies and appliances? (smart washing machine, smart tumble dryer, smart dishwasher, smart refrigerator/freezer, smart heat pump, hot water storage tank with smart charging and discharging, battery with smart charging and discharging, electric vehicle, pv, micro co-generation (micro combined heat and power), smart meter, home energy management system, home energy display. Possible answers: never heard of it, heard of it but do not understand the concept, know a little about the concept, know a lot about the concept, I own one.
Willingness to adopt SG technologies	Which of the following would you like to use in your house? Smart washing machine, smart tumble dryer, smart dishwasher, smart refrigerator/freezer, smart heat pump, hot water storage tank with smart charging and discharging, battery with smart charging and discharging, electric vehicle, pv, micro co-generation (micro combined heat and power), smart meter, home energy management system, home energy display
Motivating factors	Which of the following measures can motivate you to accept smart grids and use smart appliances?—Giving your house a more sustainable character, making your house high-tech, comparing your energy consumption to other households, sharing your results on social media, reducing your energy bill, contributing to the reliability of the grid, receiving acknowledgement for efforts, seeing the effects of your actions, reducing your CO_2 levels—(strongly motivating, motivating, slightly motivating, not motivating, I don't care).
Willingness to pay for SG technologies	How much would you be willing to invest for installing RETs or SG enabling technologies?—(€99 or less, €100–€499, €500–€999, €1000–€4999, €5000 or more, Don't know, I'm not willing to invest any money)

References

1. Edenhofer, O.; Pichs-Madruga, R.; Sokona, Y.; Seyboth, K.; Kadner, S.; Zwickel, T.; Eickemeier, P.; Hansen, G.; Schlömer, S.; von Stechow, C.; et al. (Eds.) *Renewable Energy Sources and Climate Change Mitigation: Special Report of the Intergovernmental Panel on Climate Change*; Cambridge University Press: Cambridge, UK, 2012.
2. Jacobson, M.Z.; Delucchi, M.A.; Bauer, Z.A.F.; Goodman, S.C.; Chapman, W.E.; Cameron, M.A.; Bozonnat, C.; Chobadi, L.; Clonts, H.A.; Enevoldsen, P.; et al. 100% Clean and Renewable Wind, Water, and Sunlight All-Sector Energy Roadmaps for 139 Countries of the World. *Joule* **2017**, *1*, 108–121. [CrossRef]
3. Muench, S.; Thuss, S.; Guenther, E. What hampers energy system transformations? The case of smart grids. *Energy Policy* **2014**, *73*, 80–92. [CrossRef]
4. Wissner, M. The Smart Grid—A saucerful of secrets? *Appl. Energy* **2011**, *88*, 2509–2518. [CrossRef]
5. Lovell, H. The promise of smart grids. *Local Environ.* **2019**, *24*, 580–594. [CrossRef]
6. Gyberg, P.; Palm, J. Influencing households' energy behaviour-how is this done and on what premises? *Energy Policy* **2009**, *37*, 2807–2813. [CrossRef]
7. IRENA. *A Path to Prosperity: Renewable Energy for Islands*; International Renewable Energy Agency: Bonn, Germany, 2016.
8. Smart Islands Initiative. Smart Islands Declaration: New Pathways for European Islands; Smart Islands Initiative. Available online: www.smartislandsinitiative.eu (accessed on 24 June 2021).
9. Stephanides, P.; Chalvatzis, K.J.; Li, X.; Lettice, F.; Guan, D.; Ioannidis, A.; Zafirakis, D.; Papapostolou, C. The social perspective on island energy transitions: Evidence from the Aegean archipelago. *Appl. Energy* **2019**, *255*, 113725. [CrossRef]
10. Sigrist, L.; Lobato, E.; Rouco, L.; Gazzino, M.; Cantu, M. Economic assessment of smart grid initiatives for island power systems. *Appl. Energy* **2017**, *189*, 403–415. [CrossRef]
11. Lovell, H.; Hann, V.; Watson, P. Rural laboratories and experiment at the fringes: A case study of a smart grid on Bruny Island, Australia. *Energy Res. Soc. Sci.* **2018**, *36*, 146–155. [CrossRef]
12. Jelić, M.; Batić, M.; Tomašević, N.; Barney, A.; Polatidis, H.; Crosbie, T.; Ghanem, A.D.; Short, M.; Pillai, G. Towards Self-Sustainable Island Grids through Optimal Utilization of Renewable Energy Potential and Community Engagement. *Energies* **2020**, *13*, 3386. [CrossRef]
13. Skjølsvold, T.M.; Ryghaug, M.; Throndsen, W. European island imaginaries: Examining the actors, innovations, and renewable energy transitions of 8 islands. *Energy Res. Soc. Sci.* **2020**, *65*, 101491. [CrossRef]
14. Dorotić, H.; Doračić, B.; Dobravec, V.; Pukšec, T.; Krajačić, G.; Duić, N. Integration of transport and energy sectors in island communities with 100% intermittent renewable energy sources. *Renew. Sustain. Energy Rev.* **2019**, *99*, 109–124. [CrossRef]
15. Duić, N.; da Graça Carvalho, M. Increasing renewable energy sources in island energy supply: Case study Porto Santo. *Renew. Sustain. Energy Rev.* **2004**, *8*, 383–399. [CrossRef]
16. Erdinc, O.; Paterakis, N.G.; Catalaõ, J.P.S. Overview of insular power systems under increasing penetration of renewable energy sources: Opportunities and challenges. *Renew. Sustain. Energy Rev.* **2015**, *52*, 333–346. [CrossRef]
17. Kuang, Y.; Zhang, Y.; Zhou, B.; Li, C.; Cao, Y.; Li, L.; Zeng, L. A review of renewable energy utilization in islands. *Renew. Sustain. Energy Rev.* **2016**, *59*, 504–513. [CrossRef]
18. Notton, G. Importance of islands in renewable energy production and storage: The situation of the French islands. *Renew. Sustain. Energy Rev.* **2015**, *47*, 260–269. [CrossRef]
19. Nyström, S.; Rivera, M.B.; Hedin, B.; Katzeff, C.; Menon, A.R. Challenging the image of the altruistic and flexible household in the smart grid using design fiction. In Proceedings of the Workshop on Computing within Limits, 14–15 June 2021.
20. Maldonado, E. *Final Report: Energy in the EU Outermost Regions (Renewable Energy, Energy Efficiency)*; European Commission: Brussels, Belgium, 2017.
21. Euroepan Commission. The Clean Energy for EU Islands Secretariat. Available online: https://euislands.eu/ (accessed on 15 August 2021).
22. de Lesdain, B.S. The photovoltaic installation process and the behaviour of photovoltaic producers in insular contexts: The French island example (Corsica, Reunion Island, Guadeloupe). *Energy Effic.* **2019**, *12*, 711–722. [CrossRef]
23. Morris, P.; Vine, D.; Buys, L. Critical Success Factors for Peak Electricity Demand Reduction: Insights from a Successful Intervention in a Small Island Community. *J. Consum. Policy* **2018**, *41*, 33–54. [CrossRef]
24. Martin, R. Making sense of renewable energy: Practical knowledge, sensory feedback and household understandings in a Scottish island microgrid. *Energy Res. Soc. Sci.* **2020**, *66*, 101501. [CrossRef]
25. Davidson, J. Citizenship and sustainability in dependent island communities: The case of the Huon Valley region in southern Tasmania. *Local Environ.* **2003**, *8*, 527–540. [CrossRef]
26. Mackelworth, P.H.; Caric, H. Gatekeepers of island communities: Exploring the pillars of sustainable development. *Environ. Dev. Sustain.* **2010**, *12*. [CrossRef]
27. REACT. Renewable Energy for Self-Sustainable Island Communities. Available online: https://react2020.eu/ (accessed on 20 June 2020).
28. Verbong, G.P.J.; Beemsterboer, S.; Sengers, F. Smart grids or smart users? Involving users in developing a low carbon electricity economy. *Energy Policy* **2013**, *52*, 117–125. [CrossRef]
29. Crosbie, T.; Baker, K. Energy-efficiency interventions in housing: Learning from the inhabitants. *Build. Res. Inf.* **2010**, *38*, 70–79. [CrossRef]

30. Wall, R.; Crosbie, T. Potential for reducing electricity demand for lighting in households: An exploratory socio-technical study. *Energy Policy* **2009**, *37*, 1021–1031. [CrossRef]
31. Crosbie, T.; Short, M.; Dawood, M.; Charlesworth, R. Demand response in blocks of buildings: Opportunities and requirements. *Int. J. Entrep. Sustain. Issues* **2017**, *4*, 271–281. [CrossRef]
32. Meyers, R.J.; Williams, E.D.; Matthews, H.S. Scoping the potential of monitoring and control technologies to reduce energy use in homes. *Energy Build.* **2010**, *42*, 563–569. [CrossRef]
33. Fotouhi Ghazvini, M.A.; Soares, J.; Abrishambaf, O.; Castro, R.; Vale, Z. Demand response implementation in smart households. *Energy Build.* **2017**, *143*, 129–148. [CrossRef]
34. Yadav, M.; Jamil, M.; Rizwan, M. Enabling Technologies for Smart Energy Management in a Residential Sector: A Review. In Proceedings of the Smart Cities—Opportunities and Challenges, Singapore, 21 April 2020; pp. 9–21.
35. Kloppenburg, S.; Boekelo, M. Digital platforms and the future of energy provisioning: Promises and perils for the next phase of the energy transition. *Energy Res. Soc. Sci.* **2019**, *49*, 68–73. [CrossRef]
36. Fell, M.J.; Shipworth, D.; Huebner, G.M.; Elwell, C.A. Public acceptability of domestic demand-side response in Great Britain: The role of automation and direct load control. *Energy Res. Soc. Sci.* **2015**, *9*, 72–84. [CrossRef]
37. Khan, A.R.; Mahmood, A.; Safdar, A.; Khan, Z.A.; Khan, N.A. Load forecasting, dynamic pricing and DSM in smart grid: A review. *Renew. Sustain. Energy Rev.* **2016**, *54*, 1311–1322. [CrossRef]
38. Ellabban, O.; Abu-Rub, H. Smart grid customers' acceptance and engagement: An overview. *Renew. Sustain. Energy Rev.* **2016**, *65*, 1285–1298. [CrossRef]
39. Crosbie, T.; Broderick, J.; Short, M.; Charlesworth, R.; Dawood, M. Demand Response Technology Readiness Levels for Energy Management in Blocks of Buildings. *Buildings* **2018**, *8*, 13. [CrossRef]
40. Faruqui, A.; Sergici, S. Household response to dynamic pricing of electricity: A survey of 15 experiments. *J. Regul. Econ.* **2010**, *38*, 193–225. [CrossRef]
41. Mancini, F.; Basso, L.G.; de Santoli, L. Energy Use in Residential Buildings: Impact of Building Automation Control Systems on Energy Performance and Flexibility. *Energies* **2019**, *12*, 2896. [CrossRef]
42. Batalla-Bejerano, J.; Trujillo-Baute, E.; Villa-Arrieta, M. Smart meters and consumer behaviour: Insights from the empirical literature. *Energy Policy* **2020**, *144*, 111610. [CrossRef]
43. Brent, D.A.; Friesen, L.; Gangadharan, L.; Leibbrandt, A. Behavioral Insights from Field Experiments in Environmental Economics. *Int. Rev. Environ. Resour. Econ.* **2017**, *10*, 95–143. [CrossRef]
44. Gyamfi, S.; Krumdieck, S. Price, environment and security: Exploring multi-modal motivation in voluntary residential peak demand response. *Energy Policy* **2011**, *39*, 2993–3004. [CrossRef]
45. Gyamfi, S.; Krumdieck, S.; Urmee, T. Residential peak electricity demand response—Highlights of some behavioural issues. *Renew. Sustain. Energy Rev.* **2013**, *25*, 71–77. [CrossRef]
46. Hall, N.L.; Jeanneret, T.D.; Rai, A. Cost-reflective electricity pricing: Consumer preferences and perceptions. *Energy Policy* **2016**, *95*, 62–72. [CrossRef]
47. Mert, W.; Watts, M.; Tritthart, W. Smart domestic appliances in sustainable energy systems-consumer acceptance and restrictions. In Proceedings of the ECEEE Summer Study: Act! Innovate! Deliver! Reducing Energy Demand Sustainably, Nice, France, 1–6 June 2009; pp. 1751–1761.
48. Paetz, A.-G.; Dütschke, E.; Fichtner, W. Smart Homes as a Means to Sustainable Energy Consumption: A Study of Consumer Perceptions. *J. Consum. Policy* **2012**, *35*, 23–41. [CrossRef]
49. Faruqui, A.; George, S. Quantifying Customer Response to Dynamic Pricing. *Electr. J.* **2005**, *18*, 53–63. [CrossRef]
50. Parag, Y.; Butbul, G. Flexiwatts and seamless technology: Public perceptions of demand flexibility through smart home technology. *Energy Res. Soc. Sci.* **2018**, *39*, 177–191. [CrossRef]
51. Srivastava, A.; Van Passel, S.; Laes, E. Assessing the success of electricity demand response programs: A meta-analysis. *Energy Res. Soc. Sci.* **2018**, *40*, 110–117. [CrossRef]
52. Flaim, T.; Neenan, B.; Robinson, J. Pilot Paralysis: Why Dynamic Pricing Remains Over-Hyped and Underachieved. *Electr. J.* **2013**, *26*, 8–21. [CrossRef]
53. Bamberg, S.; Möser, G. Twenty years after Hines, Hungerford, and Tomera: A new meta-analysis of psycho-social determinants of pro-environmental behaviour. *J. Environ. Psychol.* **2007**, *27*, 14–25. [CrossRef]
54. Bradley, P.; Leach, M.; Torriti, J. A review of the costs and benefits of demand response for electricity in the UK. *Energy Policy* **2013**, *52*, 312–327. [CrossRef]
55. Shipman, R.; Gillott, M.; Naghiyev, E. SWITCH: Case Studies in the Demand Side Management of Washing Appliances. *Energy Procedia* **2013**, *42*, 153–162. [CrossRef]
56. Ofgem. *Energy Demand Research Project: Final Analysis*; AECOM Building Engineering for Ofgem: Hetfordshire, UK, 2011.
57. Western Power Distribution. *SoLa Bristol SDRC 9.8 Final Report*; Western Power Distribution: Bristol, UK, 2016.
58. Bird, J. *Developing the Smarter Grid: The Role of Domestic and Small and Medium Enterprise Customers*; Report from the Customer Led Network Revolution; Nothern Power Grid: Newcastle upon Tyne, UK, 2015.
59. Carmichael, R.; Schofield, J.; Woolf, M.; Bilton, M.; Ozaki, R.; Strbac, G. *Residential Consumer Attitudes to Time-Varying Pricing*; Imperial College London: London, UK, 2014.

60. Strengers, Y. Comfort expectations: The impact of demand-management strategies in Australia. In *Comfort in a Lower Carbon Society*; Shove, E., Chappells, H., Lutzenhiser, L., Eds.; Routledge: New York, NY, USA, 2010; pp. 77–88.

61. Parag, Y. Which factors influence large households' decision to join a time-of-use program? The interplay between demand flexibility, personal benefits and national benefits. *Renew. Sustain. Energy Rev.* **2021**, *139*, 110594. [CrossRef]

62. Spence, A.; Leygue, C.; Bedwell, B.; O'Malley, C. Engaging with energy reduction: Does a climate change frame have the potential for achieving broader sustainable behaviour? *J. Environ. Psychol.* **2014**, *38*, 17–28. [CrossRef]

63. Darby, S. *The Effectiveness of Feedback on Energy Consumption: A Review for Defra of the Literature on Metering, Billing and Direct Displays*; Working Paper; Oxford Environmental Change Institute: Oxford, UK, 2006.

64. Petkov, P.; Köbler, F.; Foth, M.; Krcmar, H. Motivating Domestic Energy Conservation through Comparative, Community-Based Feedback in Mobile and Social Media. In Proceedings of the Fifth International Conference on Communities and Technologies, Brisbane, Australia, 29 June–2 July 2011; Volume 11, pp. 21–30. [CrossRef]

65. Ehrhardt-Martinez, K.; Donnelly, K.; Laitner, J. *Advanced Metering Initiatives and Residential Feedback Programs: A Meta-Review for Household Electricity-Saving Opportunities*; American Council for an Energy-Efficient Economy: Washington, DC, USA, 2010.

66. Mankoff, J.; Matthews, D.; Fussell, S.R.; Johnson, M. Leveraging Social Networks To Motivate Individuals to Reduce their Ecological Footprints. In Proceedings of the 2007 40th Annual Hawaii International Conference on System Sciences (HICSS'07), Big Island, HI, USA, 3–6 January 2007; p. 87.

67. Mankoff, J.; Fussell, S.R.; Dillahunt, T.; Glaves, R.; Grevet, C.; Johnson, M.; Matthews, D.; Matthews, H.S.; McGuire, R.; Thomson, R.; et al. StepGreen.org: Increasing Energy Saving Behaviors via Social Networks. In Proceedings of the Fourth International AAAI Conference on Weblogs and Social Media, Washington, DC, USA, 23–26 May 2010.

68. Foster, D.; Lawson, S.; Blythe, M.; Cairns, P. Wattsup? Motivating reductions in domestic energy consumption using social networks. In Proceedings of the 6th Nordic Conference on Human-Computer Interaction: Extending Boundaries, Reykjavik, Iceland, 16–20 October 2010; pp. 178–187.

69. Burchell, K.; Rettie, R.; Roberts, T.C. Householder engagement with energy consumption feedback: The role of community action and communications. *Energy Policy* **2016**, *88*, 178–186. [CrossRef]

70. Bartusch, C.; Wallin, F.; Odlare, M.; Vassileva, I.; Wester, L. Introducing a demand-based electricity distribution tariff in the residential sector: Demand response and customer perception. *Energy Policy* **2011**, *39*, 5008–5025. [CrossRef]

71. Bugden, D.; Stedman, R. Unfulfilled promise: Social acceptance of the smart grid. *Environ. Res. Lett.* **2021**, *16*, 034019. [CrossRef]

72. Bugden, D.; Stedman, R. A synthetic view of acceptance and engagement with smart meters in the United States. *Energy Res. Soc. Sci.* **2019**, *47*, 137–145. [CrossRef]

73. Parrish, B.; Heptonstall, P.; Gross, R.; Sovacool, B.K. A systematic review of motivations, enablers and barriers for consumer engagement with residential demand response. *Energy Policy* **2020**, *138*, 111221. [CrossRef]

74. Polonsky, M.J.; Vocino, A.; Grau, S.L.; Garma, R.; Ferdous, A.S. The impact of general and carbon-related environmental knowledge on attitudes and behaviour of US consumers. *J. Mark. Manag.* **2012**, *28*, 238–263. [CrossRef]

75. Flamm, B. The impacts of environmental knowledge and attitudes on vehicle ownership and use. *Transp. Res. Part D Transp. Environ.* **2009**, *14*, 272–279. [CrossRef]

76. Li, R.; Dane, G.; Finck, C.; Zeiler, W. Are building users prepared for energy flexible buildings?—A large-scale survey in the Netherlands. *Appl. Energy* **2017**, *203*, 623–634. [CrossRef]

77. Wilson, C.; Hargreaves, T.; Hauxwell-Baldwin, R. Smart homes and their users: A systematic analysis and key challenges. *Pers. Ubiquitous Comput.* **2015**, *19*, 463–476. [CrossRef]

78. Balta-Ozkan, N.; Davidson, R.; Bicket, M.; Whitmarsh, L. Social barriers to the adoption of smart homes. *Energy Policy* **2013**, *63*, 363–374. [CrossRef]

79. Hargraves, T.; Hauxwell-Baldwin, R.; Coleman, M.; Wilson, C.; Stankovic, L.; Stankovic, V.; Liao, J.; Murray, D.; Kane, T.; Hassan, T.; et al. Smart homes, control and energy management: How do smart home technologies influence control over energy use and domestic life? In Proceedings of the European Council for an Energy Efficient Economcy Conference 2015, Hyères, France, 1–6 June 2015.

80. Ford, R.; Paniamina, R. *Smart Homes: What New Zealanders Think, Have, and Want. (Project Report)*; Centre for Sustainability, University of Otago: Dunedin, New Zealand, 2016.

81. Wang, X.; Abi Ghanem, D.; Larkin, A.; McLachlan, C. How the meanings of 'home' influence energy-consuming practices in domestic buildings. *Energy Effic.* **2020**, *14*, 1. [CrossRef]

82. Nyborg, S.; Ropke, I. Constructing users in the smart grid-insights from the Danish eFlex project. *Energy Effic.* **2013**, *6*, 655–670. [CrossRef]

83. Powells, G.; Bulkeley, H.; Bell, S.; Judson, E. Peak electricity demand and the flexibility of everyday life. *Geoforum* **2014**, *55*, 43–52. [CrossRef]

84. Abi Ghanem, D.; Crosbie, T. *REACT Project Deliverable 4.1: Criteria and Framework for Participant Recruitment*; Project Report; Centre for Sustainable Engineering, Teesside University: Middlesbrough, UK, 2020.

85. Gaur, A.S.; Fitiwi, D.Z.; Curtis, J. Heat pumps and our low-carbon future: A comprehensive review. *Energy Res. Soc. Sci.* **2021**, *71*, 101764. [CrossRef]

86. Vanthournout, K.; Dupont, B.; Foubert, W.; Stuckens, C.; Claessens, S. An automated residential demand response pilot experiment, based on day-ahead dynamic pricing. *Appl. Energy* **2015**, *155*, 195–203. [CrossRef]

87. Torstensson, D.; Wallin, F. Exploring the Perception for Demand Response among Residential Consumers. *Energy Procedia* **2014**, *61*, 2797–2800. [CrossRef]
88. Reiss, P.C.; White, M.W. Household Electricity Demand, Revisited. *Rev. Econ. Stud.* **2005**, *72*, 853–883. [CrossRef]
89. Dillman, D.A.; Rosa, E.A.; Dillman, J.J. Lifestyle and home energy conservation in the United States: The poor accept lifestyle cutbacks while the wealthy invest in conservation. *J. Econ. Psychol.* **1983**, *3*, 299–315. [CrossRef]
90. Alexander, B.R. Dynamic Pricing? Not So Fast! A Residential Consumer Perspective. *Electr. J.* **2010**, *23*, 39–49. [CrossRef]
91. Shove, E.; Chappells, H. *Ordinary Consumption and Extraordinary Relationships: Utilities and Their Users*; Gronow, J., Warde, A., Eds.; Routledge: London, UK, 2013; pp. 45–58.
92. Shove, E. Converging Conventions of Comfort, Cleanliness and Convenience. *J. Consum. Policy* **2003**, *26*, 395–418. [CrossRef]
93. Warde, A. Consumption and theories of practice. *J. Consum. Cult.* **2005**, *5*, 131–153. [CrossRef]
94. Maréchal, K.; Holzemer, L. Getting a (sustainable) grip on energy consumption: The importance of household dynamics and 'habitual practices'. *Energy Res. Soc. Sci.* **2015**, *10*, 228–239. [CrossRef]
95. Christensen, T.H.; Friis, F.; Bettin, S.; Throndsen, W.; Ornetzeder, M.; Skjølsvold, T.M.; Ryghaug, M. The role of competences, engagement, and devices in configuring the impact of prices in energy demand response: Findings from three smart energy pilots with households. *Energy Policy* **2020**, *137*. [CrossRef]
96. Friis, F.; Haunstrup Christensen, T. The challenge of time shifting energy demand practices: Insights from Denmark. *Energy Res. Soc. Sci.* **2016**, *19*, 124–133. [CrossRef]
97. Michaels, L.; Parag, Y. Motivations and barriers to integrating 'prosuming' services into the future decentralized electricity grid: Findings from Israel. *Energy Res. Soc. Sci.* **2016**, *21*, 70–83. [CrossRef]
98. Nyborg, S. Pilot Users and Their Families: Inventing Flexible Pra ctices in the Smart Grid. *Sci. Technol. Stud.* **2015**, *3*, 54. [CrossRef]
99. Nicholls, L.; Strengers, Y. Peak demand and the 'family peak' period in Australia: Understanding practice (in)flexibility in households with children. *Energy Res. Soc. Sci.* **2015**, *9*, 116–124. [CrossRef]
100. Rogers, E.M. *Diffusion of Innovations*, 5th ed.; Free Press: New York, NY, USA, 2003.
101. Faiers, A.; Neame, C. Consumer attitudes towards domestic solar power systems. *Energy Policy* **2006**, *34*, 1797–1806. [CrossRef]
102. Gärling, A.; Thøgersen, J. Marketing of electric vehicles. *Bus. Strategy Environ.* **2001**, *10*, 53–65. [CrossRef]
103. Putnam, T.; Brown, D. Grassroots retrofit: Community governance and residential energy transitions in the United Kingdom. *Energy Res. Soc. Sci.* **2021**, *78*, 102102. [CrossRef]
104. Breukers, S.; Crosbie, T.; van Summeren, L. Mind the gap when implementing technologies intended to reduce or shift energy consumption in blocks-of-buildings. *Energy Environ.* **2020**, *31*, 613–633. [CrossRef]
105. Abi Ghanem, D.; Mander, S. Designing consumer engagement with the smart grids of the future: Bringing active demand technology to everyday life. *Technol. Anal. Strateg. Manag.* **2014**, *26*, 1163–1175. [CrossRef]

Article

The Impact of Detail, Shadowing and Thermal Zoning Levels on Urban Building Energy Modelling (UBEM) on a District Scale [†]

Xavier Faure *, Tim Johansson and Oleksii Pasichnyi *

Research Group for Urban Analytics and Transitions (UrbanT), Department of Sustainable Development, Environmental Science and Engineering (SEED), KTH Royal Institute of Technology, Teknikringen 10B, 100 44 Stockholm, Sweden

* Correspondence: xavierf@kth.se (X.F.); oleksii.pasichnyi@abe.kth.se (O.P.)

† This paper is an extended version of our paper published in the Proceeding of the 20th European Roundtable on Sustainable Consumption and Production, 8–10 September 2021, Graz, Austria.

Abstract: New modelling tools are required to accelerate the decarbonisation of the building sector. Urban building energy modelling (UBEM) has recently emerged as an attractive paradigm for analysing building energy performance at district and urban scales. The balance between the fidelity and accuracy of created UBEMs is known to be the cornerstone of the model's applicability. This study aimed to analyse the impact of traditionally implicit modeller choices that can greatly affect the overall UBEM performance, namely, (1) the level of detail (LoD) of the buildings' geometry; (2) thermal zoning; and (3) the surrounding shadowing environment. The analysis was conducted for two urban areas in Stockholm (Sweden) using MUBES—the newly developed UBEM. It is a bottom-up physics-based open-source tool based on Python and EnergyPlus, allowing for calibration and co-simulation. At the building scale, significant impact was detected for all three factors. At the district scale, smaller effects (<2%) were observed for the level of detail and thermal zoning. However, up to 10% difference may be due to the surrounding shadowing environment, so it is recommended that this is considered when using UBEMs even for district scale analyses. Hence, assumptions embedded in UBEMs and the scale of analysis make a difference.

Keywords: urban building energy model; UBEM; level of detail; LOD; shadowing; thermal zoning

Citation: Faure, X.; Johansson, T.; Pasichnyi, O. The Impact of Detail, Shadowing and Thermal Zoning Levels on Urban Building Energy Modelling (UBEM) on a District Scale. *Energies* **2022**, *15*, 1525. https://doi.org/10.3390/en15041525

Academic Editors: Sergio Ulgiati, Hans Schnitzer and Remo Santagata

Received: 20 January 2022
Accepted: 15 February 2022
Published: 18 February 2022

Publisher's Note: MDPI stays neutral with regard to jurisdictional claims in published maps and institutional affiliations.

1. Introduction

Buildings are responsible for one-third of the total final energy use and nearly 40% of total greenhouse gas emissions [1]. Hence, this sector is one of the key contributors to climate change and should be addressed in order to meet the 1.5 °C scenario [2]. There is a wide range of mitigation options available including the decarbonisation of supply, refurbishment of the existing building stock, and near-zero energy requirements of new buildings. However, the current speed of building energy transition is much slower than what is needed to meet national and local climate commitments [3]. New decision-making paradigms and tools are required to improve the overall efficiency of the building sector. There is an urgent need for integrated models and tools that would allow for the assessment of the benefits and deficiencies of each urban energy intervention in a holistic manner for all of involved stakeholders.

The initial uptake of city-scale building energy modelling was captured in the reviews by Swan and Ugursal [4] and Kavgic [5], which provided categorisation of the models into top–down and bottom–up, where the latter were divided into statistical and engineering. Top–down approach imposes the representation of the entire building stock as a single unit of analysis. In contrast, the bottom–up approach intends to focus on individual buildings. In their turn, statistical and engineering stand for data-driven or physics-based models being later joined by hybrid reduced-order models combining both approaches. The subsequent review by Reinhart and Davila [6] introduced the term of 'Urban Building Energy Modelling (UBEM)", which was attributed explicitly to bottom–up engineering models. This approach is different from a plain assembly of single building energy models (BEM) as it creates the automated generation of simulations based on larger amounts of structured data and a more simplistic representation of individual buildings. Most of the recent review papers have tended to focus on these types of models such as UBEMs, systematising their functional components [7], applied approaches [8], and key challenges [9]. However, a recent review by Ali et al. [10] returned to a wider scope, providing a comparative analysis of modern top–down and bottom–up urban-scale energy models.

A number of UBEM environments and tools have been developed in recent decades [11]. These include UBEMs using more detailed physics-based thermal engines such as EnergyPlus (CityBES [12], UMI [13]), simpler reduced-order models based on self-made RC networks (DIMOSIM [14], CitySim [15] or not formally named [16]), energy signatures [17], or the ISO/CEN standard method (SimStadt [18], CEA [19]). The review of UBEM cases in [20] shows that the choice of the model can be attributed to the project constraints, data and skills' availability, and, ultimately, the purpose of developed UBEM. The issue of scale has been addressed in different ways [7] including various approaches to align the created urban scale models with measured data using probabilistic calibration [21,22]. In the UBEM field, physics-based multizone dynamic models are required to evaluate design scenarios for new urban areas or carbon reduction strategies to existing building stock such as urban scale building retrofitting [6]. However, in the case of large scales, even a slight increase in resolution for one or more aspects of UBEM (e.g., spatial, temporal, scenario space) can lead to a noticeable growth in the computational burden due to the issue of dimensionality. For instance, more detailed thermal zoning will require that the higher system's dimensions are solved. In addition, another serious bottleneck for introducing higher spatial resolution to UBEM is traditionally the limited data availability.

The balance between model fidelity and accuracy is a key issue in UBEMs. Many studies have utilised archetypes (representative buildings for a group of similar buildings) to lower the number of simulations needed on a city scale [17,23,24]. A number of studies have investigated the impact of choices made when a new UBEM is set up. Three fidelity-related aspects have been regularly highlighted as having a crucial impact on the quality and applicability of the derived UBEMs, namely (a) the level of detail (LoD) of buildings' geometry [25], (b) thermal zoning [26], and (c) the shadowing effect of the surrounding environment [27]. Hence, the main value of the proposed study is in characterising the impact of these implicit assumptions on the quality of UBEMs. This contribution is expected to raise the awareness of scholars and practitioners, provide more ground-based reasons in making these modelling choices, and finally, improve the quality of decision-making based on these promising and powerful modelling tools.

This paper aimed to investigate the impact of the typical choices made at the UBEM generation stage, namely around the level of detail (LoD) of building geometries, the approach to thermal zoning, and the boundaries of the surrounding environment to be considered for shadowing. The study utilised MUBES (Massive Urban Building Energy Simulations)—a novel UBEM simulation tool presented in Section 2. The two urban districts, Minneberg and Hammarby Sjöstad (Stockholm, Sweden), used for the case study are described in Section 3. The analysis of the impact of LoD, thermal zoning, and shadowing is provided in Section 4. Finally, Section 5 summarises the paper with a discussion and our conclusions.

2. MUBES—An Open Tool for Urban Building Energy Modelling (UBEM)

2.1. UBEM Workflow

This section describes the methodology of MUBES—the new generation UBEM used for the analysis in this study. MUBES is a bottom–up physical UBEM providing a common framework for the computation of an individual buildings' energy demands in urban areas by using the data from various public data sources as inputs for dynamic building physics models. These building energy models follow a *shoebox* paradigm and are automatically generated for each individual building. These are composed of two levels: the building and the zone levels.

The building level requires building geometry with its surrounding environment, internal thermally conditioned zones/volumes delimitation, and some elements that characterize the performance of the building's envelope. The related inputs are assumed as static ones for the UBEM workflow as these are not time dependent on a yearly time basis. The zone level requires indoor elements and occupancy related inputs that have time-related impacts on energy needs. Hence, along the UBEM workflow, the latter inputs can be seen as dynamic ones. Internal heating and cooling production equipment are situated at the zone level as different production types can be present in the same building. Following the same paradigm, envelope leaks with time dependent impacts (from variables such as wind, pressure, and temperature) are also situated at the zone level. These can be differently addressed, depending on the type of zone (heated or non-heated).

The overall workflow can be described with four main steps (Figure 1): (I) data integration; (II) the generation of building models; (III) run of building energy simulations; and (IV) output and aggregation of results.

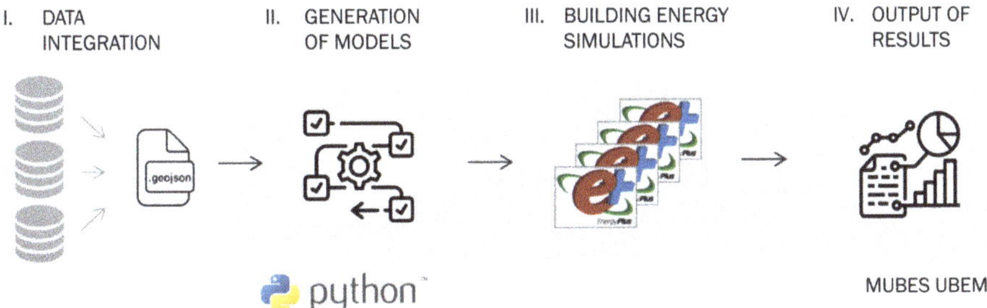

Figure 1. MUBES UBEM workflow.

The UBEM can be used in either building per building or archetype-based simulation modes. In both cases, a physics-based white box model was defined with as many elements as possible. The current version of UBEM was based on Python 3 for the structuring process and EnergyPlus 9.1 for the thermal core engine. Multithread processing was implemented for the computationally intensive processes (the generation of models and dynamic thermal simulations). While the basic function uses the eppy python package [28], a special branch was created using the geomeppy package [29] in order to enhance the thermal zoning method and enable complex building footprints to be considered. The tool with sample data is freely available under MIT license at https://github.com/KTH-UrbanT/mubes-ubem (accessed on 19 January 2022). The following subsections present the details of the used UBEM workflow for (I) data integration (Section 2.2); (II) generation of models on building (Section 2.3) and zone (Section 2.4) levels; (III) simulation options (Section 2.5); and (IV) results output (Section 2.6).

2.2. Data Integration

The process of data integration followed the Extract, Transform, Load (ETL) paradigm implemented in the Feature Manipulation Engine (FME) from SAFE software as described in [30]. The initial data sources and subsequently generated data inputs provided to MUBES UBEM are depicted in Figure 2.

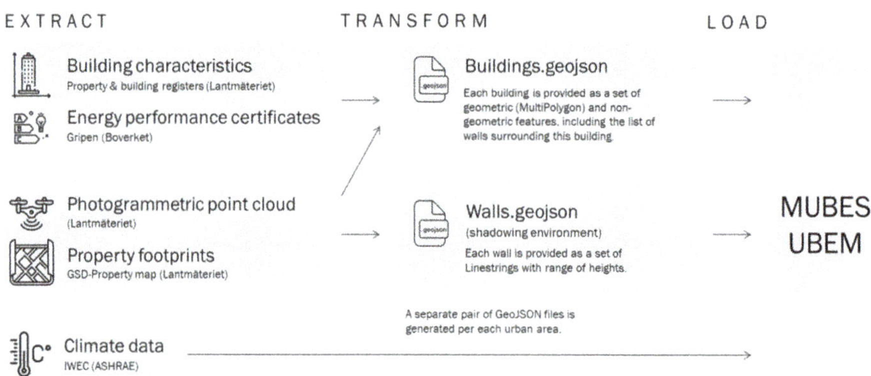

Figure 2. Data integration—primary data sources and derived data products loaded into the UBEM.

All building information was blended in FME to generate two GeoJSON files for each urban area analysed. GeoJSON is a standard geospatial data interchange format chosen due to its universality and human readability. The first file, **Buildings.geojson** contained a set of geometric (MultiPolygon) and non-geometric features, where the latter could include all attributes from the linked records for property and building cadastres (such as building purpose or form of ownership) and the energy performance certificate (EPCs) database. The second file, **Walls.geojson**, contained the geometric features (Linestrings with a range of heights) representing the shadowing environment for the whole district. Each building listed in **Buildings.geojson** was provided with a list of walls affecting the sun exposure of a particular building using the identifiers from **Walls.geojson**. This second file (**Walls.geojson**) was only needed to model the shadowing effect of the surrounding environment. In the case of larger urban areas, up to the city scale, several GeoJSON file pairs are generated.

EPCs are the essential data source in this UBEM workflow. Despite this policy instrument being universally adopted across the whole of the EU, its implementation and the quality of the resulting datasets can differ among EU member states [31]. In Sweden, EPCs are produced by independent energy experts and collected by the National Board of Housing, Building, and Planning (Boverket). EPCs are required for all larger buildings every 10 years and are based on the yearly data of the energy consumption split by different needs (space heating and cooling, domestic hot water, electricity, subdivided into collective and private areas) and energy carriers. The data origin can either be from energy supplier invoices or measured using devices especially designed for EPCs. The numeric values can be obtained either from installed meters or from the yearly collection of invoices from energy suppliers. Data gathered in EPCs also includes some useful details on the building geometry, installed equipment, occupancy type, etc.

Swedish EPCs possess a reasonable quality of data that allows them to be widely used for analysis on an urban scale [17,32]. However, they are also prone to certain problems. For instance, the time lag imposed by the methodology of the EPC data collection can result in missing effects from recent building retrofitting (be it either envelope or equipment), leading to model performance gaps. As was shown previously in [31,33], heated area attributes (area heated above 10 °C) are the key source of uncertainties in EPCs and models utilising this data source. Therefore, at the transformation stage, data from EPCs were cross-validated and enriched from other sources including building and property cadastres from Lantmäteriet and point cloud building data from the Stockholm municipality.

The following subsections describe the process for generating the energy model based on the input data provided in the main GeoJSON file.

2.3. Generation of Models—Building Level

This level is about geometry definition, thermal zoning, surrounding environment, and envelope characteristics. Each are presented separately in the following subsections.

2.3.1. Geometry Definition

Building footprints from the Swedish property map (2D) were used as the basis for the 3D model. The polygons were integrated with the EPC register by using an ETL method developed by [30,32] to obtain additional information, which was later used by the simulation engine. Each building footprint was used to clip a photogrammetric point cloud and a terrain model, which was used to calculate the median roof height and ground height of the building. This method is described further by [30,32] and was designed either to make LoD 1.2 or 1.3, considering the classification proposed in [25]. Being based on CityGML 2.0 specification ranging from LoD 0 (footprint) to LoD 4 (contains indoor features), it provides a more fine-grained specification of LoDs, specified by four sublevels for each LoD 0–3.

Building footprints from the property map can (in general) can only be used to create LoD 1.2, which can create a high deviation in building volume compared to the actual building. This is especially the case for buildings that consist of a large variety of building heights. The volume deviation can be decreased by creating LoD 1.3. This LoD does not result in a high increase in the number of surfaces compared with more detailed LoD levels [25], which is important for UBEM, as an increase in the number of surfaces for each building results in more intensive energy calculations.

To test the impact of using LoD 1.3, a method of segmentation of building footprints by different roof heights was developed. The point cloud was first cleaned by filtering 5% of the highest and lowest points and triangulating the remaining points. Triangles with high vertical slopes were kept and dissolved with their neighbours and later replaced with a centreline. Snapping was used to extend the centreline to the correct boundaries of the building footprint. The centreline was then used to cut building footprints in several parts. The median and ground height were then recalculated. Each building footprint and building part was generalised using the Douglas algorithm to minimise the number of vertices, and segment snapping was applied to remove small distances between footprints. To create the 3D model, the building footprint was set at the median ground height and was then extruded to the median roof height.

The 3D model was also used as an input to create a neighbourhood shading walls file to supply the simulation engine for shadowing computations. The buildings in the 3D model were de-aggregated into 2D lines with ground height and roof height stored as two attributes, creating a light dataset that would also be possible to re-generate later in the simulation engine. All lines were replaced by a centre point that was used to find all neighbouring points within a 250-m radius. All points were connected by a line to represent the line of sight; lines that crossed one or more buildings were filtered out. For each building, a list of the remaining walls was stored, and duplicates were removed. Finally, the walls file was created with a unique id of the walls with a corresponding wall id in the 3D buildings file. This made it possible to obtain fast and accurate neighbouring walls for a building. A low calculation time was achieved as the calculation was conducted entirely in 2D, and the height of the building and terrain was not considered, which may lead to less accurate energy calculations for some buildings.

In Section 4, the impact of the above-mentioned level of details from LoD 1.2 to LoD 1.3 is quantified for one specific district.

2.3.2. Envelope Characteristics

Since fewer elements are available for analysis at the urban scale, the envelope characteristics are in two different layers, representing the insulation effect and the inertia effect, respectively. This differentiation allows for the modelling of either lightweight or heavyweight buildings as being well insulated or not. The position of both layers can also be defined in order to capture the impact of external or internal insulation on the envelope. The definition of these lumped layers follows the resistance/capacitance paradigm for layers in a series as 1D conduction was considered in the thermal engine. Three main thermal properties are required for the layers composed of one single material, other than the thickness, namely: density (kg/m^3); thermal conductivity ($W/K/m^2$); and specific heat capacity ($J/K/kg$). Specific surface properties such as radiative properties can be defined at this stage if a specific effect is to be considered (e.g., special paint coatings or metallic surface layers). Windows are part of the envelope. The window to wall ratio (WWR) is defined as an input. Window width overlaps 95% of the zone width and the height is computed using the input given for WWR. Their energy performances are also defined in the input data.

2.3.3. Thermal Zoning

Several options for thermal zoning are implemented, from the single zone for heated and non-heated volumes up to the core and perimeter zoning option on each floor. In the case of aggregation of different floors into a single zone, the inputs are still represented at the floor level, but are then further corrected using a floor-multiplier factor, as proposed in [26]. The core and perimeter zoning option required a specific algorithm. Depending on the perimeter depth, the perimeter zone definition is automatically created starting from each edge, delimiting the core zone. The core zone definition includes a threshold for the resulting vertex's distance within three vertexes. This threshold is, by default, set to a half of the perimeter depth. This allows us to avoid having too narrow zone angles, too small edges, or too small zones. Then, for the perimeter zones, triangle zones (having a single vertex on the external polygon) are not allowed, except for the last perimeter zone definition, closing the loop over the edges of the core. Thus, perimeter zones with more than one edge in common with the core zone are allowed. The perimeter depth starts at 3 m by default and is reduced by half if any issue is encountered during the process. This algorithm was derived from the original one given in the Geomeppy package [29]. Figure 3 presents the thermal zoning options and the effect of the perimeter depth for two sample buildings.

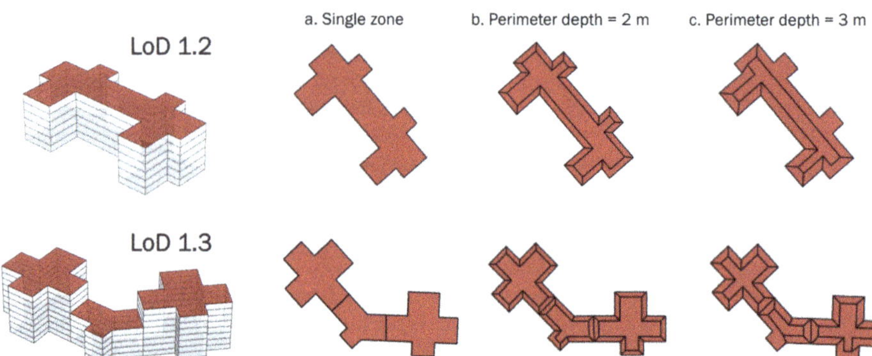

Figure 3. Thermal zoning for two sample buildings—LoD 1.2 (top) and LoD 1.3 (bottom), using three zoning strategies: single zone (**a**), and perimeter/core zoning at 2 m (**b**) and 3 m (**c**) perimeter depths.

As shown in Figure 3, non-convex zones are currently allowed, thus all external non-convex surfaces were split further into convex ones for the purpose of shortwave multireflection (see Surrounding Environment Section 2.3.4). Internal non-convex surfaces were not treated further. Internal convex ones are only needed if internal shortwave multireflection is required, which is not of concern in the case of UBEM, as no internal architecture would be available at this scale of analysis.

2.3.4. Surrounding Environment

The shadowing impact from the surrounding environment was considered for each building. Shadowing is automatically dealt with in EnergyPlus by using the *shadowing element* object. External surfaces can still receive and reflect shortwave radiation, but longwave radiation exchanges are not considered. The latter would have required the computation of the view factors between each surface before simulation, and then the use of an iterative approach to capture the heat fluxes between surfaces at each time step. Some proposals for iterative methods have been suggested by [34]. The maximum effect of a 3.6% decrease in heat needs have been observed in different locations in the U.S. In the proposed UBEM workflow, all external surfaces are sequentially considered for each building, and all visible surfaces belonging to other buildings (included in the **Walls.geojson** file introduced earlier) are reported. Then, depending on a threshold for the distance from the building's centroid, surfaces beneath the limit are viewed as shadowing surfaces in EnergyPlus. Figure 4 illustrates the distance threshold on the modelling process for one random building.

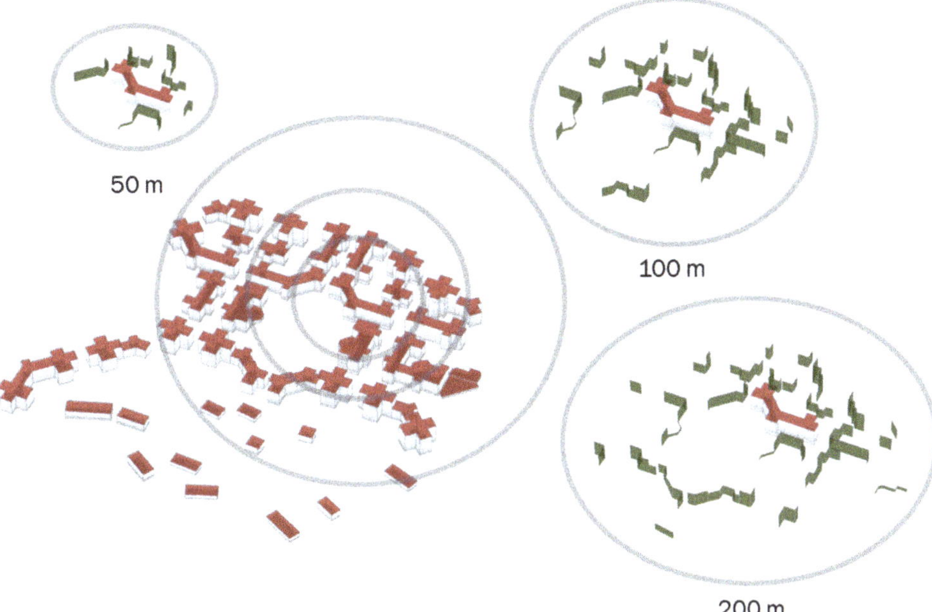

Figure 4. Shadowing effect for the environment of a random building, based on the distance thresholds (50, 100, and 200 m).

Parametric simulations for two different districts are reported in the Results (Section 4).

2.4. Generation of Models—Zone Level

This level concerns all local elements that have a time dependent impact on the zone's energy balance. Figure 5 represents the different required inputs for this level. All are presented in the following subsections.

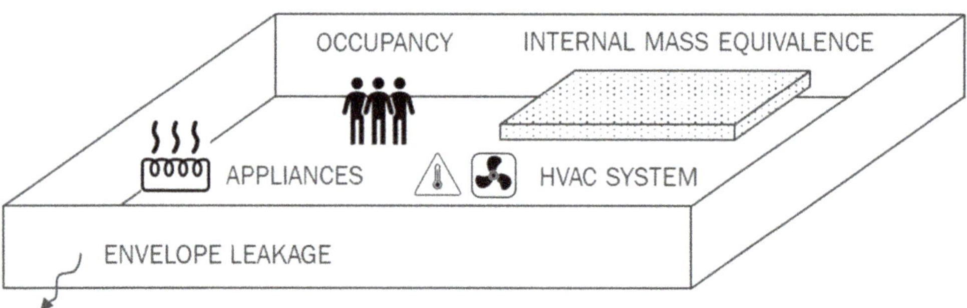

Figure 5. Model elements at the zone level.

2.4.1. Internal Mass Equivalence

The buffering effect on indoor temperature dynamics from internal furniture and partition walls is modelled through an *internal mass* object. The buffering effect is of greater importance in UBEM assumptions as the internal architecture is unknown. Indeed, all zones are defined as open spaces in which area-based elements are given as inputs. *Internal mass* object is equivalent to a material with classic thermal properties with an amount defined by weight per square meter and a link with the zone's ambient air through a surface of exchange. In the proposed UBEM, the default values are $40\ kg/m^2$ of an equivalent material with the following properties: thermal conductivity of $0.3\ W/K/m$, density of $600\ kg/m^3$, and specific heat of $1400\ J/kg/K$. The surface of exchange is twice the floor area as in [35]. The floor-multiplier is also used when thermal zoning is considered (single zones for heated and non-heated volumes).

2.4.2. Envelope Leakage

Envelope leakage is of great importance. It is influenced by the thermal gradient between indoor and outdoor conditions and the zone's height (used to compute the hydrostatic pressure gradient). Several other elements can influence related heat transfer such as stairwells, urban area density, and the building's height. The EnergyPlus infiltration model with flow coefficient enables us to consider these influenceable factors. In the proposed UBEM, a value of 0.667 was used for the pressure exponent value in the power law. In Sweden, the above listed influenceable parameters are given in the EPC's templates.

For non-heated zones, instead of the above-mentioned model, an air change rate is defined in hours per volume. This approach makes more sense for use with below ground levels without using pressure balance solvers.

2.4.3. HVAC System

In the proposed UBEM workflow, the focus is on the used-energy needs. This means that the energy carriers as well as their production and distribution efficiencies are not considered at the zone level, but are accounted for in the post-treatment and calibration stages, as will be further described. The heating, ventilation and air conditioning (HVAC) system shall thus be able to represent any kind of system embedding the indoor renewable air change rate including any potential heat recovery from it. An equivalent *shoebox* HVAC model was considered with the *Ideal Loads Air System* object, which computes, for each zone, the needed energy to match the internal temperature set point. Figure 6 presents a schematic view of such a system. A limitation can be specified for either the supplied air temperature or compensation mass flow rates of the overall supplied power. The heating and cooling supplies will, at each time step, correspond to the external needed energy for this zone to comply with its temperature setpoint. The temperature setpoint can be defined as either constant or from fixed schedules for day and night times, or even through an external file. In the proposed UBEM workflow, each zone has its own HVAC system.

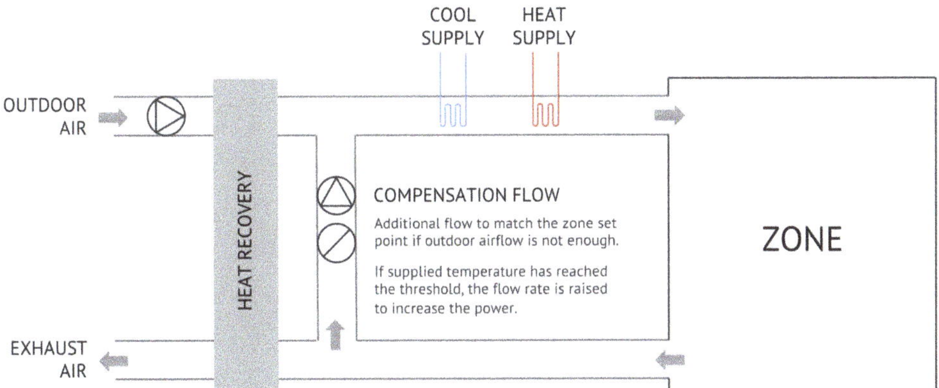

Figure 6. Model of ideal loads air system used to represent the HVAC system.

2.4.4. Occupancy

Occupancy rate can have a strong impact on the energy balance in non-residential types of buildings. In contrast to residential buildings, the number of occupants steers the air change rate and can cause significant heating effects (considering each occupant releases an average of 70 W, 15 people produce more than 1 kW of heating). The occupation density is lowest, by a large margin, in residential type buildings and does not affect the ventilation rates (the occupant's activity can if the system is demand-controlled based). Thus, for the latter type of buildings, the impact of occupancy can be embedded in the appliance's energy needs. In the proposed UBEM workflow, two options are available using either the maximum density per type of activity or hourly numbers of occupants based on random beta distribution. Scheduled timetables have also been proposed using opening and closing hours.

2.4.5. Appliances

The energy used and released by internal appliances is considered through lumped values to further decline into hourly data. Thus, a higher time resolution than just the yearly values of W/m^2 are required. Starting from a yearly value given in EPCs or other databases, the cumulative distribution of internal gain is represented by a reversed sigmoid. Equation (1) represents the cumulative distribution of internal load (CDIL) function in a regular sigmoid shape. The seasonal effect can be tuned by the slope factor γ, which would represent a greater seasonal effect for greater values. The regular sigmoid curve would represent more internal gains during the summer period (considering a starting period of the first of January), while a 6-month offset should be introduced to represent more internal gains during the winter period. A 6-month offset was thus introduced and normalised CDIL computed. Figure 7 presents the slope factor effect on the normalised CDIL. The derivative values of the CDIL were written in an external file defined as an input file of hourly watts per square meter in EnergyPlus using the *electric equipment* object in each heated zone.

$$CDIL = \frac{YearlyConsumption}{(1 + exp(-\gamma t))} \tag{1}$$

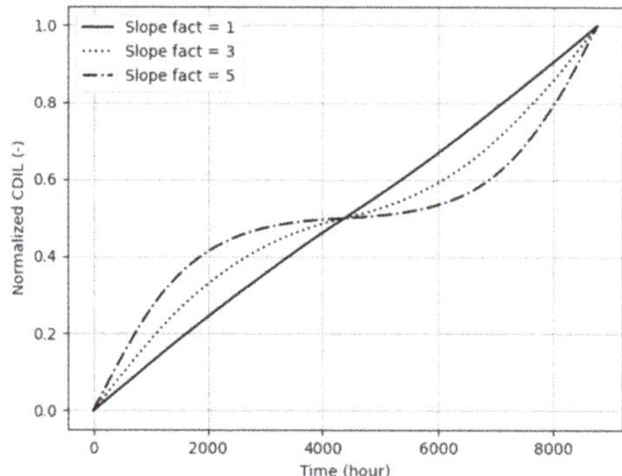

Figure 7. Internal appliance profiles.

2.5. Simulation Options

2.5.1. Domestic Hot Water

The proposed UBEM workflow assumes that domestic hot water (DHW) does not contribute to providing heat to the building. The option of modelling DHW would thus be steered by the calibration stage, using aggregated measured data for both space heating and DHW. In such cases, the related energy needs for DHW are modelled through a simple *water use equipment* object. The temperature of the hot water supply was fixed at 55 °C by default, and the temperature of hot water from the water tap was fixed at 37 °C by default, and the temperature of the cold-water supply was defined through the time series' input (or taken as constant). Together with water tap usage patterns, this results in the energy needs for DHW. As DHW might only be considered for the calibration stage (Section 2.5.2), the FMI option (Section 2.5.3) could be used to compute the water tap usage to diminish the discrepancies between the measured and simulated energy needs in non-heating periods.

2.5.2. Calibration

Despite UBEM not being a simple aggregation of BEMs, a calibration process for accurate models is still required. Even though a number of simplifications have been made in UBEMs when compared to BEMs, many inputs are still needed, which are usually associated with higher uncertainties than single BEMs. The UBEM calibration process needs to be adapted for each type of building. Hence, while missing inputs can be more or less the same for a whole sample of buildings, the calibrated inputs would definitely be different.

The probabilistic calibration has been repeatedly reported as the best fit for UBEM applications [21,22]. The iterative Bayesian process has been found to be particularly promising as it allows for the automatic adjustment of the exploring ranges for missing inputs for each building. The developed UBEM workflow is fully compatible with this technique. It provides the option of conducting numerous simulations with Latin Hypercube Sampling (LHS) for any input parameter(s) for the purpose of either sensitivity analysis or model calibration.

2.5.3. Co-Simulation Environment

The co-simulation option implemented in MUBES follows the paradigm of a functional mock-up interface (FMI) [36]. Functional mock-up units (FMU) are to be built for any model (building) that may need adjustment of one or more inputs or parameters during the simulation. All the FMUs were used in an environment dedicated to running FMUs. The FMU toolkit for EnergyPlus [37] is embedded in the MUBES UBEM workflow. FMUs can thus be automatically generated for each building in the input file. This requires the definition of specific inputs and outputs to be matched with the controlled parameters targeted in co-simulation. For the proposed UBEM tool, two examples of co-simulation are proposed using the indoor temperature setpoint and the DHW tap usage as inputs at each time step.

2.6. Output of Results

All variables available from EnergyPlus can be given as the output in the UBEM workflow. As post-treatment might be specific to each case studied, some generic methods are proposed in the UBEM, but only for the sake of illustration.

In the following sections, the above described UBEM workflow was used to make parametric simulations. The impact of the level of detail (LoD) is first highlighted, followed by thermal zoning, and the shadowing of the surrounding environment. Two different districts in Stockholm County are used for illustration. After an initial presentation of the two districts and the related database construction process, the results are presented.

3. Case Study

Two districts in Stockholm (Sweden), Minneberg and part of Hammarby Sjöstad, were considered for the impact analysis using parametric simulations (Section 4). Both districts are mainly residential, however, most of the analysed buildings include a small percentage of non-residential occupancy. Minneberg, and the analysed part of Hammarby Sjöstad, are composed of 33 and 45 buildings, respectively (Figure 8). Minneberg was developed in 1987 and is distinctive for the high homogeneity and good energy performance of its buildings (over 33 buildings, the average performance according to EPCs is 76 kWh/m^2 with a standard deviation of 10 kWh/m^2). Hammarby Sjöstad is world-famous as one of the first environmental districts with ambitious energy targets [38]. However, the selected part belongs to the earliest stage of its development (2000–2003), containing more diverse architecture solutions with a noticeably higher variance of building energy performance (over 45 buildings, the average performance according to EPCs is 114 kWh/m^2, with a standard deviation of 43 kWh/m^2). Only Minneberg was used for the analysis of the level of detail (LoD) impact, while both districts were analysed for the impacts of thermal zoning and the surrounding shadowing environment on the thermal energy demand intensity (TEDI) for space heating. The earlier described UBEM workflow (Figure 1) and input data (Figure 2) were used in both cases.

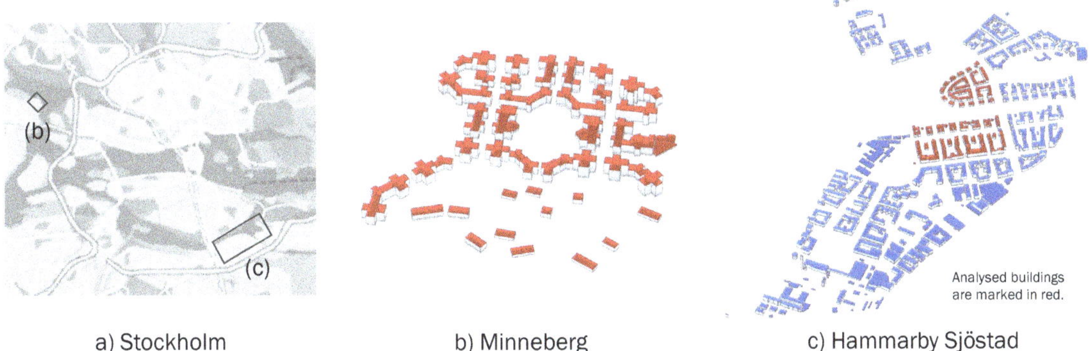

Figure 8. The two urban areas in Stockholm (**a**) considered in this study—(**b**) Minneberg and (**c**) Hammarby Sjöstad. The buildings analysed are marked in red.

4. Results

In this section, the results of parametric simulations are presented to analyse the impact of the level of detail (LoD) (Section 4.1, the thermal zoning impact (Section 4.2), and the shadowing impact of the surrounding environment (Section 4.3). Two different districts in Stockholm municipality (Section 3) were used for the purpose of illustration. For all simulations, the climate of Stockholm Arlanda airport was used from the IWEC typical year database from ASHRAE [39].

4.1. The Impact of Level of Detail (LoD)

Minneberg district was used to investigate the impact of the level of detail (LoD) of the building geometry. Two levels of detail were analysed: LoD 1.2 and LoD 1.3. For this analysis, 23 out of 33 buildings were used as other buildings were not available in the LoD 1.2 format. The application of the two LoDs resulted in two different geometries generated for each building, as depicted for the two sample buildings in Figure 9.

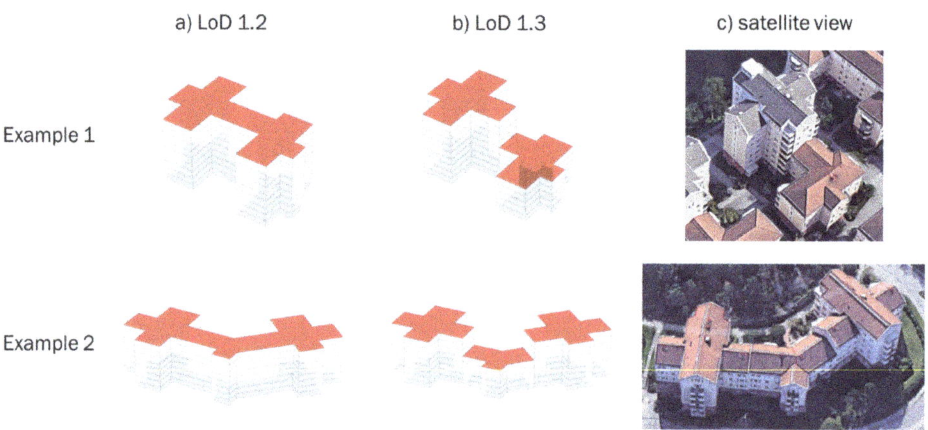

Figure 9. Two sample buildings represented with (**a**) LoD 1.2 and (**b**) LoD 1.3 models in comparison to (**c**) a satellite view.

The calculated thermal energy demand was normalised using the heated area to compensate for the difference in total heated area. However, the changes in the surface of the external envelope and the solar gains emerging from the choice of LoD still led to different heat gains and losses. Figure 10 presents the deviation between the thermal energy demand intensity (TEDI) for LoD 1.3 and the reference of LoD 1.2 versus the change in shape factor (the ratio between the envelope surface area facing outwards and its volume) induced by upgrading from LoD 1.2 to LoD 1.3. The results show that even though most discrepancies remained below 4%, some buildings demonstrated more than 10% greater heat needs for LoD 1.3 than LoD 1.2. The two largest changes were observed in the case of buildings 9 and 10 where the shape factor increase was nearly 20%. Thus, at the UBEM scale, keeping LoD 1.2 could lead to a 10% extra discrepancy of TEDI for some buildings. However, for the overall considered district (23 buildings), the difference remained below 1% (ΔTEDI = 0.76%). Hence, using a higher level of detail might be irrelevant for some larger scale UBEMs targeted at lower spatial resolution. At the same time, making the extra effort by using LoD 1.3 can be worth it in the case of building calibration or analysing the impact of ECMs, as in this case, the identified 10% extra TEDI would result in a skewed definition of calibrated building parameters or wrongly estimated energy savings.

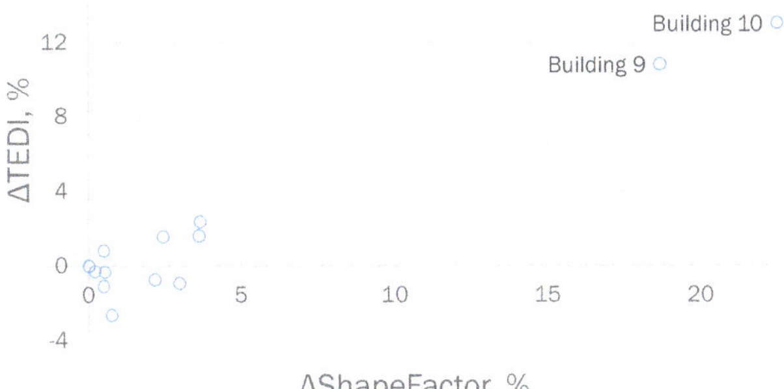

Figure 10. Relative change of shape factor and thermal energy demand intensity (TEDI) from the upgrade of the level of detail (LoD) for buildings in the Minneberg district, from *LoD 1.2* to *LoD 1.3* (LoD 1.2 serves as a reference).

4.2. Impact of Thermal Zoning

This subsection presents, for the two districts described above, the impact of different *thermal zoning* resolutions. Figure 11 presents the different options available in the UBEM for a simple building: (**a**) single zone for heated and non-heated volumes, (**b**) single zone per floor, (**c**) core-perimeter zones for heated and non-heated volumes, and (**d**) core-perimeter zones per floor.

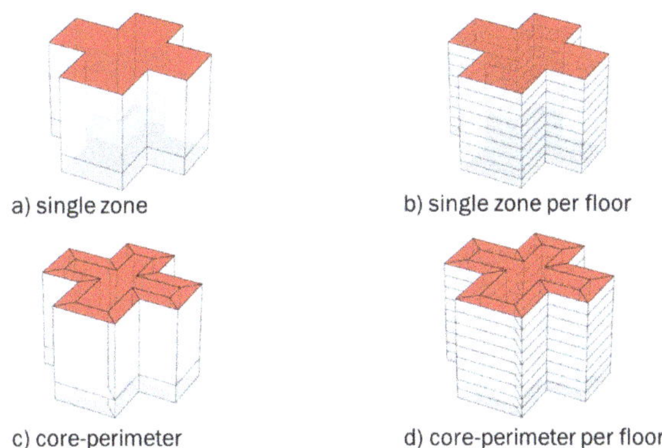

Figure 11. Four thermal zoning approaches, applied to a sample building consisting of three regular floors and two basement floors: (**a**) single zone for heated and non-heated volumes, (**b**) single zone per floor, (**c**) core-perimeter zones for heated and non-heated volumes, and (**d**) core-perimeter zones per floor.

The paradigm of floor multiplier was applied for options (**a**) and (**c**). The core and perimeter (**c, d**) zone definition followed the algorithm presented above (Section 2.3.3). All elements other than the thermal zones remained the same within the different simulation setups presented below. The impact of thermal zoning is characterised by the change in TEDI. Figure 12 presents the distributions of absolute *(left)* and relative *(right)* discrepancies along the four zoning options, with (**b**) *(single zone per floor)* as the reference. The same trends were observed for the three geometry cases (one district with LoD 1.2 and two with LoD 1.3). The configuration with *single zone* (**a**) remained close, with a minor underestimation of TEDI, to the configuration (**b**) with *single zone per floor*. The *core and perimeter zone* approach (**c, d**) increases TEDI by a small amount, keeping the same difference between configuration (**c**) and (**d**) as between (**a**) and (**b**), with a very minor underestimation of TEDI when aggregating the different floors into one volume (**c**). These results match the findings of similar studies reported earlier [26]. The highest *relative* difference applies for the buildings with the lowest consumption, while the highest *absolute* difference was observed for the buildings with the highest consumption.

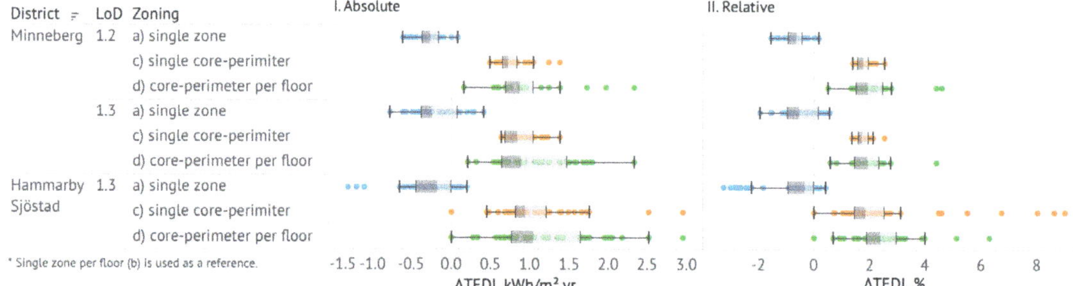

Figure 12. Absolute (**I**) and relative (**II**) change in thermal energy demand intensity (TEDI) for buildings in Minneberg (LoD 1.2 and LoD 1.3) and Hammarby Sjöstad (LoD 1.3) districts with single zone (a), single core−perimeter (c), and core-perimeter per floor (d) zoning applied. Single zone per floor (b) was used as a reference.

Table 1 provides the calculated changes of TEDI across different LoDs and thermal zoning approaches at a district scale. These values suggest that, similarly to the varying LoDs, different thermal zoning approaches might lead to the same results and are not worthy of interest for analysis made on a district scale. As a great deal of extra time is required when using a *core and perimeter zone on each floor* (**d**), *one zone per floor* (**b**) zoning can be suggested as the default choice for UBEM studies.

Table 1. Change in the total thermal energy demand intensity (TEDI) due to different zoning approaches (Figure 11) at the district scale.

		Total TEDI Difference, % Single Zone per Floor (b) Is Used as a Reference.		
District	**LoD**	**(a) Single Zone**	**(c) Single Core-Perimeter**	**(d) Core-Perimeter Per Floor**
Minneberg	1.2	−0.5	1.7	2.1
	1.3	−0.3	1.7	2.0
Hammarby Sjöstad	1.3	−0.4	2.0	2.3

4.3. Impact of Surrounding Shadowing Environment

This subsection explores the impact of the threshold distance, beyond which the shadowing effect of surrounding buildings is not considered. The distance was defined as presented earlier in the model workflow (Section 2.3.4). Parametric simulations were conducted for all buildings in the two case districts with a fixed LoD (1.3) and thermal zoning *(b, one zone per floor)* configuration.

The buildings' performances, estimated as TEDI, were obtained for each building and aggregated at the district scale. The TEDI factor represents the ratio of TEDI for each shadowing distance over the maximum TEDI computed for all shadowing distances. As expected, there was an evident dependency of the shadowing effect from the surrounding environment. Figure 13 shows that on a *building level* (**i**), greater shadowing areas resulted in higher TEDIs. While this held true for both districts, the aggregated results at the district level were quite different. Only 5% TEDI difference was observed for Minneberg at the district scale, while 12% TEDI difference was computed for Hammarby Sjöstad.

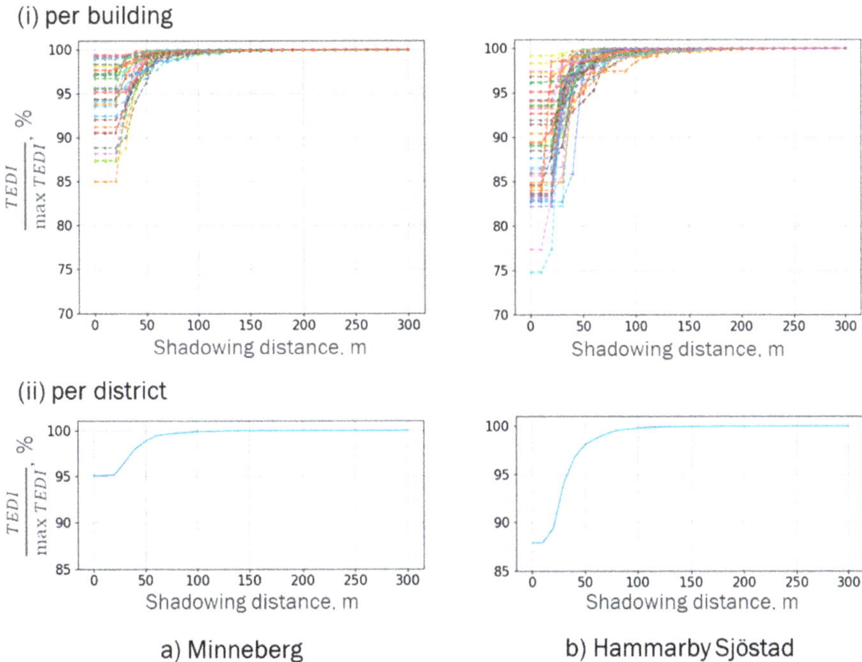

Figure 13. Impact of shadowing environment limited by a distance threshold on thermal energy demand intensity (TEDI) for each building (top) and the entire district (bottom) in (**a**) Minneberg and (**b**) Hammarby Sjöstad districts. Maximum TEDI was used as the reference.

These results allow us to characterize the magnitude of the effect of certain thresholds for shadowing environments on TEDI. At a district scale, differences below 2% could be achieved by including all shadowing surfaces within 50 m from the building's centroid. Furthermore, surfaces farther than 150 m did not seem to have any effect at the district level. At the same time, these results suggest that a threshold of 200 m should be kept for analysis at the building level.

5. Conclusions and Discussion

This paper has presented MUBES—a new simulation tool for urban building energy modelling (UBEM). This tool can be used for a number of applications including: (a) analysis of the current energy performance of an existing building stock at a district or city scale; (b) mapping the system effects from the large scale roll-out of retrofitting actions; (c) generation of a calibrated sample of simulations that can further be used to compensate for missing data; and (d) analysis of various operation strategies for the building stock on a district scale that could improve the overall performance of the urban energy system (including power distribution grid or district heating network).

MUBES UBEM follows a physics-based paradigm using a Python-based framework as an environment for the generation and management of simulations and EnergyPlus as a core thermal engine. To enable analysis of the impact of the level of detail (LoD), the geometry definition with photogrammetric point cloud method was conducted at the data integration stage. The developed UBEM workflow generates models for building energy performance simulations building-by-building and runs simulations at the district scale in a fully automated way. Input data are provided through a GeoJSON file containing both geometric (polygons for all building's external surfaces) and non-geometric properties for each building integrated from several data sources. At its core, the workflow follows a *shoebox* paradigm with ideal HVAC system, and in this way provides additional robustness for the further expansions required for intaking input data in other formats.

The developed simulation tool was used to investigate the impact of three aspects that can affect the performance of UBEMs on a district/urban scale: (1) the level of detail (LoD) for input building geometries; (2) thermal zoning approach; and (3) the shadowing effect of the surrounding environment. Following the analysis of these phenomena for the two case districts in Stockholm, the subsequent conclusions can be drawn:

Level of detail (LoD). A change in the LoD from 1.2 to 1.3 resulted into quite distinctive shape factors (0–20%) for some buildings, leading to a noticeable (0–13%) impact on the thermal energy demand intensity (TEDI) for space heating at the building scale. At the same time, for a district scale analysis, given a certain level of homogeneity of the analysed district, a more detailed LoD 1.3 might not be required. For instance, in the case of the studied district of Minneberg, the overall TEDI difference (ΔTEDI) at the district scale remained below 1%, despite a change of over 10% for some buildings. Hence, as use of LoD 1.3 may require extra effort in data collection, LoD 1.2 could be seen as sufficient for district scale analysis. On the other hand, bottom-up physical models are required to accurately compute the impact of energy conservation measures that are to be estimated. Thus, as these impacts might be less accurately estimated with LoD 1.2, it would still be recommended to use LoD 1.3 if available.

Thermal zoning. The analysis of various thermal zoning approaches has mostly confirmed earlier studies. Particularly, the overall ΔTEDI at the district scale has remained below 5%, despite a more pronounced effect for some buildings. The analysis showed that a single zone option for heated and non-heated volumes should be avoided, which is in line with recommendations from existing standards. At the same time, a compromise of having one zone per floor was still found to be acceptable. For higher buildings, the merging of middle floor zones while keeping bottom and top floor zones separate could be worthy of further investigation.

Surrounding shadowing environment. Up to 12% of ΔTEDI could be attributed to the change in the shadowing environment in the case of two districts with quite different types of building geometries, with a monotone increase in TEDI along with the increase in the shadowing distance threshold. At the district scale, limited effects (below 2%) were observed for the nearest shadowing environment (up to 50 m). Furthermore, surfaces farther than 100 m did not have any profound effect at the district scale for both studied areas. At the building scale, the limited effects' threshold rose to 150 m. However, as extra computing time is negligible, the authors would advise keeping 200 m for all simulations.

We conclude that the analysed modeller assumptions embedded in UBEMs have a distinct impact on the UBEMs' outcome and suggest promoting more explicit documentation of these choices in upcoming UBEM studies.

Author Contributions: Conceptualisation, X.F., T.J. and O.P.; Methodology, X.F., T.J. and O.P.; Software, X.F. and T.J.; Validation, X.F.; Formal analysis, X.F., T.J. and O.P.; Investigation, X.F., T.J. and O.P.; Resources, X.F., T.J. and O.P.; Data curation, X.F. and T.J.; Writing—original draft preparation, X.F.; Writing—review and editing, X.F., T.J. and O.P.; Visualisation, X.F. and O.P.; Supervision, X.F.; Funding acquisition, O.P. All authors have read and agreed to the published version of the manuscript.

Funding: This research was funded by the Swedish Energy Agency (Energimyndigheten) via the E2B2 research programme, project nos. 40846-2 and 46896-1.

Institutional Review Board Statement: Not applicable.

Informed Consent Statement: Not applicable.

Data Availability Statement: The presented UBEM platform and sample input dataset for the district of Minneberg are provided open source under MIT license at https://github.com/KTH-UrbanT/ mubes-ubem (accessed on 19 January 2022). The raw data utilised in the study were obtained from Swedish public bodies (Boverket and Lantmäteriet) and are limited to use within particular research projects.

Acknowledgments: We express gratitude to Boverket and Lantmäteriet for providing the data extracts from the EPC database 'Gripen', building and property cadastre data and photogrammetric building data, respectively. The computations were tested using resources provided by the Swedish National Infrastructure for Computing (SNIC) at SNIC Science Cloud (SSC) partially funded by the Swedish Research Council through grant agreement no. 2018-05973.

Conflicts of Interest: The authors declare no conflict of interest.

References

1. Global Status Report for Buildings and Construction 2019—Analysis. Available online: https://www.iea.org/reports/global-status-report-for-buildings-and-construction-2019 (accessed on 8 November 2021).
2. Rogelj, J.; Shindell, D.; Jiang, K.; Fifita, S.; Forster, P.; Ginzburg, V.; Handa, C.; Kobayashi, S.; Kriegler, E.; Mundaca, L.; et al. *Mitigation Pathways Compatible with 1.5 °C in the Context of Sustainable Development*; Intergovernmental Panel on Climate Change: Geneva, Switzerland, 2018; 82p.
3. Masson-Delmotte, V.; Pörtner, H.-O.; Skea, J.; Zhai, P.; Roberts, D.; Shukla, P.R.; Pirani, A.; Pidcock, R.; Chen, Y.; Lonnoy, E.; et al. *An IPCC Special Report on the impacts of Global Warming of 1.5 °C above Pre-Industrial Levels and Related Global Greenhouse Gas Emission Pathways, in the Context of Strengthening the Global Response to the Threat of Climate Change, Sustainable Development, and Efforts to Eradicate Poverty*; Intergovernmental Panel on Climate Change: Geneva, Switzerland, 2018; 630p.
4. Swan, L.G.; Ugursal, V.I. Modeling of end-use energy consumption in the residential sector: A review of modeling techniques. *Renew. Sustain. Energy Rev.* **2009**, *13*, 1819–1835. [CrossRef]
5. Kavgic, M.; Mavrogianni, A.; Mumovic, D.; Summerfield, A.; Stevanovic, Z.; Djurovic-Petrovic, M. A review of bottom-up building stock models for energy consumption in the residential sector. *Build. Environ.* **2010**, *45*, 1683–1697. [CrossRef]
6. Reinhart, C.F.; Cerezo Davila, C. Urban building energy modeling—A review of a nascent field. *Build. Environ.* **2016**, *97*, 196–202. [CrossRef]
7. Ferrando, M.; Causone, F.; Hong, T.; Chen, Y. Urban building energy modeling (UBEM) tools: A state-of-the-art review of bottom-up physics-based approaches. *Sustain. Cities Soc.* **2020**, *62*, 102408. [CrossRef]
8. Johari, F.; Peronato, G.; Sadeghian, P.; Zhao, X.; Widén, J. Urban building energy modeling: State of the art and future prospects. *Renew. Sustain. Energy Rev.* **2020**, *128*, 109902. [CrossRef]
9. Hong, T.; Chen, Y.; Luo, X.; Luo, N.; Lee, S.H. Ten questions on urban building energy modeling. *Build. Environ.* **2020**, *168*, 106508. [CrossRef]
10. Ali, U.; Shamsi, M.H.; Hoare, C.; Mangina, E.; O'Donnell, J. Review of urban building energy modeling (UBEM) approaches, methods and tools using qualitative and quantitative analysis. *Energy Build.* **2021**, *246*, 111073. [CrossRef]
11. Sola, A.; Corchero, C.; Salom, J.; Sanmarti, M. Simulation Tools to Build Urban-Scale Energy Models: A Review. *Energies* **2018**, *11*, 3269. [CrossRef]
12. Hong, T.; Chen, Y.; Lee, S.H.; Piette, M. CityBES: A Web-based Platform to Support City-Scale Building Energy Efficiency. In Proceedings of the 5th International Urban Computing Workshop, San Francisco, CA, USA, 13 August 2016.
13. Reinhart, C.F.; Dogan, T.; Jakubiec, J.A.; Rakha, T.; Sang, A. UMI—An urban simulation environment for building energy use, daylighting and walkability. In Proceedings of the 13th Conference of International Building Performance Simulation Association, Chambery, France, 25–28 August 2013; pp. 476–483.
14. Garreau, E.; Abdelouadoud, Y.; Herrera, E.; Keilholz, W.; Kyriakodis, G.-E.; Partenay, V.; Riederer, P. DIstrict MOdeller and SIMulator (DIMOSIM)—A dynamic simulation platform based on a bottom-up approach for district and territory energetic assessment. *Energy Build.* **2021**, *251*, 111354. [CrossRef]
15. Robinson, D.; Haldi, F.; Leroux, P.; Perez, D.; Rasheed, A.; Wilke, U. CITYSIM: Comprehensive Micro-Simulation of Resource Flows for Sustainable Urban Planning. 2009. Available online: https://www.semanticscholar.org/paper/CITYSIM%3A-Comprehensive-Micro-Simulation-of-Resource-Robinson-Haldi/0fb6af269aef7d6c69123e5aa8d778aa98ddd834 (accessed on 7 February 2022).
16. Lundström, L.; Akander, J. Bayesian Calibration with Augmented Stochastic State-Space Models of District-Heated Multifamily Buildings. *Energies* **2020**, *13*, 76. [CrossRef]
17. Pasichnyi, O.; Wallin, J.; Kordas, O. Data-driven building archetypes for urban building energy modelling. *Energy* **2019**, *181*, 360–377. [CrossRef]
18. Nouvel, R.; Brassel, K.-H.; Bruse, M.; Duminil, E.; Coors, V.; Eicker, U.; Robinson, D. SimStadt, a new workflow-driven urban energy simulation platform for CityGML city models. In Proceedings of the International Conference CISBAT 2015 Future Buildings and Districts Sustainability from Nano to Urban Scale, Lausanne, Switzerland, 9–11 September 2015. [CrossRef]
19. Fonseca, J.A.; Nguyen, T.-A.; Schlueter, A.; Marechal, F. City Energy Analyst (CEA): Integrated framework for analysis and optimization of building energy systems in neighborhoods and city districts. *Energy Build.* **2016**, *113*, 202–226. [CrossRef]
20. Ang, Y.Q.; Berzolla, Z.M.; Reinhart, C.F. From concept to application: A review of use cases in urban building energy modeling. *Appl. Energy* **2020**, *279*, 115738. [CrossRef]
21. Sokol, J.; Cerezo Davila, C.; Reinhart, C.F. Validation of a Bayesian-based method for defining residential archetypes in urban building energy models. *Energy Build.* **2017**, *134*, 11–24. [CrossRef]

22. Wang, C.-K.; Tindemans, S.; Miller, C.; Agugiaro, G.; Stoter, J. Bayesian calibration at the urban scale: A case study on a large residential heating demand application in Amsterdam. *J. Build. Perform. Simul.* **2020**, *13*, 347–361. [CrossRef]
23. Cerezo Davila, C.; Reinhart, C.F.; Bemis, J.L. Modeling Boston: A workflow for the efficient generation and maintenance of urban building energy models from existing geospatial datasets. *Energy* **2016**, *117*, 237–250. [CrossRef]
24. Cerezo, C.; Sokol, J.; AlKhaled, S.; Reinhart, C.; Al-Mumin, A.; Hajiah, A. Comparison of four building archetype characterization methods in urban building energy modeling (UBEM): A residential case study in Kuwait City. *Energy Build.* **2017**, *154*, 321–334. [CrossRef]
25. Biljecki, F.; Ledoux, H.; Stoter, J. An improved LOD specification for 3D building models. *Comput. Environ. Urban Syst.* **2016**, *59*, 25–37. [CrossRef]
26. Chen, Y.; Hong, T. Impacts of building geometry modeling methods on the simulation results of urban building energy models. *Appl. Energy* **2018**, *215*, 717–735. [CrossRef]
27. Nikoofard, S.; Ugursal, V.I.; Beausoleil-Morrison, I. Effect of external shading on household energy requirement for heating and cooling in Canada. *Energy Build.* **2011**, *43*, 1627–1635. [CrossRef]
28. Philip, S. Eppy: Scripting Language for E+ Idf Files, and E+ Output Files. Available online: https://github.com/santoshphilip/eppy (accessed on 11 January 2022).
29. Bull, J. Geomeppy: Geometry Editing for E+ Idf Files. Available online: https://github.com/jamiebull1/geomeppy (accessed on 11 January 2022).
30. Johansson, T.; Olofsson, T.; Mangold, M. Development of an energy atlas for renovation of the multifamily building stock in Sweden. *Appl. Energy* **2017**, *203*, 723–736. [CrossRef]
31. Pasichnyi, O.; Wallin, J.; Levihn, F.; Shahrokni, H.; Kordas, O. Energy performance certificates—New opportunities for data-enabled urban energy policy instruments? *Energy Policy* **2019**, *127*, 486–499. [CrossRef]
32. Johansson, T.; Vesterlund, M.; Olofsson, T.; Dahl, J. Energy performance certificates and 3-dimensional city models as a means to reach national targets—A case study of the city of Kiruna. *Energy Convers. Manag.* **2016**, *116*, 42–57. [CrossRef]
33. Mangold, M.; Österbring, M.; Wallbaum, H. Handling data uncertainties when using Swedish energy performance certificate data to describe energy usage in the building stock. *Energy Build.* **2015**, *102*, 328–336. [CrossRef]
34. Luo, X.; Hong, T.; Tang, Y.-H. Modeling Thermal Interactions between Buildings in an Urban Context. *Energies* **2020**, *13*, 2382. [CrossRef]
35. Hong, T.; Lee, S.H. Integrating physics-based models with sensor data: An inverse modeling approach. *Build. Environ.* **2019**, *154*, 23–31. [CrossRef]
36. Functional Mock-up Interface. Available online: https://fmi-standard.org/ (accessed on 15 December 2021).
37. Introduction—FMU Export User Guide. Available online: https://simulationresearch.lbl.gov/fmu/EnergyPlus/export/ (accessed on 15 December 2021).
38. Pandis Iverot, S.; Brandt, N. The development of a sustainable urban district in Hammarby Sjöstad, Stockholm, Sweden? *Environ. Dev. Sustain.* **2011**, *13*, 1043–1064. [CrossRef]
39. Weather Data Center. Available online: https://www.ashrae.org/technical-resources/bookstore/weather-data-center (accessed on 15 December 2021).

Article

Roadmap to Neutrality—What Foundational Questions Need Answering to Determine One's Ideal Decarbonisation Strategy

Stefan M. Buettner

EEP—Institute for Energy Efficiency in Production, University of Stuttgart, 70569 Stuttgart, Germany; stefan.buettner@eep.uni-stuttgart.de; Tel.: +49-711-970-1156

Abstract: Considering increasingly ambitious pledges by countries and various forms of pressure from current international constellations, society, investors, and clients further up the supply chain, the question for companies is not so much whether to take decarbonisation action, but what action and by when. However, determining an ideal mix of measures to apply 'decarbonisation efficiency' requires more than knowledge of technically feasible measures and how to combine them to achieve the most economic outcome: In this paper, working in a 'backcasting' manner, the author describes seven aspects which heavily influence the composition of an 'ideal mix' that executive leadership needs to take a (strategic) position on. Contrary to previous studies, these aspects consider underlying motivations and span across (socio-)economic, technical, regulatory, strategic, corporate culture, and environmental factors and further underline the necessity of clarity of definitions. How these decisions influence the determination of the decarbonisation-efficient ideal mix of measures is further explored by providing concrete examples. Insights into the choices taken by German manufacturers regarding several of these aspects stem from about 850 responses to the 'Energy Efficiency Index of German Industry'. Knowledge of the status quo, and clarity in definitions, objectives, time frames, and scope are key.

Keywords: decarbonisation; climate neutrality; industrial energy saving; strategic decision making; net-zero; road mapping; energy efficiency; ideal mix; sustainability strategy; energy efficiency index

Citation: Buettner, S.M. Roadmap to Neutrality—What Foundational Questions Need Answering to Determine One's Ideal Decarbonisation Strategy. *Energies* **2022**, *15*, 3126. https://doi.org/ 10.3390/en15093126

Academic Editors: Sergio Ulgiati, Hans Schnitzer and Remo Santagata

Received: 8 February 2022
Accepted: 22 April 2022
Published: 25 April 2022

Publisher's Note: MDPI stays neutral with regard to jurisdictional claims in published maps and institutional affiliations.

1. Introduction

1.1. Background

Ahead of the United Nations' Climate Conference COP26 in Glasgow, a vast array of severe weather incidences across the globe—floods, storms, droughts, increase in temperature, melting ice shelves, etc. [1], underlined the warnings presented by various bodies [2–5]. The latter stress that significant action is required by policymakers to still be able to limit global warming to less than 2 °C, ideally 1.5 °C, above pre-industrial levels, as agreed in the Paris Climate Agreement [6].

The Intergovernmental Panel on Climate Change (IPCC), the International Energy Agency (IEA), the German Energy Agency, and many other bodies have published reports, roadmaps, and scenarios [4,7,8] on actions necessary to meet the set target. The pace of environmental change suggests that actions should be taken sooner rather than later to keep the required action trajectory manageable and maintain the ability to meet the target. Nonetheless, unforeseen situations, such as the conflict between Russia and the Ukraine and the linkages to energy-dependency, can further increase the urgency of decarbonisation and switching to renewables [9]. In fact, events of an imminent magnitude can trigger stakeholders to societally endorsed changes of policy and concerted action in a time of crisis. For example, this was the case with the COVID-19 pandemic and also with the nuclear reactor catastrophe in Fukushima, which led Germany to move away from nuclear energy and announce the *Energiewende* [10].

Ahead of COP26, an increasing number of countries have reacted by declaring their ambitions for net-zero emissions in line with the requirement to submit updated intended nationally determined contributions (INDCs) to the United Nations Framework Convention on Climate Change. According to the Climate Action Tracker [11], "over 140 countries had announced or are considering net-zero targets, covering 90% of global emissions". Net-zero means that emissions remaining after reduction efforts are balanced out through offsetting (i.e., via carbon sinks or compensatory projects) [12]. While many countries pledge to reach net-zero emissions by 2050, some aim at reaching this goal earlier (e.g., Germany 2045), some later (e.g., China 2060). Moreover, whereas some countries target carbon neutrality, others target climate neutrality (e.g., European Union 2050) [13]. Carbon neutral only refers to carbon-dioxide emissions (CO_2), whilst climate neutral includes other emissions such as methane, etc. How countries aim to achieve these goals, however, remains vastly vague.

As setting a target never automatically leads to its achievement or even further actions, it is crucial to equip policymakers with the insights needed (on how) to achieve net-zero in actuality and effectively. Looking at the demands faced by governments to fight and prioritize climate change, it may seem like it is up to the governments alone to mitigate climate change. However, by *direct* action, governments only account for the emissions of their immediate actions and on their premises. Conversely, they have *indirect* influence on the emissions of their entire economy through regulatory measures and policies. These may include bans, minimum requirements, mandatory actions, the provision of infrastructure, incentives, and subsidies.

Typically, most emissions are caused by energy generation and key economic sectors, such as transport, industry, housing, and agriculture [14,15]. Therefore, achieving climate change targets essentially comes down to getting these sectors to reduce their emissions, usually with the aforementioned set of policy measures.

Specifically, the challenge is to identify which set of measures is effective and economic to decarbonise which part of the economy. Instructive measures have proven impactful in the past (i.e., minimum standards, phasing-out of incandescent light bulbs, etc.) [16,17]. Nonetheless, given that achieving net-zero requires emissions to be cut or removed across the board, it is necessary that individual and intrinsic actions are as broad as possible. Hence, it is essential to find effective means to trigger such intrinsic wish in stakeholders to reduce emissions, in other words, convincing them to 'buy-in'. This way, rather than avoiding regulations and trying to find loopholes, stakeholders proactively look for means with which they can succeed in meeting their self-set targets.

Two key challenges arise: Firstly, one has to identify means that successfully trigger the (intrinsic) decision to decarbonise and, secondly, to provide those who have taken this decision (or are at least contemplating to) with the means to decarbonise effectively.

As stakeholders are principally aware of their own operations, they have a good chance finding ways to reduce their emission footprint. The cumulative proactive efforts of stakeholders then allow governments to shift attention from the spot-policing of compliance (with instructed policies) to ensuring a suitable environment for stakeholders to be able to decarbonise (i.e., planning capacities, generation and transmission infrastructure, support mechanisms). Furthermore, potential gaps in stakeholder ambitions to meet the countries' goals can then be addressed.

1.2. Industrial Sector of High Relevance for Achieving Net-Zero—But How to Get Started?

One of the most relevant groups in the energy transition is the industrial sector. Not only does it account for a large proportion of most countries' energy consumption, but also for associated energy- and process-related emissions [18–22]. Furthermore, this sector determines the shape, performance, and durability, as well as the energy and resource consumption of goods during production and service life, but also their repairability, recyclability, and how and where the required raw materials are sourced. Hence, the industrial sector influences all other sectors by controlling product and machinery characteristics as well as their modes of operation (e.g., power stations, turbines, transmission infrastructure equipment, vehicles, materials for new buildings and retrofits, machinery, electronics, clothing, or furniture). These decisions largely determine the embedded emissions of all produce, a factor which is rapidly gaining in importance. This is further underlined by both the 'Sustainable Product Initiative', which is developed by the European Commission at present, including "requirements on mandatory sustainability labelling and disclosure of information to consumers on products along value chains" [23], and a 'Resource Passport' for buildings that the German government plans to introduce, along with reshaping its support programmes from purely considering energy-related characteristics to the whole lifecycle footprint [24].

Therefore, decarbonisation in manufacturing can be considered a critical enabler to the question of how to achieve carbon or climate neutrality on a country-wide level and beyond. A growing number of studies thus explore pathways for deep decarbonisation, particularly of energy-intensive industries. According to Nurdiawati et al. [25] (p. 2), many of these studies "focused much on the technological pathways and less on the supportive enabling reforms that would facilitate their uptake". Bauer et al. [26] explore pathways for decarbonising four emission-intensive sectors, even moving beyond direct emissions to also considering value-chain and end-consumers emissions. Bataille et al. [27] (p. 1) present an "integrated [policy] strategy for a managed transition" in energy intensive industries, also including technology options, and Rissman et al. [28] review policy options, sociological, technological, and practical solutions in detail.

These studies address decarbonisation of industry from either a policy, a supply-side, or technology perspective—often with a focus on energy intensive industries—but are short of giving corporate stakeholders (irrespective of their company's energy intensity) *concrete* advice on how to get started from an individual company's perspective. Similarly, studies such as the one by Johnson et al. [29] analyse and compare *national* roadmaps for decarbonising the heavy industry on a global scale, alongside factors such as ambition, financial effort, and mitigation measures. Nevertheless, this approach again leaves a gap when it comes to company-tailored advice.

Consultancies and advisories fill this gap insufficiently. While they generally indicate which steps have to be taken by a stakeholder to shape a decarbonisation roadmap from a company perspective [30–32], they either do not go into sufficient detail, or do not address the prerequisite, qualifying steps, notably those of strategic decision making. These, such as the motivation leading to the decision to decarbonise, however, often have serious implications on the shape of an 'ideal' decarbonisation strategy and how it can be implemented effectively.

An effective way to develop decarbonisation roadmaps could involve applying approaches from the backcasting framework literature. This concept, established by Robinson [33], refers to a strategy where stakeholders/policymakers set up a target (energy consumption/emissions) and work backwards from this target to reach it in the future. This framework is widely applied in designing emission-reduction pathways. In this context, a new strand of the scenario literature includes a focus on low-carbon scenario road mapping. As part of this new literature, Hughes and Strachan find "that low carbon scenarios tend to focus either on qualitative, social trend-based approaches to developing futures, or on purely technological, engineering-based views of an energy system" [34] (p. 46). In particular, technologically focused studies, such as Bataille et al. [35] and Manders et al. [36], often operate within a 'backcasting' framework explained by Holmberg and Robèrt [37]. However, they argue that road mapping the future is always, to some extent, hampered by uncertainty and that therefore the system level, as well as the actor and the technology level, must be considered. Thus, one may argue, that due to the uncertainty and inaccuracy of existing studies and roadmaps, they remain low in their ability to give *concrete* advice.

Having said that, studies that not only focus on either technology, individual, social, or system level are still rare. Similarly, there is a lack of studies that take into account the whole industrial/manufacturing sector instead of only focusing on its energy-intensive parts. Closing this gap, and thus contributing to effective decarbonisation roadmaps, is the aim of this article.

1.3. The Issue: Enabling Corporate Stakeholders to Decarbonise Effectively

The present article addresses this gap by answering the following research question: What foundational questions matter and need answering to provide practical guidance to corporate stakeholders on how to shape an effective and tailored decarbonisation strategy?

Derived from professional practice and applying a mix-methods approach based on data gathered via the Energy Efficiency Index of German Industry (EEI) [38], this work addresses underlying motivations and spans across (socio-)economic, technical, regulatory, strategic, corporate culture, and environmental factors. It further underlines the necessity of a mutual understanding, clarity, and communication of definitions and targets.

Plenty of companies have already made pledges related to emission reductions. However, these companies constitute only a small proportion of the global manufacturing industry, even though they might be big in size individually. Nonetheless, to achieve net-zero on a societal level, it is not sufficient to address the largest emitters only, but to find ways to reach at best all emitters. Specifically, it is crucial to get their 'buy-in', irrespective of their emission intensity or size, and empower them (and the communities they are embedded in) to take action.

Tackling these challenges, this work aims at aiding executive leadership, as well as other company functions relevant to the transition, in shaping their pathway to net-zero effectively. It further provides insights to policy makers, service providers, financiers, and the general public on (often not obvious) obstacles, needs for support, and infrastructure, as well as interdependencies along the process. Several of the general principles may also apply and, therefore, prove to be helpful to other sectors, state actors, communities, or individuals.

The motivation for this article partly arose out of a meeting with a company invested in advancing energy efficiency, but which had not yet seen the point in decarbonisation. Following an explanation of why it is in their best interest to take decarbonisation seriously (by highlighting a series of external pressure points), the manager expressed the belief that immediate action was necessary. To brief the company's CEO, the manager then enquired what aspects the executive leadership of a manufacturing company needs to consider to shape an effective and economic strategy. Although the analysis may generally be broadly applicable to many stakeholder-types, the author focuses on (predominantly manufacturing) companies that have taken the decision to decarbonise or contemplate whether to do so.

Following a backcasting approach, this article provides an overview of seven foundational questions that need answering to enable a general understanding, as well as to provide practical guidance on how to shape an effective and tailored decarbonisation strategy. The results demonstrate that clarity in definitions, objectives, timeframes, and scope, as well as a thorough understanding of the status quo and the technically feasible options, are key. In light of changing emission and energy prices, as well as the goal of ensuring resilience against external shocks, digital solutions, and an adjusted approach to economic viability calculations are needed to help with keeping such a strategy ideal.

2. Methods and Materials

As discussed earlier, previous studies about decarbonisation road mapping tend to focus either on the system (national roadmaps) or on the individual level (specific sectors). Furthermore, they tend to concentrate either on policy or technological factors. This article digs a bit deeper by taking most of these factors into consideration and combining them, thus eventually requiring a combination of qualitative and quantitative elements.

The associated methodology applied by the author is a backcasting method, as described by Robinson [33] (p. 339), that is adjusted for the context of company decision-makers and the goal of decarbonisation. The resulting seven individual steps take inspiration from the six steps originally proposed by Robinson but differ in their shape and nature. 'Backcasting' in this context means working backwards from the desired outcome to the ingredients that need to be obtained or taken into account to reach that future. It is thus an explicitly normative approach [33] (p. 337).

In an iterative process, starting in May 2019, the author analysed manufacturing companies' stand towards decarbonisation with a particular focus on local decarbonisation efforts, notably around energy efficiency.

The qualitative element of the analysis of companies' actions, ambitions, and intentions is based on primary sources. Direct conversations with companies allow for a first-hand understanding of their viewpoints and needs. The businesses consulted were manufacturing companies that are either clients in energy efficiency or decarbonisation projects, participate in the Energy Efficiency Index of German Industry (EEI), seek guidance on the topic or partook in events concerning industrial decarbonisation. In addition, business press, newspaper articles, press releases, and pledges from companies, as well as feedback received in context of public speeches and outcomes from expert discussions have been taken into account. These kinds of observations promise to shed light on aspects concerning willingness and efforts to decarbonise.

Afterwards, these observations were tested quantitatively within the framework of the Energy Efficiency Index of German Industry (EEI) to confirm the anecdotal evidence and assess the actual progress of decarbonisation. Introduced in 2013, EEI surveys German manufacturing companies of all sizes, energy intensities, and across 27 sub-sectors twice a year. It aims at gaining an understanding of companies' stands, expectations, plans, opinions, experiences towards energy efficiency, and increasingly also decarbonisation. EEI data is gathered applying a mixed-methods approach combining online (ca. 10%) and telephone surveys (ca. 90%) [38].

An iterative process was applied to deepen the understanding of interdependencies and elements that are the foundational ingredients that enable—or hold back—decarbonisation. Whenever the EEI uncovered a relevant finding, the next data collection, after pre-testing, was utilised to drill deeper. In total, evidence from five data collections is considered in the context of this article (cf. Table 1).

Table 1. Energy Efficiency Index of German Industry (EEI) datasets referred to [39–43].

EEI Data Collection	Data Collection Period	Observations
2019/2	October–November 2019	915
2020/1	May 2020	863
2020/2	October–November 2020	884
2021/1	April–May 2021	717
2021/2	November 2021–January 2022	865

To provide a general overview, a series of EEI questions of the past five data collections were identified to illustrate selected aspects: (a) whether companies plan to decarbonise, and (b) if so, by when. What level of ambition they have for (c) 2025 and (d) for 2030 and (e) optimising for which dimension(s). Based on (f), what motivation they do so, and (g) what weight different determinants have in deciding for decarbonisation measures. Beside the area of observation (h), EEI explores the increasing relevance of product carbon footprints (i). The awareness of companies' emission footprint (j), along with knowledge about energy consumption and type (k, l) and energy saving potentials (m), are explored to assess companies' knowledge of their status quo [39–43].

3. Results

Before making a decision, one often considers the implications and repercussions of that decision. Nevertheless, even after a thorough consideration it is not unlikely that an aspect that significantly impacts the overall ambition is overlooked—unless one has succeeded in a very similar or identical undertaking before. Decarbonising one's business is to some extent like building a house for the first time. After completion, one has learnt much about what to do better or differently the next time. Nonetheless, in many instances one only builds one house (if any). Roughly the same applies in the case of decarbonisation: once it is achieved—however (in-)efficiently—there is rarely a situation where one does it again from scratch (unless a company has multiple sites and started with a pilot one or offers the experience as a service to others). Again, similar to one's house, there remains the prospect of continuous optimisation. While some improvements might be incremental, other interventions would require significant interference if at all possible (for example switching from a radiator-based heating system to underfloor heating to allow the installed heat pump to serve the home with heat more efficiently [44]). Setting a clear target to be reached in the future and being aware of multidimensional factors, which might influence how it is reached, is the ultimate goal for a successful decarbonisation strategy.

Therefore, it is of high relevance—to stakeholders of any sector—to *find answers to seven foundational questions*, ideally before, but at least simultaneously to taking action. Only the response to these questions allows one to determine one's ideal decarbonisation strategy, or to make an informed decision whether to go ahead and act, or even to openly pledge to take action.

(1) Terminology;
(2) Optimisation variable;
(3) Level of ambition;
(4) Area of observation;
(5) Motivation and needs;
(6) Priorities;
(7) Status quo.

Based on the responses to these, it is then possible to derive (a) general, and (b) specific routes of action to determine a decarbonisation strategy suiting one's situation, goals, and opportunities. Making use of (c) digitalisation and (d) a modified form of economic viability calculations allows one to find one's ideal roadmap to neutrality and to adjust it dynamically to changing environments.

Why these seven, one could argue. Essentially, every one of them is guided by the notion of what could go wrong (or has gone wrong elsewhere), what could reduce the efficiency and/or effectiveness of a decarbonisation strategy, and how this can be avoided.

As mentioned when discussing the backcasting framework and also as explained by Rissmann et al., "the best practice in designing efficient industrial operations is to analyse the entire process by working "backwards" from the desired application to the energy consuming-equipment" [28] (p. 16). Transposed to the context of this article, the "desired application" reflects the desired outcome.

In this context, however, the outcome needs to be further specified as decarbonisation can be understood differently, achieved differently, and should be pursued differently, if it is to address different motivations or to consider different priorities. Therefore, as Rissman et al. stated referring to increasing efficiency of industrial systems and processes, "design should be an integrative process that accounts for how each part of the system affects other parts." [28] (p. 16).

In this article, "design" refers to the preliminary steps (i.e., strategic considerations) that need to be taken, typically by executive leadership, to allow them, and subsequently their company to shape and pursue an effective and efficient roadmap to neutrality.

Other than the practical "design layers" that describe step by step the "how" of increasing efficiency [28] (p. 16) [45], the seven foundational questions address the "what", "where", "by when", and "why", as well as the "how". Nevertheless, they apply on a more strategic than a specifically practical level.

The following sections will provide a more detailed explanation of the seven dimensions (Sections 3.1–3.7), followed by an overview of how they guide implementation in general and more individually (Sections 3.8 and 3.9), as well as steps to make and keep a strategy ideal (Sections 3.10 and 3.11).

3.1. Terminology

The foundation of an effective decarbonisation strategy, as of any other work in any other area, is to establish mutual understanding and clarity across all stakeholders involved regarding the terms used and how they are understood. Otherwise, misunderstandings or misperceptions will lead to either unnecessary action being taken or, worse, essential actions being overlooked.

Buettner [46] points out that a key reason for the frequent mixed-up between carbon- and climate neutrality is that, while CO_2-equivalents (CO_2-eq.) are the 'currency' to measure greenhouse gas (GHG) emissions adversely affecting our climate, the suffix "-eq" (standing for equivalents) gets easily lost on the way. This is particularly the case in oral or simplified conversation and correspondence.

Apart from this, it is further possible that the difference between carbon neutrality, climate neutrality, and environmental neutrality itself is not clear. However, this unclarity has fortunately been decreasing over the past three years. In short, climate neutrality exceeds the ambition of carbon neutrality by also addressing methane and all other gases that have a warming potential for the atmosphere (GHGs), such as nitrous oxide and hydrofluorocarbons. Environmental neutrality reaches even further and addresses all other gases and substances that have a negative impact on the environment (such as particulate matter and sulphur dioxide, cf. Figure 1) [46].

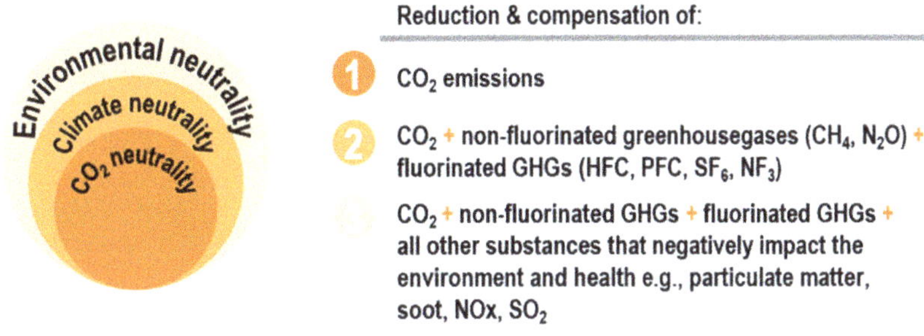

Figure 1. Defining different neutralities and what is needed to achieve them [46,47].

This frequent lack in clarity regarding definitions can also be observed beyond private sector stakeholders, in the public sector, in politics, public discussion and in media, for instance when reporting on targets: The German business paper Handelsblatt and the New York times diverge over the target set by Japan in late 2020. According to Handelsblatt [48], Japan is aiming for climate neutrality, while the New York Times [49] reports carbon neutrality to be the target. Without the means to retrieve the information from the original source in the language of origin, one will not know which neutrality is being targeted by Japan.

Therefore, establishing clarity of the target dimension and how it is being defined is crucial [46] for all stakeholders involved in the process (i.e., within a company), thus making it the *first success criterion* to any kind of net-zero pledge.

3.2. Optimisation Variable

Even if the terminology is commonly understood, a strategy can only be effective if it serves achieving a clearly defined objective, in this instance one or multiple target dimensions that serve as variable(s) that are optimised for [50]. In context of emission reduction optimisation, common variables are (not exhaustive):

(a)   Reduction of energy consumption (reduces emissions);
(b) CO_2-neutrality (usually includes reduction of **a**);

(c) Climate neutrality (includes **a** and **b** and is policy goal of, e.g., EU and Germany);

(d) Environmental neutrality (includes **a**, **b** and **c**).

For stakeholders in general, but also for a company in particular, it makes sense to pursue pragmatic pathways to effectively achieve what is needed. However, it is also relevant to observe the legislator's target setting, notably its target dimension. If climate neutrality is the country's target, policies are very likely tailored to serve this goal and companies are well-advised to take this into consideration rather than looking only at a subset of this dimension (e.g., carbon neutrality).

Even though the optimisation variables **a–d** are not mutually exclusive, the Energy Efficiency Index of German Industry (EEI) observed in its second data collection 2020 [41] that the 834 participating manufacturers on average optimise their companies towards two target dimensions. This suggests that within a further reaching optimisation variable, they also aim at optimising for (at least) one of its components in particular:

Most companies (58%) optimise towards a reduction of energy demand, second most (53%) for the reduction of CO_2-emissions. The fact that just over a third of companies indicate they want to optimise for GHG reductions (36%) or overall environmental impacts (36%) leads to the surmise that GHG reductions or, in other words, the means to reach climate neutrality remain abstract in the industrial context. This stands in opposition to the fact that climate neutrality has been the known target of both Germany and the European Union (EU) at the time of the data collection (cf. Figure 2) [41].

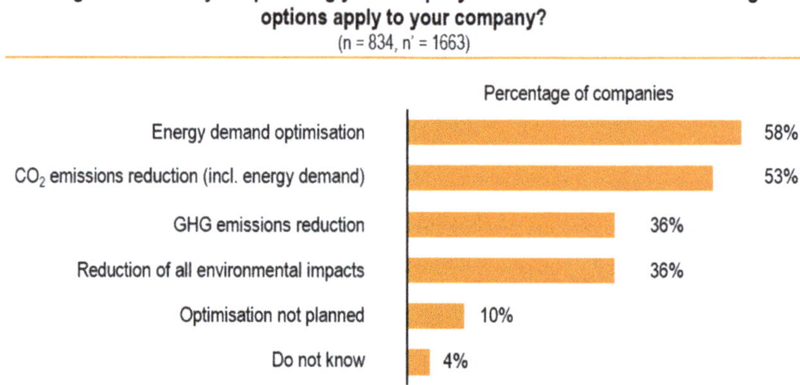

Figure 2. Target dimensions companies optimise towards [41].

Addressing the potential issue of climate neutrality being rather complex due to some of its hard to identify and quantify sub-components (e.g., nitrous oxide and hydrofluoro-carbon.), the United Nations Economic Commission for Europe's Task Force on Carbon Neutrality is pursuing an in-between target dimension: Carbon neutrality plus methane reduction (and hydrogen) in its carbon neutrality project [51] (para 17) [52]. An agreement to reduce global methane emissions in context of COP26, counting more than 100 countries to date [53], indicates the notion of 'carbon neutrality +' to be tangible for those that find it difficult to commit to the further reaching climate neutrality goal.

After awareness of terminologies, determining the optimisation variable(s) as target dimensions and overarching goals that stakeholders are aiming to work and orient their forthcoming actions towards is thus the *second success criterion* on the path to net-zero.

3.3. Level of Ambition

The choice of target dimension (e.g., carbon or climate neutrality) only provides a limited indication of the level of ambition, as it remains unclear by when it is to be achieved. Very timely target years usually suggest a high level of ambition, whereas far into the future targets indicate either a very cautious regime, limited means to reach the goal earlier or simply lacking ambition. The German energy provider RWE plans to become climate neutral two years after the scheduled German coal phase-out—in 2040 [54]. Very timely target years, however, often significantly depend upon compensatory measures rather than actual emission reductions [55].

Clarity on the level of ambition is only achieved when it is also determined (a) *by when* the goal should be achieved and (b) *what percentage reduction* of the target dimension this is set to be. The latter is of high relevance, as there are scenarios in which a 100% reduction either cannot be achieved or simply is not the goal. This is the case if the target dimension is energy consumption, or if proportions of the energy- or process-related emissions cannot be avoided through reduction, substitution, or other alike means. In such cases, it could be attempted to balance remaining emissions through offsets (e.g., compensation) to manage a 'net-zero' instead of the 'actual zero' state in respect to their target dimension. Nevertheless, several stakeholders the author works with object to compensatory projects by principle and exclude these from their feasible set of decarbonisation measures, thus excluding themselves from the option of reaching 'net'-zero.

Beyond defining an ambition in terms of the finish line, it makes sense to also consider *interim milestones* to ensure the target can be met and potentially unpopular interventions are not being postponed to the future. Moreover, interim milestones ensure that the trajectory required to achieve the target is the same as the actual trajectory and adjustments are made if necessary. While there is no requirement to determine interim goals for companies, it is logical to do so in terms of year and level of achievement by then.

Many countries have set milestones for (at least) 2030 [56] (p. 41). As thorough assessments by these countries into the state of play are to be expected, it makes sense for stakeholders operating in these countries to define a milestone that ideally is already following the country's target for the respective year(s), too. The cases of Germany and the Netherlands being successfully sued at their constitutional courts over insufficient short- to medium-term action towards their 2030 targets underlines that additional interim milestones and, if necessary additional actions could be of relevance [57,58]. This is also why the outcome of the Glasgow Climate Conference COP26 encourages revisiting the current level of action, status, and subsequent tightening of pledges in shorter cycles than originally agreed upon in the Paris Climate Agreement (Art. 4 (9)) [6,59]. The current crisis, which has led to a desire in many European countries to quickly reduce dependence on fossil fuel imports, adds an additional and concrete urgency [9].

Nonetheless, countries can only succeed in meeting their (climate) goals, if they get the individual emitters, notably across building, transport, and industrial sectors, to reduce their (energy- and process-related) emissions.

Looking at the ambitions of German manufacturing, 59% of the 852 companies participating in the EEI in autumn 2019, ahead of the COVID-19 pandemic, indicated they plan to achieve net-zero. Of these 488, two thirds aim to have met this goal already before or by 2025 (cf. Figure 3). Peaking numbers in 2025, 2030, 2035, 2040, and 2045 (highlighted in yellow) suggest that semi-decades are chosen by many companies as their target years or at least milestones. The data further suggests that a vast majority of companies participating in the EEI prefer taking substantial immediate or at least short-term action [39].

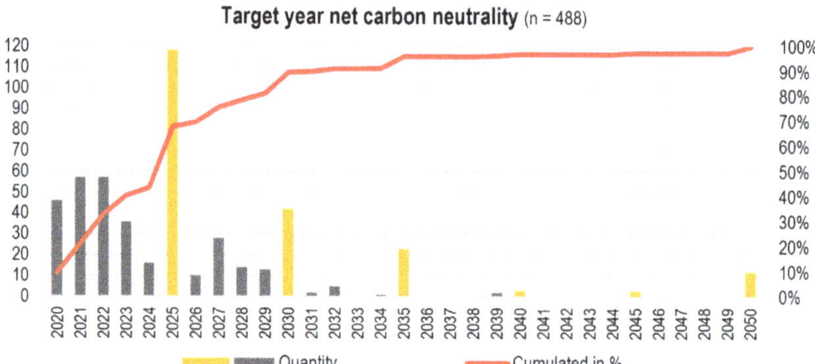

Figure 3. By when do companies plan to reach net carbon neutrality [38,39]?

Taking the likely impact of the COVID-19 pandemic into consideration and addressing the apparently important milestone year 2025, the first iteration of EEI in 2020 found that the 611 participating companies on average, and based on 2019 figures, aim (on average) at reducing their GHG emissions by 22.1% by then [40]. Asking for their 2030 ambitions at the time when the enhanced target of the European Commission for 2030 was being discussed (autumn 2020), 415 companies participating in the second data collection of EEI in 2020 expressed to aim (on average) for a 26.4% GHG reduction (based on 2019) [41]. This data confirms that (at least participating) companies consider substantial short-term action, accounting for more than 80% of what is planned for the whole decade, to happen within its first half. The arising curve of ambition appears to follow a path similar to limited growth functions, whereas policy action is often perceived to follow the opposite path of an exponential growth curve slowly growing towards 2030 and then taking up pace. The action gap arising from this/from what stakeholders need to enable them to meet their goals, and the current impact of policy, is explored further by Buettner et al. [60].

The level of ambition—the combination of target dimension, percentage-goal, and due date—can either be 'simply' determined by stakeholders, or be set once 'all cards are on deck', meaning all relevant (limiting) factors and potentials, feasible measures, as well as their costs are known. Irrespective of when exactly this decision is taken, setting and announcing a level of ambition is the *third success criterion* on the path to net-zero.

3.4. Area of Observation

In the context of target setting, the area of observation, or the 'system barrier' is not always clear and obvious. Like the necessity to establish clarity of definitions, it is necessary to define to what the set target dimension and level of ambition refer.

This leads to three questions that need to be considered by stakeholders.

(1) Does the target apply to one site, multiple sites, or all sites of the stakeholder, or only to those in countries where some form of CO_2-levy is operational or considered. Does it only apply to sites in selected countries, e.g., Germany? Intuitively, it would be understandable if stakeholders prioritise those sites where there is an elevated levy-induced 'incentive' to take action, respectively those where the enabling environment makes it easier to succeed when taking action. From the author's practical experience, companies often initially focus on one site, or sites within their home country and then, when actions prove to be successful, they gradually expand beyond both geographically and in terms of efforts taken on the initial site.

(2) Are we referring to emissions and energy use in relation to this site/these sites only and, if so, including or excluding the corporate vehicle fleet (Scope 1 + 2). Or does the ambition go beyond the direct and indirect emissions that are under quasi-direct control of the stakeholder? Such Scope 3 emissions arise indirectly from one's action but are often

outwit direct control, and include business travel, the workforce's commute and additional emissions arising along up- and downstream supply chains (cf. Figure 4) [61].

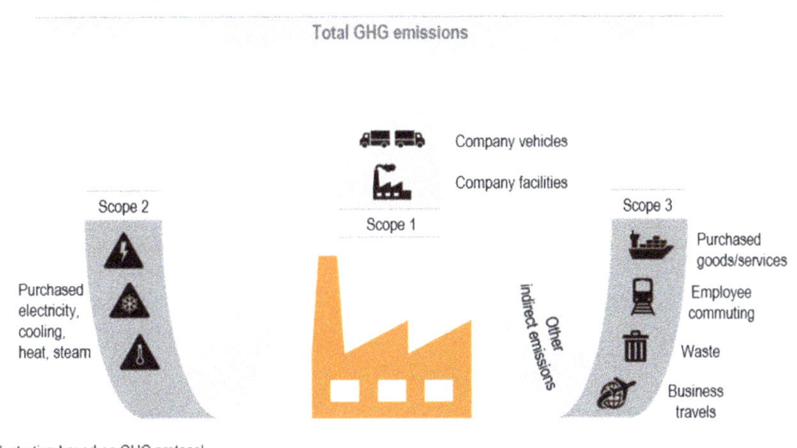

Figure 4. Carbon footprint assessment [38,52].

To the author's experience, most companies initially only address their energy-related emissions (Scope 2), as well as emissions directly arising from their work (Scope 1), due to the complexities of addressing Scope 3 emissions. Complexities arise predominantly out of potential double counting: Scope 1 emissions of one company might be Scope 3 emissions of another company [61,62]. Currently under investigation by EEI in its second data collection 2021, the interim analysis suggests that 77% of the 848 (846, 843) companies responding to this question strive to address Scope 1 emissions or have done so successfully already, 78% target Scope 2 and 75% Scope 3 emissions (cf. Figure 5). Further analysis of the new data will allow an examination of whether companies on average only address Scope 3 after a head start on Scope 1 and 2. The interim analysis suggests so: the progress is furthest in respect to Scope 2, followed by Scope 1 and with a substantial gap in Scope 3, which is understandable, as Scope 2 is 'easiest' to achieve by optimising energy supply contracts [43].

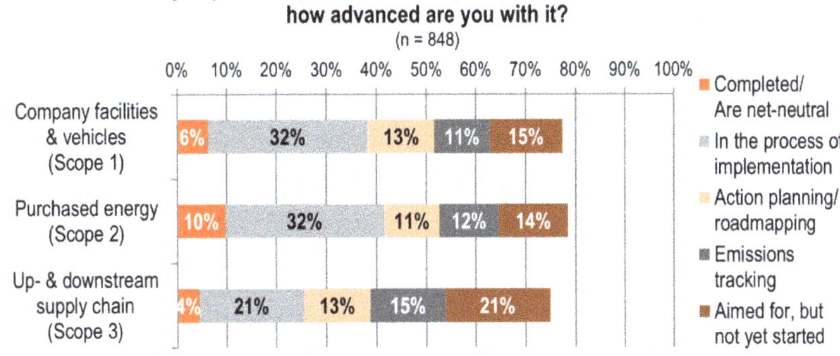

Figure 5. Companies' plans and current state in respect to Scope 1–3 [43].

(3) Approaching emissions from a different, a product angle: are only those emissions considered up to the point when (a) a product leaves the premises or arrives at the customer/the shop? Or is the additional emission footprint of the product (b) arising during its useful life, or (c) even until it is fully disassembled and recycled of relevance, too? Particularly in the automotive industry (b), this is of high relevance to meet the European Union's requirement on new vehicles to not exceed 95 g of CO_2-eq per km on fleet average to avoid being fined 95 Euros per gram and vehicle exceeding the average [63]. Considering the large footprint carried by the manufacture of lithium-ion batteries, but also steel, aluminium etc., manufacturers such as Volkswagen work to sell their electric vehicles with a net-zero footprint at the point of handover [64]. A significant undertaking, as many end products' Scope 3 emissions make up more than 75°% of the overall "Product Carbon Footprint" (PCF)—82% in the automotive industry [65] (p. 9).

The automotive industry is not the only sector where PCFs are increasingly found. The chemical giant BASF announced the assessment of the carbon footprint of all its products, as well [66]. Interim analysis of the EEI's second data collection in 2021 suggest that 37% of 829 companies responding to this question consider the PCF until the point of handover, 13% until the end of useful life, and 21% until the product is fully recycled/disposed of. However, 29% do not consider their products' PCF at all at this point. In total, almost 71% of companies work to offer products with a 'net-zero' footprint in one form or another, at least in respect to the point of handover (cf. Figure 6) [43], which is a good move in context of the EU's sustainable product legislative initiative mentioned earlier [23].

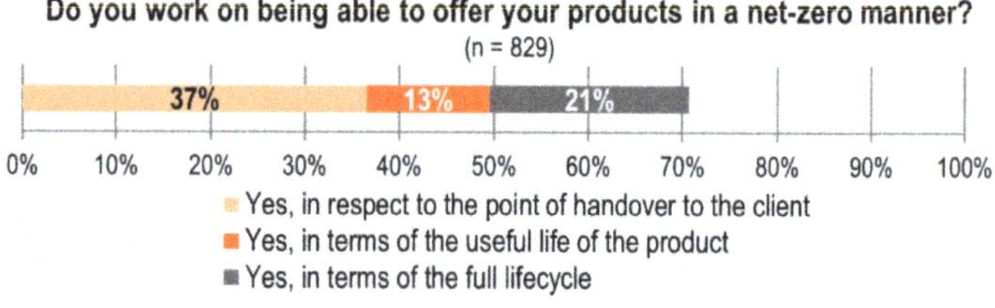

Figure 6. Companies' goals in relation to their products' carbon footprint [43].

As the bandwidth and efforts required largely differ depending on what system barriers are being set, defining the area of observation, the spatial, as well as the scope of reduction, constitutes the *fourth success criterion* to reach net-zero.

3.5. Motivation and Needs

Beyond the somewhat technical questions of what, by when, and how far, it is of critical relevance to explore *why* decarbonisation is sought. What is the underlying motivation of the executive leadership and the stakeholder for pursuing net-zero? Motivation plays a large role in determining one's ideal strategy and mix of measures, as elements that are of high internal (e.g., corporate culture) or external (e.g., legislation) relevance may be emphasised over a purely technical composition of measures. The motivation also determines how the topic of decarbonisation is embedded in the stakeholder's overall strategy.

Common motivators include (not exhaustive) [67]:

- Requirements of the upstream supply-chain;
- Requirements of investors/shareholders;
- Image improvement: display leadership and innovativeness;
- Image improvement: attracting and retaining skilled personnel;
- Pursuing societal responsibility and corporate culture;
- Meeting societal expectations;
- Demands from policymakers and meeting legal requirements;
- Long-term economic advantages, including building up competency;

- Risk reduction regarding external shocks, such as energy price and acquisition and emission costs;
- Ensuring security of supply arising from (micro-) outages.

As Buettner and König [67] outline analysing these motivators, there is an increasing pressure to take action, triggered by both, but not only, investors and up-stream supply-chains. The latter has just been confirmed by EEI [43]: around a third of 836 participating companies are facing emission-related contractual demands from their upstream supply-chain. Image is not only of relevance to remain able to sell one's products but also to attract and retain scarce skilled personnel. The steeply increasing price of (a) CO_2 within the European Emission trading system (EU ETS, currently at 96 EUR/tCO_2-eq, [68]), (b) electricity, and (c) gas are an increasing cause of concern among stakeholders [69–72], even more so since Ukraine was attacked.

As decarbonisation measures that best address the various motivators can differ widely, getting a clear picture of the main motivator(s) for the decision to act constitutes the *fifth success criterion* on the path to net-zero.

3.6. Priorities

While answering the question of why, when, and what is the essential foundation of determining a roadmap to neutrality, the latter can only succeed if further decision criteria are being determined. These criteria are needed to rank and filter feasible measures simultaneously or after scoring how well these measures address the key motivators. Decision criteria include (not exhaustive) [73]:

- Level of investment;
- Investment cost per tonne of CO_2-eq. avoided;
- Emission cost savings (absolute or relative to invest);
- Image effect through visible measures;
- Expected increase in productivity
- Technical aspects and risks (complexity and difficulty level);
- Disruption of operations (cross-cutting-, support processes or core processes);
- Implementation competence (experience with type of measure or access to personnel with necessary skills);
- Impact on company valuation
- Payback time (including emission-related opportunity costs of inaction);
- Availability of required material and equipment (supply bottlenecks).

Analysing data of the EEI [40], Buettner and König found that economic factors such as absolute and relative level of investment have the highest priority as decision criterion [73]. Given that, they also found that technical aspects are the third most frequently mentioned primary decision criterion, having asked 787 companies. They further identified that the aggregate findings diverge significantly when assessing the top three decision factors from a company size, energy intensity, or sub-sectorial perspective. In context of the GHG reduction target, looking at the primary decision factor only, implementation competence stands out (cf. Figure 7). Either companies setting a particularly ambitious GHG reduction target (understandably) look particularly at their implementation competence when deciding which action(s) to pursue *or*, companies that have (access to) implementation competence (are able) to set more ambitious targets. At least, these two readings appear to play a role for the upper two quartiles of companies illustrated in Figure 7, indicating 'implementation competence' to be their primary decision criterion when selecting of individual measures, as the median GHG reduction target is at the same level as for the other primary decision criteria [40].

As the criteria according to which measures are vetted for feasibility and ranked have a significant impact on how the set of individual measures of a decarbonisation roadmap will look like, deciding upon the top three determinants or their ranking order is the *sixth success criterion* on the way to net-zero.

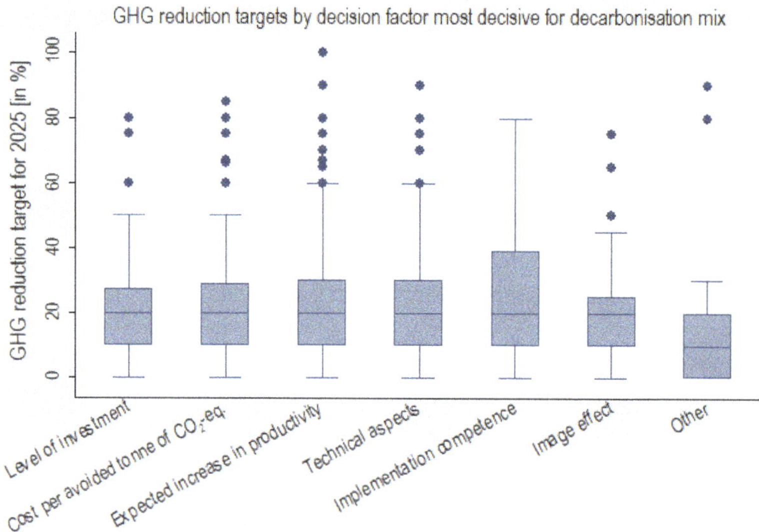

Figure 7. GHG reduction targets by decision factor most decisive for decarbonisation mix [40].

3.7. Status Quo

While the first six success criteria are largely a strategic and economic decision to be taken by the stakeholder, they are still insufficient to derive a successful decarbonisation strategy. Determining one's ideal decarbonisation strategy and subsequently a concrete set of measures is dependent on knowing about where one stands right now-the status quo. As simple as determining the status quo sounds, it requires a thorough assessment across various dimensions:

(a) *What has already been done?* How is the state of the sites, machinery, and equipment? Are there any obvious low hanging fruits?

(b) *What intervention is approaching anyway?* This can be replacement investments, a restructuring of the production line, process, or product range.

(c) *How 'safe' is the site in its existence?* This is of relevance if investing into high efficiency technologies that are pricey to acquire but promise large relative energy and emission savings. If the (non-environmental) sustainability of the business model or production technology is questionable it might, however, not make sense to invest large sums at that site.

(d) *What is the current energy consumption per type of fuel and site, and what are the energy and process-related emissions in respect to the target dimension and area of observation?* Based on this information, stakeholders will know where they are starting from and potentially also where interventions might promise the biggest impact per effort taken.

(e) *What are the local conditions?*

    o   Are there undeveloped areas or available roof spaces? For instance, for on-site generation of renewable energy, energy storage, or heat recovery systems.

    o   How are the climatic conditions? This includes temperature range (e.g., for air/air heat pumps or air conditioning needs and level of insulation), solar radiation (to harness solar energy), wind and air corridors (to apply micro wind generation or use passive ventilation), adjacent waters (for micro-hydro or air/water heat pumps), geology (regarding earthquake risk and for geothermal energy including air/ground heat pumps) and environmental protection zones (e.g., limited development due to protected species or drinking water protection areas).

o   How is the surrounding infrastructure? Is there access to overland power lines, proximity to wind farms, solar parks, or hydro power stations? Are there nearby plots of land that would be suitable for these technologies (for off-premises self-generation)?

o   Who is in the neighbourhood? This is primarily the proximity to entities with whom a symbiotic relationship could be built, typically a sender or recipient of secondary energy or secondary raw materials either on the stakeholders' site (i.e., pre-heating of processes), the industrial estate or in the borough (i.e., feeding waste heat into district heating grid, as Aurubis does for Hamburg's Hafencity [74]). Here, it also plays a role how ambitious the local authority is, as well as the state, region, and country the site is located in and further, whether there are support- and co-funding schemes or other support-mechanisms in place to benefit from or to reduce the overall investments.

According to EEI, about half of participating companies have not been aware of their energy- or process-related emission footprint at the time of participation (cf. Figure 8) [41].

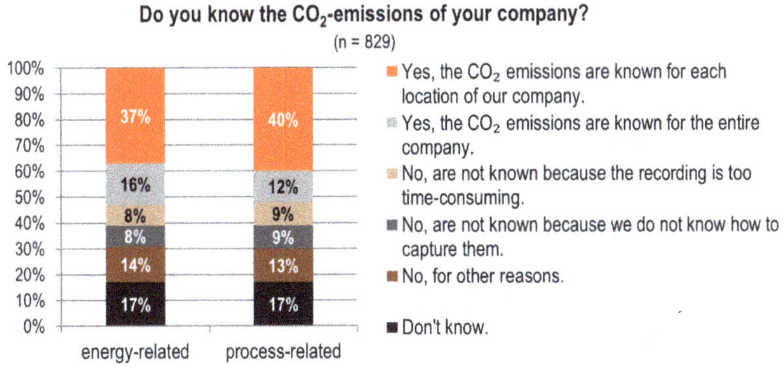

Do you know the CO$_2$-emissions of your company?

(n = 829)

Figure 8. Companies' knowledge of energy- and process-related CO$_2$-emissions [41].

Apart from lighting, the majority of companies were also not aware of their percentage energy saving potentials of the cross-cutting technologies they use (cf. Figure 9) [42].

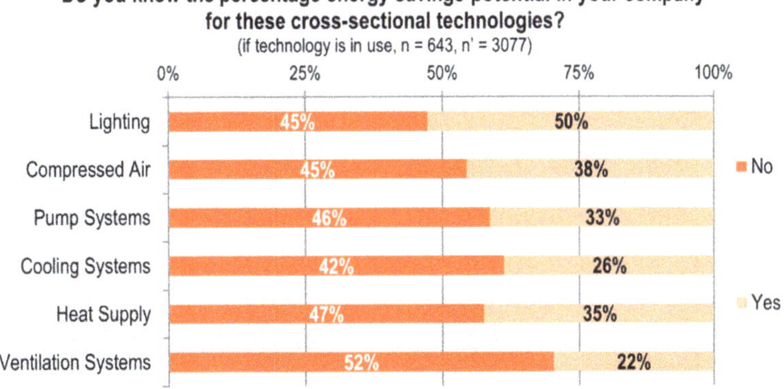

Figure 9. Companies' knowledge of their percentage energy savings potential in cross-cutting technologies they use [42].

More than four out of ten companies were unaware of what proportion of their energy is used for heating and cooling (cf. Figure 10). The latter are, in contrast to electricity, rather immobile, harder to electrify, and difficult to decarbonise, but they offer great potentials for waste energy utilisation, which 22% of participating companies do not harness at all (cf. Figure 11) [42].

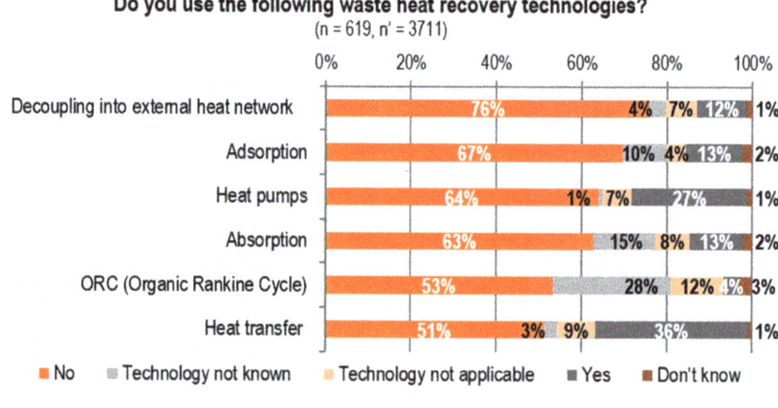

Figure 10. Companies' knowledge of share of energy used for heating and cooling [42].

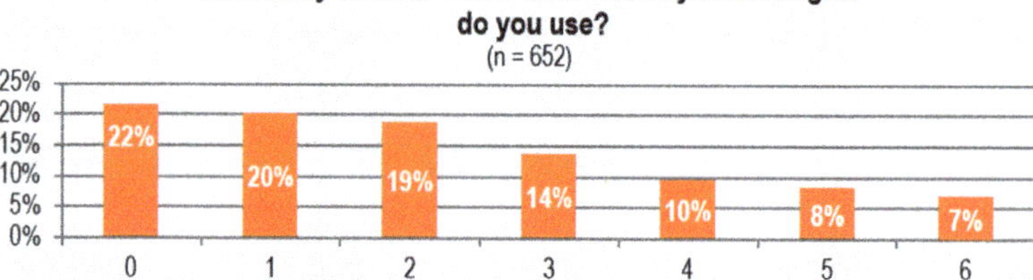

Figure 11. Waste heat recovery technologies used [42].

Acquiring a fair understanding of the status quo, the foundation the road to neutrality is built on, is the starting point of all further steps and hence the *seventh success criterion*.

With the answers to these seven foundational questions, spanning across economic, technical, strategic, principled, and geo-spatial dimensions, it is then feasible for stakeholders to derive both *general* (Section 3.8) and *specific* ways (Section 3.9) forward.

3.8. General

Building on the answers to the seven foundational questions, it is now necessary to determine the proportion to which the goal is to be achieved through measures that can be *implemented locally* and measures that are to be *implemented externally or by others*.

As described by Buettner et al. [38], *internal measures* can include:

- Reduction of energy consumption (and of the connected load) through energy efficiency measures, including utilising waste energy and passive resources such as passive ventilation or solar gains.
- Reduction of process-related or process-induced emissions, for instance by substituting (metallurgical) coke with green hydrogen in steel production, by identifying alternative chemical transformation pathways that are less emission intensive but lead to an equivalent outcome, or by developing more resource efficient processes and products that require a smaller proportion of emission intensive ingredients (e.g., cement clinker in the cement industry).
- Self-generation of renewable energies and their storage, such as solar-, wind-, hydro- or geothermal energy, including means of flexibilising the energy demand.

External measures are all other measures, such as:

- Acquisition of renewable energy (e.g., electricity, hydrogen, biomass, biogas, district heating).
- Procurement of (intermediate) products, raw materials, services, and mobility that have a net-zero emission footprint—either directly acquired on the market or via requirements set for suppliers.
- Offsetting emissions through projects (e.g., afforestation or efficiency-replacement programs through one's own products—comparable to a self-initiated scrappage scheme).
- Offsetting through purchase of certificates.
- Acceptance of the payment of emission charges (in this case 'net-zero' is out of reach in most scenarios).

Carbon capture and storage or utilisation (CCU/CCS/CCUS) is an additional measure, but it does not avoid the emergence of emissions, it only prevents them from being emitted into the atmosphere. While emissions are captured locally (*internal* measure), their further treatment can take place locally as well as elsewhere (*external* measure) [47]. A vast range of studies (such as Cresko et al. [45] and Rissmann et al. [28]) provide further and concrete detail on internal and external measures.

To determine the sequence of measures and the split between local and external measures, both the prioritisation procedures and the scoping outcomes are instrumental as the potential effect of individual measures, investment cost, complexity, payback time, and other key performance indicators will differ and need to be weighted.

It needs to be stated that the split will change over time and with progressing implementation. Bosch, for instance, announced in May 2019 that it would reach carbon neutrality by 2020 [55]. This was only feasible by launching activities in all areas. As local measures could not all be implemented within such a short period, the coverage gap was addressed through offsetting via climate protection projects and the acquisition of green energy. With the progressing implementation of local measures, these external measures can be melted down to a degree until the optimal constellation for net-zero carbon emission is reached. In the meantime, Bosch has changed to the political target dimension of the European Union, climate neutrality, and clarified that succeeding in their original area of observation (Scope 1 + 2), they are now working on Scope 3 [75].

Unless addressed when responding to the seven foundational questions, it is essential to make the decision of whether the tool of compensation through projects or certificates is within one's toolkit. Offsetting does allow reaching net-zero in an expedited manner at the additional cost of the certificates/for the projects—literally buying time until emission saving measures implemented locally take an effect. Accepting emission costs until these can be avoided 'naturally' is the alternative. In the author's experience, several companies rule out compensation as an instrument of their decarbonisation toolkit, as they consider it cheating, since it does not help them reach actual zero emissions. Furthermore, they may wish to avoid the repercussions if such projects are found to be dubious or faulty, or simply want to work towards zero 'naturally' [76–78].

3.9. Specific

Beyond the general types of measures described in the previous sub-chapter, there are further interventions, very specific to the situation of a stakeholder and their status quo, that present an opportunity to take a *technology leap* on the way to shape a net-zero business model. This is to replace existing machinery and equipment with innovative cutting-edge ones that also allow for capitalisation on the opportunities presented by automation, digitalisation, and machine learning. This can, for instance, be control systems that adjust the source of energy, storage, and a range of energy flexibility means by the current availability and price of clean energy, including virtual storage [79]. Another example is factory operation systems that report machine data to a central dashboard in a plug-n-play manner. Similar to the interoperability of "Internet of Things" (IoT) devices in more recent smart home systems or computer operating systems, they adjust to different form factors via drivers built around a core operating system [80]. Other studies also highlight the growing importance of digitalisation in other areas of sustainable business performance, such as cloud-manufacturing, recyclability, and circular economy [81].

In addition to this, Sustainability Key Performance Indicators (KPIs) can be defined based on the decisions made until this point to allow strategic management to monitor the progress on and effectively pursue the road to net-zero, but also as a basis for sustainability reporting [82,83].

3.10. Economic Viability

Buettner and Wang [47] point out that in the context of decarbonisation it is necessary to reconsider traditional economic viability calculations to assess the economic performance of technically feasible measures. The traditional model does neither account for increasing energy costs, nor for the increasing costs of inaction in the format of emission pricing (the price within the EU Emission Trading System (EU ETS), for instance, has risen by over 50% between 1 November 2021 and 1 February 2022 [68]). Further, a short payback time is often a key decision criterion due to various reasons, including business cycles, useful life of machinery, etc. However, in the context of decarbonisation, it makes sense to look for the best constellation for the respective milestone or target year.

To apply this, all types of measures remaining up to this point are to be assessed based on their economic merits, including emission costs avoided, and then weighted and scored as defined by the stakeholder. Simplified, the resulting ranking order constitutes the ideal configuration at that very point of time. 'Simplified', as some measures might depend on each other, are not compatible or only unleash their highest efficiency if applied in a bundle.

3.11. Dynamic Adjustment to Changing Environments

As energy prices and emission costs change over time, the ideal configuration changes over time, as well. To keep one's optimal decarbonisation strategy up to date with energy and emission price developments, it is advisable to make the ranking table of measures described in Section 3.10 dynamically respond to such changes. This is of particular relevance, as these cost-changes can have a significant impact on the ranking order of potential measures in a multiple year timeframe. As described by Buettner and Wang [47], building on energy and emission cost schedules and forecasts, it is then feasible to optimise the mix of measures based on specific milestone or target years, or a combination of these, respectively.

Combining all of the factors discussed in this chapter result in a focus-, situation-, priority- and specificity-driven approach, which is a very individual puzzle that changes its configuration over time.

4. Conclusions and Discussion

Within this article, the author illustrated how the methods applied lead to an understanding of how everything is connected. Using the backcasting method, he provides a step-by-step overview of seven foundational aspects that require attention, thereby helping decision makers in shaping a successful and effective decarbonisation strategy.

Even though the general approach towards what needs to be done may be similar to approaches applied by others, this roadmap to neutrality differs by (a) taking the perspective of an executive decision maker on the demand-side and (b) going a level deeper, where most other approaches indicate what needs to be done, either in general from a system or country level [28,29], or on a micro level (i.e., technical optimisation options and procedures) [25,26,45]. In addition, where existing approaches outline technical roadmaps [45] or indicate what must be done but not always how [30–32,84] and stop short of putting it into context, the approach presented explains the underlying strategic aspects that need to be considered beforehand. Firstly, such considerations raise awareness of the implications of decisions (to be) made and, secondly, ensure the ability to take decarbonisation actions in the best manner and interest of the company. Finally, this approach differs in its methodology by combining qualitative and quantitative data, which (a) allows one to validate learnings from individual cases on a much wider basis, and (b) to interpret broad quantitative findings in context, as sometimes multiple readings appear plausible.

Determining one's ideal decarbonisation strategy, associated decarbonisation roadmap and range of concrete measures essentially comes down to considering one's situation, priorities and motivations, and focus. With these points—addressed by the seven success criteria—one's specific puzzle of measures falls into place.

As shown in the step-by-step approach, clarity regarding the terminology of the target dimension (e.g., carbon vs. climate neutrality; Section 3.1) and the optimisation variables, inferable from this target dimension (Section 3.2), are the first two steps. This is important, as a target can only be set and achieved effectively if it is clearly defined, and ideally is also in line with general country- and regional-level goals. Given this, the level of ambition (Section 3.3) needs to be clarified, as it goes beyond the previously mentioned dimensions, including time-targets and reduction goals. Here, it may also make sense to establish interim milestones to assess progress in smaller steps. Next, stakeholders should define the area of observation and the system barriers (Section 3.4), as well as the scope of emission reductions. This includes the chosen sites the company intends to decarbonise (spatial) and the scopes of emission—scope 1, 2, and 3—that are supposed to be reduced. Besides these rather technical decisions, the identification of one's intrinsic motivation to decarbonise can also be crucial (Section 3.5). Such motivators can reach from purely economic rationales and legal requirements to reputational issues and social responsibility. Being clear about their motivations, stakeholders also need to formulate their priorities, which serve as criteria for the implementation of measures (Section 3.6). Data from EEI shows that, on average, companies rank investment level highest and that the ranking largely depends on company size and energy-intensity level. Finally, yet importantly, it is essential that companies know their starting point—their status quo (Section 3.7). Only those who are aware of their fundamentals can hope to effectively build on them. This includes current levels of energy consumption and emissions but also many other factors, such as surrounding infrastructure and climatic conditions.

After one has fulfilled all of the abovementioned points, further decisions on whether to take external (e.g., acquisition of renewable energy) or internal (e.g., reduction of energy consumption) measures to reach the target need to be made (Section 3.8). Deciding on whether to count on compensation measures or not is part of this process. More specific decisions on which measures to take depend on the individual situation of a company (Section 3.9).

Nevertheless, it remains to be underlined that the road to net-zero does not end with meeting the (milestone-)targets set within time. Like reaching one's ideal weight, it is one challenge to reach it, and another one to keep it. The ideal mixture of measures to maintain it is likely to change with time, situation, and environment.

An adjusted form of economic viability assessment (Section 3.10), as well as a continuous adjustment to current prices, availabilities, changing environments and policies (Section 3.11) will ease the challenge of keeping the decarbonisation strategy and associated mix of measures ideal over time.

Data of the Energy Efficiency Index of German Industry illustrated that a significant proportion of manufacturers participating in the survey are already on a good path. However, the remaining companies need to be picked up, and much work remains to be done across all areas looked at to successfully transition to a net-zero economy and to keep it net-zero.

Even though most of the evidence was gathered from German manufacturers and reflects the situation in Germany, it can be argued that the seven foundational questions are likely to remain valid irrespective of geography or culture. In contrast, the answers to the seven questions are likely to be different depending on those factors. Therefore, the currently ongoing data collection via the Energy Efficiency Barometer of Industry and the exchange with bodies, stakeholders, and companies in other geographies is of particular interest. Whether the seven questions can be also applied to areas outside the industry remains to be assessed by further analysis.

Funding: The article processing charge was funded by the Deutsche Forschungsgemeinschaft (DFG) in the funding programme Open Access Publishing.

Institutional Review Board Statement: Not applicable.

Informed Consent Statement: Informed consent was obtained from all subjects involved in the study.

Data Availability Statement: The data presented in this study are available on request from the corresponding author. The data are not publicly available due to privacy issues.

Acknowledgments: The underlying research in the form of the Energy Efficiency Index of German Industry (EEI, #EEIndex) would not have been possible without the continuous support of the Karl-Schlecht-Foundation and the Heinz und Heide Dürr Foundation, as well as the companies participating and the network of partners of the EEI, including those reviewing and supporting in progress of developing this paper and the EEI data collection process, notably Samuel Wörz, Frederick Vierhub-Lorenz, Marina Gilles, Anabel Reichle, Ole Pfister, and Werner König, the decarbonisation team and the team of student researchers. In Germany, evidence is usually collected each April/May and October/November (www.eep.uni-stuttgart.de/eei; accessed on 21 April 2022); the #EEBarometer runs all year round in nine further languages across 88 countries (www.eep.uni-stuttgart.de/eeei; accessed on 21 April 2022). The summarised results and recordings of briefings on the results can also be found there. All conclusions, errors, or oversights are solely the responsibility of the author.

Conflicts of Interest: The author declares no conflict of interest. The funders had no role in the design of the study; in the collection, analyses, or interpretation of data; in the writing of the manuscript; or in the decision to publish the results.

References

1. McClean, D. Earth Day: 2020 Saw a Major Rise in Floods and Storms. UNDRR. 2021. Available online: https://www.undrr.org/news/earth-day-2020-saw-major-rise-floods-and-storms (accessed on 29 December 2021).
2. Renouf, J.S. Making sense of climate change—The lived experience of experts. *Clim. Chang.* **2021**, *164*, 14. [CrossRef]
3. Ripple, W.J.; Wolf, C.; Galetti, M.; Newsome, T.M.; Green, T.L.; Alamgir, M.; Crist, E.; Mahmoud, M.I.; Laurance, W.F. The Role of Scientists' Warning in Shifting Policy from Growth to Conservation Economy. *BioScience* **2018**, *68*, 239–240. [CrossRef]

4. Masson-Delmotte, V.; Zhai, P.; Pörtner, H.O.; Roberts, D.; Skea, J.; Shukla, P.R.; Pirani, A.; Moufouma-Okia, W.; Péan, C.; Pidcock, R.; et al. An IPCC Special Report on the Impacts of Global Warmin of 1.5 °C above Pre-Industrial Levels and Related Global Greenhouse Gas Emission Pathways, in the Context of Strengthening the Global Response to the Threat of Climate Change, Sustainable Development, and Efforts to Eradicate Proverty. IPCC. 2019. Available online: https://www.ipcc.ch/site/assets/uploads/sites/2/2019/06/SR15_Full_Report_High_Res.pdf (accessed on 31 January 2022).

5. Masson-Delmote, V.; Zhai, P.; Piani, A.; Connors, S.L.; Péan, C.; Berger, S.; Caud, N.; Chen, Y.; Goldfarb, L.; Gomis, M.I.; et al. Climate Change 2021—The Physical Science Basis—Summary for Policymaker. IPCC. 2021. Available online: https://www.ipcc.ch/report/ar6/wg1/downloads/report/IPCC_AR6_WGI_SPM_final.pdf (accessed on 31 January 2022).

6. United Nations. Paris Agreement. 2015. Available online: https://unfccc.int/sites/default/files/english_paris_agreement.pdf (accessed on 28 December 2021).

7. IEA. World Energy Outlook 2021. 2021. Available online: https://iea.blob.core.windows.net/assets/4ed140c1-c3f3-4fd9-acae-789a4e14a23c/WorldEnergyOutlook2021.pdf (accessed on 28 December 2021).

8. Jugel, C.; Albicker, M.; Bamberg, C.; Battaglia, M.; Brunken, E.; Bründlinger, T.; Dorfinger, P.; Döring, A.; Friese, J.; Gründing, D.; et al. dena-Leitstudie Aufbruch Klimaneutralität. Deutsche Energie-Agentur GmbH. 2021. Available online: https://www.dena.de/fileadmin/dena/Publikationen/PDFs/2021/Abschlussbericht_dena-Leitstudie_Aufbruch_Klimaneutralitaet.pdf (accessed on 28 December 2021).

9. The Economist. How to Escape the Bear Market: Europe Reconsiders Its Energy Future. 2022. Available online: https://www.economist.com/business/2022/03/05/europe-reconsiders-its-energy-future (accessed on 23 March 2022).

10. Fairley, P. Germany Folds. *IEEE Spectr.* **2011**, *48*, 47. [CrossRef]

11. Climate Action Tracker. CAT Net Zero Target Evaluations. 2021. Available online: https://climateactiontracker.org/global/cat-net-zero-target-evaluations/ (accessed on 31 December 2021).

12. Net Zero Climate. What Is Net Zero? Available online: https://netzeroclimate.org/what-is-net-zero/ (accessed on 31 December 2021).

13. Net Zero Tracker. Net Zero Tracker Beta. 2021. Available online: https://zerotracker.net/ (accessed on 31 December 2021).

14. Lamb, W.F.; Wiedmann, T.; Pongratz, J.; Andrew, R.; Crippa, M.; Olivier, J.G.J.; Wiedenhofer, D.; Mattioli, G.; Khourdajie, A.A.; House, J.; et al. A review of trends and drivers of greenhouse gas emissions by sector from 1990 to 2018. *Environ. Res. Lett.* **2021**, *16*, 073005. [CrossRef]

15. IEA. Global Energy Review: CO_2 Emissions in 2020. 2021. Available online: https://www.iea.org/articles/global-energy-review-co2-emissions-in-2020 (accessed on 28 December 2021).

16. Kim, Y.J.; Brown, M. Impact of domestic energy-efficiency policies on foreign innovation: The case of lighting technologies. *Energy Policy* **2019**, *128*, 539–552. [CrossRef]

17. Houde, S.; Spurlock, C.A. Minimum Energy Efficiency Standards for Appliances. Old and New Economic Rationales. *Econ. Energy Environ. Policy* **2016**, *5*, 65–84. Available online: https://www.jstor.org/stable/26189506 (accessed on 28 December 2021). [CrossRef]

18. Umweltbundesamt. Energieverbrauch 2020 nach Sektoren und Energieträgern. 2021. Available online: https://www.umweltbundesamt.de/bild/endenergieverbrauch-2020-nach-sektoren (accessed on 10 December 2021).

19. Umweltbundesamt. Energiebedingte Treibhausgas-Emissionen 1990–2019. 2021. Available online: https://www.umweltbundesamt.de/sites/default/files/styles/800w400h/public/medien/384/bilder/2_abb_energiebed-thg-emi_2021-06-02.png (accessed on 10 December 2021).

20. Umweltbundesamt. Entwicklung der Treibhausgasemissionen in Deutschland 2010–2019. 2021. Available online: https://www.umweltbundesamt.de/sites/default/files/medien/421/bilder/1_entwicklung_der_treibhausgasemissionen_in_deutschland_0.jpg (accessed on 10 December 2021).

21. IEA. Global CO_2 Emissions by Sector, 2019. 2021. Available online: https://www.iea.org/data-and-statistics/charts/global-co2-emissions-by-sector-2019 (accessed on 31 January 2022).

22. IEA. Key World Energy Statistics 2021. 2021. Available online: https://www.iea.org/reports/key-world-energy-statistics-2021 (accessed on 31 January 2022).

23. European Commission. Sustainable Product Policy & Ecodesign. Available online: https://ec.europa.eu/growth/industry/sustainability/sustainable-product-policy-ecodesign_en (accessed on 30 March 2022).

24. Herz, C. Neuer Ressourcenpass für Gebäude: Was jetzt auf Mieter und Eigentümer Zukommt. Handelsblatt. 2022. Available online: https://www.handelsblatt.com/finanzen/immobilien/immobilien-neuer-ressourcenpass-fuer-gebaeude-was-jetzt-auf-mieter-und-eigentuemer-zukommt/28051264.html (accessed on 27 March 2022).

25. Nurdiawati, A.; Urban, F. Towards Deep Decarbonisation of Energy-Intensive Industries: A Review of Current Status, Technologies and Policies. *Energies* **2021**, *14*, 2408. [CrossRef]

26. Bauer, F.; Hansen, T.; Nilsson, L.J. Assessing the feasibility of archetypal transition pathways towards carbon neutrality—A comparative analysis of European industries. *Resour. Conserv. Recycl.* **2022**, *177*, 106015. [CrossRef]

27. Bataille, C.; Åhman, M.; Neuhoff, K.; Nilsson, L.J.; Fischedick, M.; Lechtenböhmer, S.; Solano-Rodriquez, B.; Denis-Ryan, A.; Stiebert, S.; Waisman, H.; et al. A review of technology and policy deep decarbonization pathway options for making energy-intensive industry production consistent with the Paris Agreement. *J. Clean. Prod.* **2018**, *187*, 960–973. [CrossRef]

28. Rissman, J.; Bataille, C.; Masanet, E.; Aden, N.; Morrow, W.R.; Zhou, N.; Elliott, N.; Dell, R.; Heeren, N.; Huckestein, B.; et al. Technologies and policies to decarbonize global industry: Review and assessment of mitigation drivers through 2070. *Appl. Energy* **2020**, *266*, 114848. [CrossRef]

29. Johnson, O.W.; Mete, G.; Sanchez, F.; Shawoo, Z.; Talebian, S. Towards Climate-Neutral Heavy Industry: An Analysis of Industry Transition Roadmaps. *Appl. Sci.* **2021**, *11*, 5375. [CrossRef]

30. South Pole. Become Climate Neutral. 2021. Available online: https://www.southpole.com/sustainability-solutions/become-climate-neutral (accessed on 31 December 2021).

31. Climate Neutral. How It Works. 2021. Available online: https://www.climateneutral.org/how-it-works (accessed on 31 December 2021).

32. Fashion Cloud. So Wird Dein Unternehmen Klimaneutral. 2021. Available online: https://fashion.cloud/de/so-wird-dein-unternehmen-klimaneutral/ (accessed on 31 December 2021).

33. Robinson, J.B. Energy backcasting A proposed method of policy analysis. *Energy Policy* **1982**, *10*, 337–344. [CrossRef]

34. Hughes, N.; Strachan, N.; Gross, R. The structure of uncertainty in future low carbon pathways. *Energy Policy* **2013**, *52*, 45–54. [CrossRef]

35. Bataille, C.; Waisman, H.; Colombier, M.; Segafredo, L.; Williams, J. The Deep Decarbonization Pathways Project (DDPP): Insights and emerging issues. *Clim. Policy* **2016**, *16*, 1–6. [CrossRef]

36. Mander, S.L.; Bows, A.; Anderson, K.L.; Shackley, S.; Agnolucci, P.; Ekins, P. The Tyndall decarbonisation scenarios—Part I: Development of a backcasting methodology with stakeholder participation. *Energy Policy* **2008**, *36*, 3754–3763. [CrossRef]

37. Holmberg, J.; Robèrt, K.-H. Backcasting—A framework for for strategic planning. *Int. J. Sustain. Dev. World Ecol.* **2000**, *7*, 291–308. [CrossRef]

38. Buettner, S.M.; Schneider, C.; König, W.; Mac Nulty, H.; Piccolroaz, C.; Sauer, A. How do German manufacturers react to the increasing societal pressure for decarbonisation? *Appl. Sci.* **2022**, *12*, 543. [CrossRef]

39. EEP. *Der Energieeffizienz-Index der Deutschen Industrie. Umfrageergebnisse 2. Halbjahr 2019*; Institut für Energieeffizienz in der Produktion, Universität Stuttgart: Stuttgart, Germany, 2019. Available online: https://www.eep.uni-stuttgart.de/eei/archiv-aeltere-erhebungen/ (accessed on 21 April 2022).

40. EEP. *Der Energieeffizienz-Index der Deutschen Industrie. Umfrageergebnisse 1. Halbjahr 2020*; Institut für Energieeffizienz in der Produktion, Universität Stuttgart: Stuttgart, Germany, 2020. Available online: https://www.eep.uni-stuttgart.de/eei/archiv-aeltere-erhebungen/ (accessed on 21 April 2022).

41. EEP. *Der Energieeffizienz-Index der Deutschen Industrie. Umfrageergebnisse 2. Halbjahr 2020*; Institut für Energieeffizienz in der Produktion, Universität Stuttgart: Stuttgart, Germany, 2020. Available online: https://www.eep.uni-stuttgart.de/eei/archiv-aeltere-erhebungen/ (accessed on 21 April 2022).

42. EEP. *Der Energieeffizienz-Index der Deutschen Industrie. Umfrageergebnisse 1. Halbjahr 2021*; Institut für Energieeffizienz in der Produktion, Universität Stuttgart: Stuttgart, Germany, 2021. Available online: https://www.eep.uni-stuttgart.de/eei/archiv-aeltere-erhebungen/ (accessed on 21 April 2022).

43. EEP. *Der Energieeffizienz-Index der Deutschen Industrie. Umfrageergebnisse 2. Halbjahr 2021*; Institut für Energieeffizienz in der Produktion, Universität Stuttgart: Stuttgart, Germany, 2022. Available online: https://www.eep.uni-stuttgart.de/eei/archiv-aeltere-erhebungen/ (accessed on 21 April 2022).

44. Weber, R. Wärmepumpen als Klimaretter? 2021. Available online: https://www.daserste.de/information/wirtschaft-boerse/plusminus/sendung/waermepumpe-bewertung-100.html (accessed on 30 December 2021).

45. Cresko, J.; Shenoy, D.; Liddell, H.P.; Sabouni, R. *Chapter 6: Innovating Clean Energy Technologies in Advanced Manufacturing September 2015. Quadrennial Technology Review 2015*; U.S. Department of Energy: Washington, DC, USA, 2015; pp. 180–225. Available online: https://www.energy.gov/sites/prod/files/2017/03/f34/qtr-2015-chapter6.pdf (accessed on 21 April 2022).

46. Buettner, S.M. Framing the Ambition of Carbon Neutrality, UNECE Task Force on Industrial Energy Efficiency. In *Group of Experts on Energy Efficiency*; GEEE-7/2020/INF.2; United Nations Economic Commission for Europe: Geneva, Switzerland, 2020; Available online: https://www.unece.org/fileadmin/DAM/energy/se/pdfs/geee/geee7_Sept2020/GEEE-7.2020.INF.2_final_v.2.pdf (accessed on 8 February 2022).

47. Buettner, S.M.; Wang, D. A Pathway to Reducing Greenhouse Gas Footprint in Manufacturing: Determinants for an Economic Assessment of Industrial Decarbonisation Measures. EEP—Institute for Energy Efficiency in Production, University of Stuttgart, Stuttgart, Germany. 2022; *unpublished*.

48. DPA. Japans neuer Ministerpräsident peilt Klimaneutralität bis 2050 an. Handelsblatt. 2020. Available online: https://www.handelsblatt.com/politik/international/energiewende-japans-neuer-ministerpraesident-peilt-klimaneutralitaet-bis-2050-an/26307902.html (accessed on 30 December 2021).

49. Dooley, B.; Inoue, M.; Hida, H. Japan's New Leader Sets Ambitious Goal of Carbon Neutrality by 2050. *The New York Times*, 2020. Available online: https://www.nytimes.com/2020/10/26/business/japan-carbon-neutral.html (accessed on 30 December 2021).

50. Watkins, M.D. Demystifying Strategy: The What, Who, How, and Why. Harvard Business Review. 2007. Available online: https://hbr.org/2007/09/demystifying-strategy-the-what#:~|{:text=A%20good%20strategy%20provides%20a,prioritize)%20to%20achieve%20desired%20goals (accessed on 4 February 2022).

51. Group of Experts on Cleaner Electricity Systems. Framework for Attaining Carbon Neutrality in the United Nations Economic Commission for Europe (ECE) Region by 2050. ECE/ENERGY/GE.5/2020/8 Rev.1. United Nations Economic and Social Council, 2020. Available online: https://unece.org/fileadmin/DAM/energy/se/pdfs/CES/ge16_2020/ECE_ENERGY_GE.5_2020_8_rev1.pdf (accessed on 7 February 2022).

52. G2G. G2G Talk: An Interview with Olga Algayerova. How Is the UN Economic Commission for Europe (UNECE) Working with Countries to Attain Carbon Neutrality in the ECE Region? 2021. Available online: https://www.c2g2.net/c2gtalk-olga-algayerova/ (accessed on 31 December 2021).

53. U.S. Department of State. United States, European Union, and Partners Formally Launch Global Mathane Pledge to Keep 1.5C within Reach. 2021. Available online: https://www.state.gov/united-states-european-union-and-partners-formally-launch-global-methane-pledge-to-keep-1-5c-within-reach/ (accessed on 30 December 2021).

54. Müller, B. So Will RWE bis 2040 Klimaneutral Werden. Süddeutsche Zeitung. 2019. Available online: https://www.sueddeutsche.de/wirtschaft/rwe-klimaneutral-2040-1.4622120 (accessed on 4 February 2022).

55. Mazzei, A. Climate Action: Bosch to Be Carbon Neutral Worldwide by 2020. Bosch. 2019. Available online: https://www.bosch-presse.de/pressportal/de/en/climate-action-bosch-to-be-carbon-neutral-world-wide-by-2020-188800.html (accessed on 4 February 2022).

56. Jeudy-Hugo, S.; Lo Re, L.; Falduto, C. Understanding Countries Net-Zero Emissions Targets. OECD, 2021. Available online: https://www.oecd.org/officialdocuments/publicdisplaydocumentpdf/?cote=COM/ENV/EPOC/IEA/SLT(2021)3&docLanguage=En (accessed on 7 February 2022).

57. Bundesverfassungsgericht. Constitutional Complaints against the Federal Climate Change Act Partially Successful. 2021. Available online: https://www.bundesverfassungsgericht.de/SharedDocs/Pressemitteilungen/EN/2021/bvg21-031.html (accessed on 31 December 2021).

58. Spier, J. 'The "Strongest" Climate Ruling Yet': The Dutch Supreme Court's Urgenda Judgment. *Neth. Int. Law Rev.* **2020**, *67*, 319–391. [CrossRef]

59. Evans, S.; Gabbatiss, J.; McSweeney, R.; Chandrasekhar, A.; Tandon, A.; Viglione, G.; Hausfather, Z.; You, X.; Goodman, J.; Hayes, S. COP26: Key outcomers agreed at the UN climate talks in Glasgow. Carbon Brief Staff. 2021. Available online: https://www.carbonbrief.org/cop26-key-outcomes-agreed-at-the-un-climate-talks-in-glasgow (accessed on 31 December 2021).

60. Buettner, S.M.; Döpp, J.; König, W.; Strauch, L. Increasing the Voltage—Equencing Decarbonisation with Green Power & Efficiency. EEP—Institute for Energy Efficiency in Production, University of Stuttgart, Stuttgart, Germany. 2022; *unpublished*.

61. Greenhouse Gas Protocol. FAQ. World Resources Institute. Available online: https://ghgprotocol.org/sites/default/files/standards_supporting/FAQ.pdf (accessed on 30 December 2021).

62. Rajan, A. The Daunting Task of Decarbonising Investment Portfolios. Financial Times. 2020. Available online: https://www.ft.com/content/ecdf1231-93d6-4819-a32a-a49a7a25fcea (accessed on 31 January 2022).

63. European Environment Agency. CO_2 Performance of New Passenger Cars in Europe. 2021. Available online: https://www.eea.europa.eu/ims/co2-performance-of-new-passenger (accessed on 30 December 2021).

64. Volkswagen. Volkswagen Way to Zero. 2021. Available online: https://www.volkswagen.de/de/marke-und-erlebnis/waytozero.html (accessed on 31 December 2021).

65. World Economic Forum. Net-Zero Challenge: The Supply Chain Opportunity. 2021. Available online: https://www3.weforum.org/docs/WEF_Net_Zero_Challenge_The_Supply_Chain_Opportunity_2021.pdf (accessed on 6 February 2022).

66. Haupt, C. BASF Ermittelt CO_2-Fußabdruck aller Verkaufsprodukte. Wirtschaftspresse. 2020. Available online: https://www.basf.com/global/de/media/news-releases/2020/07/p-20-260.html (accessed on 31 December 2021).

67. Buettner, S.M.; König, W. Looking behind decarbonisation—What pressure points trigger action? In Proceedings of the European Council for an Energy Efficient Economy ECEEE Summer Study, digital, 7–11 June 2021; pp. 345–354. Available online: https://www.eceee.org/library/conference_proceedings/eceee_Summer_Studies/2021/3-policy-finance-and-governance/looking-behind-decarbonisation-what-pressure-points-trigger-action/ (accessed on 21 April 2022).

68. Ember. Daily Carbon Prices. 2022. Available online: https://ember-climate.org/data/carbon-price-viewer/ (accessed on 7 February 2022).

69. Stratmann, K. Erste Unternehmen Kapitulieren vor Hohen Gas- und Strompreisen—Was Die Politik Nun Tun Will. Handelsblatt. 2022. Available online: https://www.handelsblatt.com/politik/deutschland/strom-und-gas-erste-unternehmen-kapitulieren-vor-hohen-gas-und-strompreisen-was-die-politik-nun-tun-will/27935982.html (accessed on 31 January 2022).

70. Flauger, J.; Kersting, S. Der Strompreis-Schock: Industrie Zahlt So Viel Wie Seit Einem Jahrzehnt Nicht Mehr. Handelsblatt. 2021. Available online: https://www.handelsblatt.com/unternehmen/energie/energiewirtschaft-der-strompreis-schock-industrie-zahlt-so-viel-wie-seit-einem-jahrzehnt-nicht-mehr/27360246.html (accessed on 31 January 2022).

71. DIHK. DIHK-Umfrage: Unternehmen Leiden Unter Hohen Strom- und Gaspreisen. 2021. Available online: https://www.dihk.de/de/themen-und-positionen/wirtschaftspolitik/energie/dihk-umfrage-unternehmen-leiden-unter-hohen-strom-und-gaspreisen-61652 (accessed on 31 January 2022).

72. Rieger, J. Insolvent Trotz Guter Auftragslage. SWR. 2022. Available online: https://www.tagesschau.de/wirtschaft/unternehmen/gaspreise-insolvenzen-unternehmen-101.html (accessed on 31 January 2022).

73. Buettner, S.M.; König, W. Determining the ideal mix—(finding out) what range of measures is best for one's business? In Proceedings of the ACEEE Summer Study, digital, 13–15 July 2021; Available online: https://www.aceee.org/sites/default/files/pdfs/ssi21/panel-2/Buettner.pdf (accessed on 8 February 2022).

74. Aurubis, A.G. CO_2-Free Heat Supply for Hamburg's Hafen City District. Available online: https://www.wire-tradefair.com/cgi-bin/md_wiretube/lib/pub/tt.cgi?oid=2377518&lang=2 (accessed on 6 February 2022).

75. Bosch. Unternehmensweiter Umweltschutz: Klimaneutralität Seit 2020. 2022. Available online: https://www.bosch.com/de/nachhaltigkeit/umwelt/ (accessed on 31. January 2022).
76. Economento. Greenpeace Wirft VW Fehlende CO_2-Kompensation bei ID.—Elektroautos vor. 2020. Available online: https://ecomento.de/2020/09/30/greenpeace-wirft-vw-fehlende-co2-kompensation-bei-elektroautos-vor/ (accessed on 31 January 2022).
77. Greenfield, P. Carbon Offsets Used by Major Airlines Based on Flawed System, Warn Experts. The Guardian. 2021. Available online: https://www.theguardian.com/environment/2021/may/04/carbon-offsets-used-by-major-airlines-based-on-flawed-system-warn-experts (accessed on 31 January 2022).
78. AL Ghussain, A. The Biggest Problem with Carbon Offsetting Is That It Doesn't Really Work. Greenpeace. 2020. Available online: https://www.greenpeace.org.uk/news/the-biggest-problem-with-carbon-offsetting-is-that-it-doesnt-really-work/ (accessed on 31 January 2022).
79. Federal Ministry of Education and Research. How the Kopernicus Project Synergie Helps Industry Match Its Electricity Demand to the Supply. Available online: https://www.kopernikus-projekte.de/en/projects/synergie (accessed on 31 January 2022).
80. FABOS. FABOS-Open, Distributed, Real-Time Capable and Secure Operation System for Production. 2022. Available online: https://www.fab-os.org/en/ (accessed on 28 January 2022).
81. Agrawal, R.; Wankhede, V.A.; Kumar, A.; Upadhyay, A.; Garza-Reyes, J.A. Nexus of circular economy and sustainable business performance in the era of digitalization. *Int. J. Product. Perform. Manag.* **2021**, *71*, 748–774. [CrossRef]
82. Wang, D.; Buettner, S.M. KPIs in CSR reporting as a vehicle for climate neutrality. In Proceedings of the European Council for an Energy Efficient Economy ECEEE Summer Study, digital, 7–11 June 2021; Available online: https://www.eceee.org/library/conference_proceedings/eceee_Summer_Studies/2021/3-policy-finance-and-governance/kpis-in-csr-reporting-as-a-vehicle-for-climate-neutrality/ (accessed on 21 April 2022).
83. Hristov, I.; Chirico, A. The Role of Sustainability Key Performance Indicators (KPI's) in Implementing Sustainable Strategies. *Sustainability* **2019**, *11*, 5742. [CrossRef]
84. The Economist Applied. How Do You Make Your Company Carbon-Neutral? 2020. Available online: https://applied.economist.com/articles/how-do-you-go-carbon-neutral (accessed on 28 January 2022).

Article

The Thermoeconomic Environment Cost Indicator ($i_{ex\text{-}TEE}$) as a One-Dimensional Measure of Resource Sustainability

Sobhy Khedr [1], Melchiorre Casisi [2,*] and Mauro Reini [1]

[1] Department of Engineering and Architecture, University of Trieste, 34127 Trieste, Italy; sobhy.khedr88@gmail.com (S.K.); reini@units.it (M.R.)

[2] Polytechnic Department of Engineering and Architecture, University of Udine, 33100 Udine, Italy

* Correspondence: melchiorre.casisi@uniud.it

Abstract: This paper presents a conceptual development of sustainability evaluation, through an exergy-based indicator, by using the new concept of the Thermoeconomic Environment (TEE). The exergy-based accounting methods here considered as a background are Extended Exergy Accounting (EEA), which can be used to quantify the exergy cost of externalities like labor, monetary inputs, and pollutants, and Cumulative Exergy Consumption (CExC), which can be used to quantify the consumption of primary resources embodied in a final product or service. The new concept of bioresource stock replacement cost is presented, highlighting how the framework of the TEE offers an option for evaluating the exergy cost of products of biological systems. This sustainability indicator is defined based on the exergy cost of all resources directly and indirectly consumed by the system, the equivalent exergy cost of all externalities implied in the production process and the exergy cost of the final product.

Keywords: resource sustainability; exergy; exergy cost accounting; exergy cost of biological resources

Citation: Khedr, S.; Casisi, M.; Reini, M. The Thermoeconomic Environment Cost Indicator ($i_{ex\text{-}TEE}$) as a One-Dimensional Measure of Resource Sustainability. *Energies* **2022**, *15*, 2260. https://doi.org/10.3390/en15062260

Academic Editors: Sergio Ulgiati, Hans Schnitzer and Remo Santagata

Received: 13 December 2021
Accepted: 11 March 2022
Published: 19 March 2022

Publisher's Note: MDPI stays neutral with regard to jurisdictional claims in published maps and institutional affiliations.

1. Introduction

It can be noted that an effort is under development in the literature to define resource sustainability indicators based on thermodynamic quantities, in particular on exergy. Exergy is widely recognized as a proper tool for evaluating the resources required by energy systems [1–5] or by technological production systems in general [6–11]. In addition to the basic exergy analysis, an exergy cost accounting must be implemented [12–15]. When dealing with complex, multi-component, energy systems with both direct and indirect exergy consumption, exergy cost accounting is required for obtaining a certain product flow. Furthermore, when the goal is to assess the impact or the sustainability of the production system, the actual primary exergy resources directly and indirectly available for the production system itself must be considered. The expectation is that an exergy-based sustainability indicator could encompass, in a one-dimensional figure, various aspects of sustainability, or even all of them.

This paper first summarizes different exergy-based cost accounting approaches presented in the literature, highlighting the effort to include in the analysis a progressively more complete picture of the indirect effects and externalities of the production process, which may affect the sustainability of the process itself. Then, an extension of the previous cost accounting method is presented, based on the concept of TEE. This is a consistent ultimate boundary of exergy cost accounting, where various exergy reservoirs of limited content are immersed in the zero-exergy matrix, as shown in Figure 1, but they remain separated from it because of some confinement constraints. Starting from this very simple but meaningful framework, the concept of bioresource stock replacement (BSR) cost is precisely defined, allowing the exergy cost evaluation of all biological resources used as production process inputs. Introducing the concept of the BSR cost does not need any arbitrary hypothesis, or cost allocation rules not consistent with the input/output framework [16] that is a characteristic of the great majority of the cost accounting approaches presented in the literature. Moreover, it is consistent with the replacement cost of mineral resources presented recently by Valero and Valero [17], as an extension of the thermoecological cost introduced by Szargut [15,18]. Finally, the Thermoeconomic Environment Cost (TEEC) is presented, and, on this basis, a new exergy sustainability indicator is easily defined, and its potential as a one-dimension measure of resource sustainability is discussed.

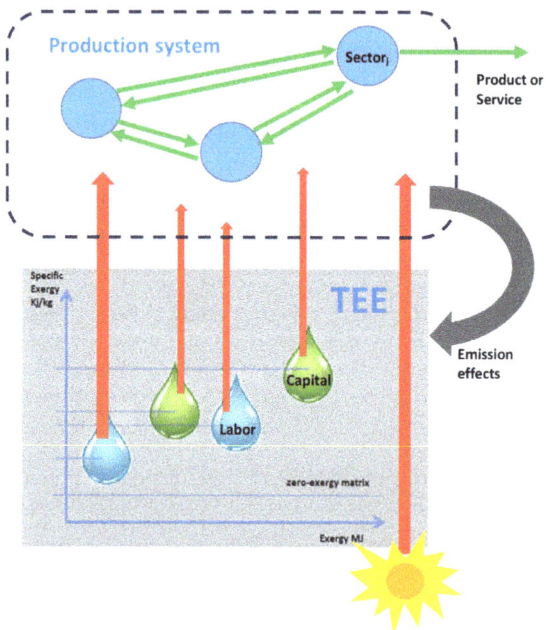

Figure 1. A qualitative representation of the flows involved in the TEEC evaluation.

An Outlook to Some Sustainability Indices in the Literature

A critical summary of the previous effort for identifying exergy-based sustainability indicators can be found in Kharrazi et al. [19], where the authors highlight the limitations of two approaches based on thermodynamics in defining a proper sustainability index: Emergy Synthesis and Exergy Analysis. In particular, agreeing with Kharrazi et al., the sustainability index proposed by Emergy Synthesis [20,21] allows us to highlight important measurements of sustainability, but it does not consider any limit to the minimization of input emergy consumption, implicitly assuming that a reduction is always possible and desirable, as is a wider usage of renewable resources. In addition, the emergy sustainability index is defined as a ratio [21] where the product of the yield and the input renewable resources is the numerator, whilst the sum of the capital invested plus the input non-renewable resources, both multiplied by the capital invested, is the denominator. The reasons for such a definition are not immediately evident. Moreover, its physical meaning is not clear, beyond the idea that a higher product yield and a higher renewable input at constant non-renewable and capital resource consumption is a more sustainable condition for a certain system.

Kharrazi et al. [19] also recognize that recent methods based on exergy cost accounting (like the EEA [22–25]) attempt to unify capital investment, human labor, and environmental resources into a common exergetic description. Nevertheless, they note that, in the exergy literature, no sustainability index similar to the one defined by the Emergy Synthesis model have been presented. In fact, the latter not only considers the strict (and arbitrary) control volume of the analyzed system but also attempts to consider the direct and indirect effects of system activities. On the other hand, emergy cannot be obtained from a straightforward input-output approach. Instead, a peculiar algebra must be used, which implies a non-conservative nature of the emergy itself.

Various exergy indexes claim to express sustainability [26], but they mainly relay the exergy efficiency concept without prescribing a specific control volume or a pre-defined origin of the supply chains that feed the considered production process or component. This is the case, for instance, of the Depletion number (D_P) and the sustainability index (SI) shown by Rosen [27]. D_P is the complement to one part of the exergy efficiency model, and SI is the inverse, i.e., they convey the same information as the exergy efficiency model itself.

A different definition of the exergy sustainability index is used in [28], as the ratio of the exergy of the products and the exergy content of waste flows. Another index was defined by Dewulf et al. [29] as the fraction of renewable energy in the total input (named the exergy renewability indicator) and the ratio of the input exergy and the sum of the same input exergy plus the expected exergy consumption for a complete abatement of harmful wastes from the process (named the environmental compatibility).

All these indices do make sense, but the direct and indirect effects of system activities, outside the control volume of the system itself are not systematically investigated, they are simply supposed to be proportional to the exergy of some input or output flow.

In the following, the most relevant properties expected in a sustainability index are presented by critically combining and integrating the requirements highlighted in [30–32]:

a.   It must be expressed by a—possibly simple—numeric expression and produce results that can be unambiguously ranked within two opposite limits.

b.   It must be calculated based on intrinsic properties of both the process (the system that it refers to) and of the (local or global) environment.

c.   It must be normalized in some sense, so that it may be used to compare different systems, different environmental conditions, different scenarios and/or different time series for the same community.

d.   It must be calculated based on an unambiguous, reproducible method under a well-defined set of fundamental assumptions.

e.   It must comply with the accepted laws of physics.

2. Exergy Cost Accounting for Assessing Sustainability

Generally speaking, when dealing with complex, multi-component, energy systems with both direct and indirect consumptions for obtaining a certain product, an exergy cost accounting must be implemented besides the basic exergy analysis. Exergy cost accounting definition requires:

a.   cost allocation rules input/output algebra by Leontief [16] are widely accepted, but other cost allocation rules may be found in the literature [13,33,34].

b.   clear limits for the control volume, where the start of the exergy supply chains of the system is located, and where the unit exergy costs of all inputs crossing the limits must be known [35].

Some other additional conditions must be considered to use exergy cost as a sustainability indicator. The actual primary exergy resources must be identified, and the exergy cost of polluting emissions must be evaluated. There is wide agreement about cost allocation rules and in practice, all exergy-based costs must be allocated to the product. There is still some investigation to avoid what is called double accounting when a multi-products system must be analyzed or some other constraints occur. The conservative nature of the cost flow through the energy conversion system is important if the aim is to quantitatively evaluate the impact on the primary resources of goods or services, and not to obtain only meaningful indicators. Moreover, to assess sustainability, it is important to indicate the impact affecting the resources available at the present moment, not in the distant past time, so the time scale must be defined properly. Even if the cost allocation rules are defined and consistent with conservative cost balances of all control volumes, the ultimate boundaries play an important role in exergy cost accounting and must be consistent with assessments of the impact in primary exergy resources of a product and service. The reference environment used in the basic exergy analysis cannot be identified with these ultimate boundaries of the exergy cost accounting analysis [36]. Because it is perfectly homogenous, its temperature and pressure cannot be modified, and it cannot be affected in any way by its interaction with the production system being considered.

If the goal is to assess sustainability, and not merely to obtain a rational comparison among products or technologies, the additional conditions may be summarized as:

a. The actual primary exergy resources must be identified, considering both renewable and non-renewable resources.
b. The impact affecting the resources available at the present moment, not that in a distant past time, must be assessed.
c. All exergy costs related to polluting emissions must be evaluated, besides the exergy costs of the inputs.

It must be recognized that the EEA defined by Sciubba [22–24,37] has achieved an important advance in this direction. It measures the amount of primary exergy absorbed by a system throughout its life cycle, without any special attention to biological resources, which are accounted-for in their exergy content. In addition to material and energy flows received directly and indirectly from nature (where all the supply chains start), EEA includes externalities for capital investments, human labor and polluting emissions, the latter being calculated on a remediation basis, similarly to Dewulf et al. [29]. The exergy cost of the products, as well as the exergy efficiency of a process or a region [38,39] calculated through the EEA approach, are certainly not limited to the only strict (and arbitrary) control volume of the analyzed system.

2.1. Definition of the Thermoeconomic Environment (TEE)

The TEE was introduced as a consistent ultimate boundary for the exergy cost accounting, with the following objects [40]:

- Overcoming the drawbacks of the reference environment used in basic exergy analysis. Some of these drawbacks are that the reference environment has no resources of energy or raw materials that are required to be consumed to obtain a specific process or product. The reference temperature cannot be modified, which means that some phenomena, like global warming [41], cannot be accounted-for. In addition, the reference environment is not affected by any polluting emissions from the production system.
- Defining a framework consistent with the formulations of CExC [42] and EEA.
- CExC and EEA are milestones of the effort of including in the exergy accounting analysis a progressively more complete picture of the indirect effects and externalities of the production process. Some of the ideas developed in those approaches will be used in the following to define the proposed sustainability indicator.
- Suggesting new options for a consistent exergy cost definition of all resources. As will be evident in the following, the framework of the TEE may help the definition of a proper exergy cost for the effect of polluting emissions from the production process, or for the indirect destruction of resources, including living biomass.

The TEE is defined as a set of reservoirs, where different kinds of natural resources are confined. All of them are surrounded by the zero-exergy matrix (the dead state). Each available resource has a specific exergy content greater than zero, as shown in Figure 2.

From the previous definition, it can be easily inferred that the TEE is not too big to be modified by the interactions with the production processes because the amount of exergy in each reservoir is limited and because the confined conditions of the reservoirs can be compromised. In addition, to consider some real-world phenomena like the periodic oscillations of the availability of solar energy or global warming, which is increasing as a consequence of GHG emissions, it must be recognized that even the zero-exergy matrix may change its temperature $T°$ and composition.

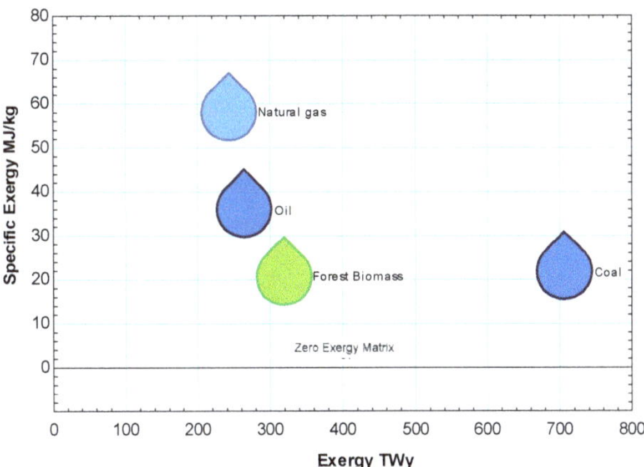

Figure 2. A partial semi-qualitative representation of the TEE.

2.2. Chemical Exergy Calculation

The zero-exergy matrix can be defined as the reference state model introduced by Szargut [43,44]. It is based on the identification of a set of reference substances whose specific chemical exergies can be determined as concentration exergy with respect to an ideal mixture of gas at $T°$, $P°$. The chemical exergies of all other substances in the TEE can be calculated by considering reversible chemical reactions. In this way, crude fossil fuels and other mineral resources are not obtained as confined inside reservoirs, but they may be better regarded as obtained all together, i.e., mixed in a condition that may be identified as the Thanatia planet introduced by Valero and Valero [17]. Notice also that additional exergy must be consumed to obtain the resources in a confined way, as they are found in real-world mines or as they are regarded as being inside the TEE.

The specific exergy costs of each available resource inside the TEE are the basis of the accounting: the specific exergy costs of each reservoir may be considered equal to 1, consistent with the hypotheses of the EEA and CExC models. This expresses the idea that a certain exergy stock of non-renewable resources is available in the TEE at the present moment, together with a set of exergy flows of renewable resources (including the renewable parts of all partially renewable reservoirs).

2.3. The Exergy Cost of Mineral Resources

If the dynamic process allowing exergy accumulation inside the reservoirs can be neglected, the assumption of the specific exergy costs of each reservoir being equal to 1 may be correct even if a larger amount of exergy was required in the distant past. For instance, this may have been the case when the accumulation process was very slow compared with the duration of the considered production process, such as in the case of natural fossil fuel, or other mineral reservoirs.

On the other hand, if a non-negligible dynamic exists inside the TEE, exergy extraction from a certain reservoir may produce additional exergy destruction in other reservoirs. In this second case, two options can be identified:

- To extend the supply chain describing the indirect consumption of resources.
- To define a set of unit exergy costs, not equal to one, which is regarded as equivalent to the mechanism of additional exergy resource destruction.

In 2011, [45,46] Valero introduced the exergy replacement costs (ERC) and the model of Thanatia to assess the concentration exergy of mineral resources based on their scarcity in nature [17].

In the TEE language, the proposals by Valero and Valero may be re-formulated by stating that, in the Thanatia planet, the confining constraints of all reservoirs were destroyed, all minerals are mixed, and they have all reacted with the zero-exergy matrix. Then, the ECR is the exergy required to produce a reservoir of a certain mineral resource, from the conditions defined for the Thanatia planet, by using real-world, irreversible technologies, as shown in Figure 3. This methodology was introduced to assess the concentration exergy of mineral resources based on their scarcity in nature. The combination of the ERC concept with the Thermo-Ecological Cost method (TEC) originally proposed by Szargut, allows us to assess products considering the exergy associated with the consumption of non-renewable resources extracted directly from nature, taking their scarcity into account.

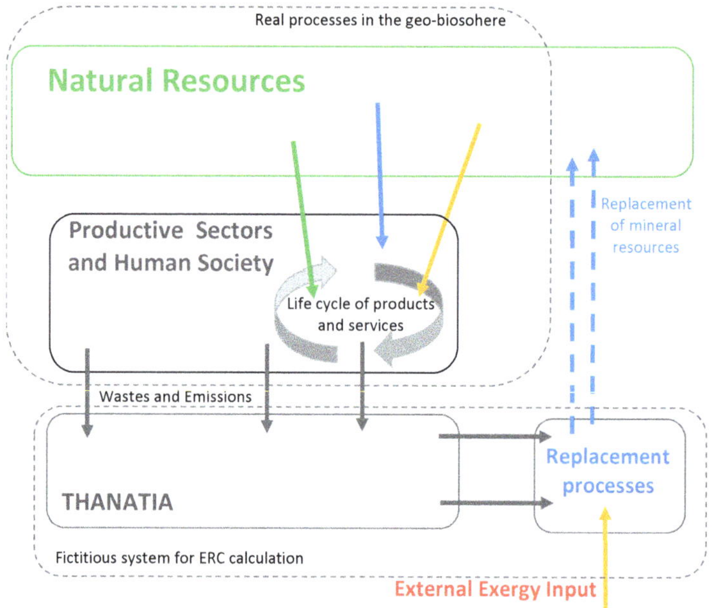

Figure 3. Cradle-to-grave-to-cradle process for calculating the exergy replacement cost (ERC).

The only exergy input external to the geo-biosphere is solar energy (and possibly tidal and geothermal energy). Therefore, the ERC can be properly understood as the cost that should be paid to consider a non-renewable resource as if it were completely renewed using solar energy, i.e., as if it were renewable on a human time scale, like solar energy itself. It is worth noting that this interpretation makes the ERC of mineral resources and renewable energy in the input of a generic production process homogeneous, so that they may be added together without inconsistency.

2.4. The Exergy Equivalent of Capital and Human Work

In the EEA, externalities can be assigned "equivalent exergy values", under a set of assumptions [25,47]. The more recent proposal by Rocco and Colombo [48] may be regarded as an attractive alternative, since it was directly derived from the input/output algebra by Leontief [8]. In this approach, the interactions among the sectors of the whole production system are described by the monetary magnitudes usually adopted in the economic analysis. Then, the exergy evaluation of each flow in the model is obtained by considering the exergy of all the inputs coming from the environment and feeding the sectors (the nodes) of the production network. In this way, the exergy equivalent of capital has not been evaluated explicitly and, if it is evaluated a posteriori, the result may be different for the different production sectors considered.

As far as the exergy evaluation of human work is concerned, the approach suggested by Rocco and Colombo [48] is a direct extension of their exergy input/output analysis. The human labor production sector is embodied as an additional sector, without the need for any arbitrary assumptions, as schematized in Figure 4. This sector supplies the required human labor to all the others (only two big sectors, final goods production, and intermediate goods production are shown in the Figure) and receives from the final goods production sector all the necessary inputs for human labor production. Obviously, additional information, such as the quantitative evaluation of the inputs required by the human labor activities from each one of the other sectors and, the human working hours required by each of them, is required.

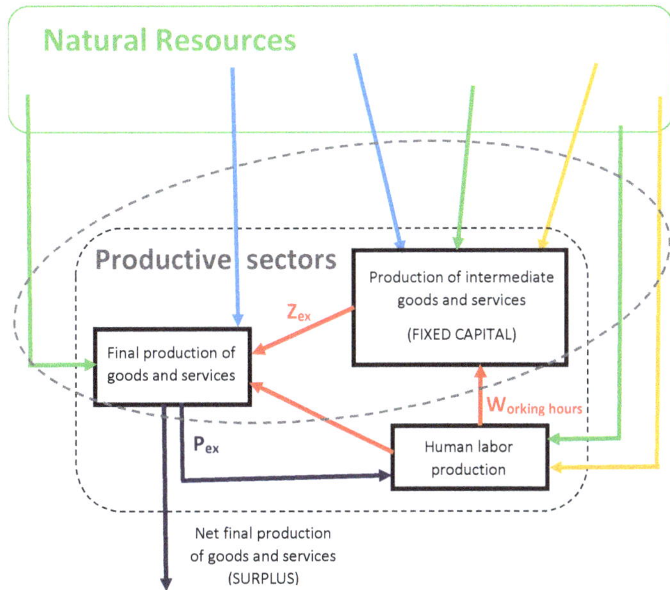

Figure 4. Schematic of the sub-system introduced by Rocco for the internalization of human labor in embodied energy analysis. Adapted from [41].

2.5. The Exergy Cost of Products of Biological Systems

The frame of the TEE offers an option for evaluating the exergy cost of products of biological systems. As shown in Figure 5, a fictitious extension of the system, with the function of replacing the stock of the bioresource reservoir, was considered analogously with the replacement processes considered in the definition of the ERC of mineral resources. The object is to stay as close as possible to the latter methodology. Unfortunately, the ultimate waste produced by the use of biological substances include carbon dioxide, water, and very few other elements, so the replacement processes of the original resources (forests, agriculture fields, ecosystems, etc.) cannot be defined based on actual technology. The methodology is then adapted as follows.

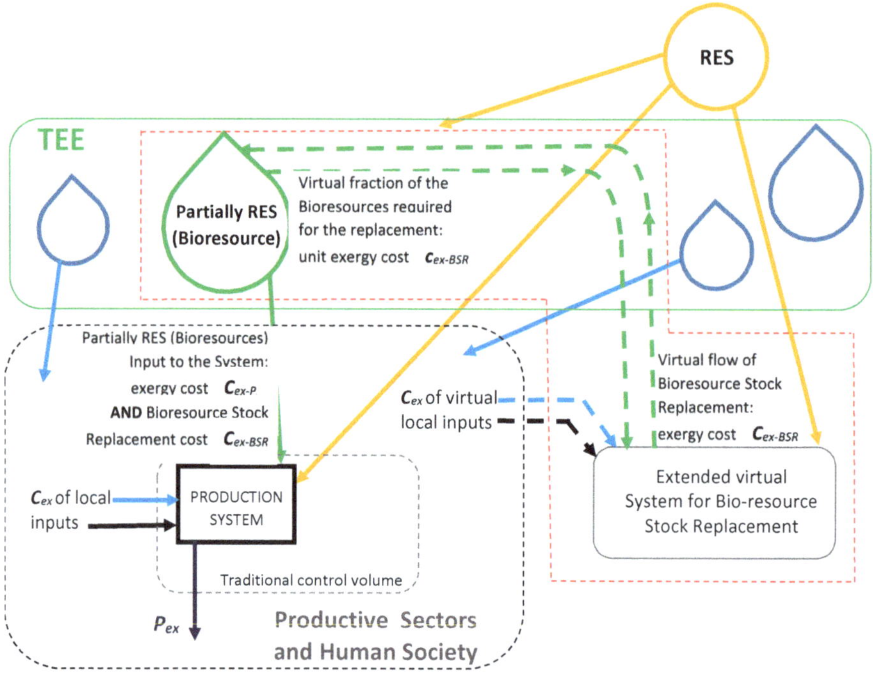

Figure 5. The concept of bioresource stock replacement cost.

If the bioresource is consumed (or indirectly destroyed) at an extraction rate (β) lower than its natural growth rate in the reservoir (α), the stock is not affected, and the input to the production system is regarded as completely renewable (specific exergy costs equal to 1). If instead the extraction rate is greater than the growth rate ($\beta > \alpha$), the fictitious extension of the system must cultivate the ecosystem to replace the original stock. The BSR cost (C_{ex-BSR}) can thus be calculated because all the input costs of the extension of the real system are known:

- Solar energy and other renewable resources have a unit exergy cost equal to one.
- All non-renewable resources can be evaluated at their exergy cost, including indirect consumption and the ERC of the mineral resources.
- The capital and human work can be evaluated at their exergy equivalent, via one of the methodologies previously outlined.

Notice that even a fraction (ρ) of the bioresources considered must be virtually extracted to be used as an input to the extended system for bioresource replacement. This is because the living substance cannot be obtained from the products of the economic sectors with actual technologies without using some living input. The unit exergy cost of this flow must be regarded as the same as the BSR, without introducing any problem in the calculation of the latter, based on the usual rules of exergy cost accounting. This assumption is equivalent to considering a bifurcation of the virtual flow of BSR into two parts, one for the actual replacement and one for recirculating the input required by the virtual system. This cost allocation rule in bifurcating flows must be regarded as a well-consolidated result in the field of Thermoeconomics [49]. Moreover, it can be easily noted from Figure 5 that the BSR cost can be inferred from the cost balance of the sub-system inside the dotted red line, without the need to explicitly know the cost of the bioresource recirculated as an input to the virtual system. In fact, the unit cost of the bioresource consumed by the production system, disregarding the stock replacement, is known to be is equal to its chemical exergy (consistently with EEA).

It is worth noting that, if the extended system for BSR is considered, the differential equation governing the bioresource stock decline (Equation (1)) is replaced by differential Equation (2):

$$\frac{dM}{dt} = (\alpha - \beta)M_0 \tag{1}$$

$$\frac{dM}{dt} = [\alpha(1 - \rho) - \beta - \rho]M_0 + \frac{dR}{dt} = 0 \tag{2}$$

where dR/dt is the flow of bioresource replacement allowing a constant value M_0 of the bioresource inside the reservoir to be maintained. Therefore, it can be easily obtained that:

$$\frac{dR}{dt} = M_0[\beta - \alpha + \rho(1 + \alpha)] \tag{3}$$

$$\rho = \frac{(\beta - \alpha)}{(k - \alpha)} \tag{4}$$

where ρ, in the last equation, can be more properly understood as the fraction of the whole M, where a growth rate $(k > \beta > \alpha)$ must be obtained thanks to the additional local inputs coming from the productive sectors and additional renewable energy resources. The last two terms, evaluated at their proper exergy cost, constitute the BSR cost of the partially renewable input consumed by the production system.

2.6. The Exergy Evaluation of Polluting Emissions

To assess polluting emissions, the exergy remediation cost has been suggested in the literature for both the CExC and the EEA models. In the EEA model, the direct and indirect exergy consumption during the overall system operation and to support system decommissioning are considered, consistent with the LCA approach [23]. The ecological cost of the polluting emissions is calculated on a remediation basis by introducing a fictitious extension of the system, where the waste treatment process has been completed, as shown in Figure 6.

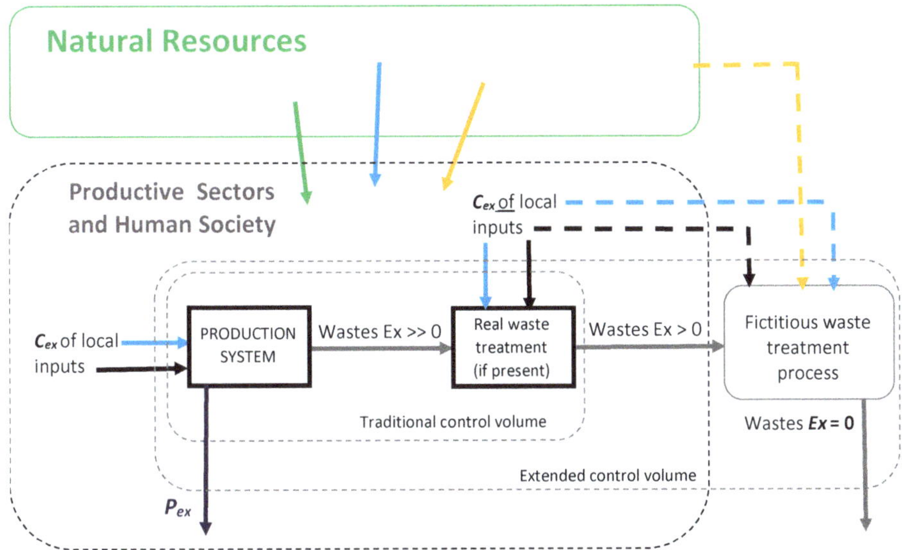

Figure 6. The extended control volume for exergy cost evaluation, following the EEA.

It is worth noting that the remediation cost for neutralizing the chemical and physical exergy of waste (the cost actually incurred plus the virtual cost) may be the same whether or not waste treatment strategies are applied. Treatments are required to convert all wastes into a flow with temperature and composition similar to those of the zero-exergy matrix of the TEE, but the non-incurred part of the cost is calculated, inside the extension of the system. The cost for the actual treatment is generally even higher because real processes are less efficient than virtual ones. The result is that a highly polluting plant may appear to be less resource-consuming (more sustainable) than a plant that obtains the same product cleanly.

In an alternative approach, the actual exergy cost of polluting emissions can be defined, in the frame of the TEE, as the real exergy stock depletion produced by the polluting emissions, caused by:

- The destruction of the confine restrictions of reservoirs.
- The variation in the zero-exergy matrix temperature or composition.
- The dilution of substances inside the reservoirs, reducing their concentration.
- The indirect destruction of the (living) biomass stock inside the reservoirs.

In this way, virtuous plants, which effectively include emission neutralization systems, may have a specific exergy cost of their products lower than polluting plants, highlighting that the former requires less consumption of resources (i.e., they are more sustainable).

3. The Thermoeconomic Environment Cost and the Exergy Resource Sustainability

Combining all previous considerations, the TEEC can be calculated as follows:

$$C_{ex-P} = \sum C_{ex-RES} + \sum (C_{ex-PRS} + C_{ex-BSR}) + \sum (C_{ex-NRS} + C_{ex-Rep}) \tag{5}$$

where:

- C_{ex-P} is the TEEC of the product P.
- C_{ex-RES} is the exergy cost of the product P, taking into account only RES.
- C_{ex-PRS} and C_{ex-NRS} are the exergy costs of the product P, taking into account only partially RES, or non-RES, respectively,
- C_{ex-BSR} and C_{ex-Rep} are the exergy costs of the product P, taking into account only the exergy BSR cost of partially RES, or the ERC of mineral non-RES, respectively.

As shown in Figure 7, the flows extracted from the TEE must include both the direct inputs and all the other real exergy stock depletions in the whole TEE, because of the polluting emissions. In this context, an exergy-based sustainability indicator easily arises, named i_{ex-TEE}, as the ratio between the exergy cost (calculated ignoring the ERC of non-RES and the BSR cost) in the numerator and the total exergy TEEC (calculated taking all terms into account) in the denominator.

$$i_{ex-TEE} = \frac{\sum C_{ex-RES} + \sum C_{ex-PRS} + \sum C_{ex-NRS}}{C_{ex-P}} \tag{6}$$

The exergy resource sustainability index i_{ex-TEE} is equal to one in the ideal case, where all direct and indirect consumptions are in the form of RES, while it is internal to the 0–1 interval in all real cases, where both RES and non-RES are consumed.

The index i_{ex-TEE} may approach one only if resources with a very low ERC or BSR cost are consumed, i.e., if all non-RES or partially RES possibly consumed are non-rare. It is worth noting that the recycling of materials reduces the value of both the ERC of the mineral resources and the BSR cost in this model, increasing the value of the proposed exergy-based sustainability indicator.

The index i_{ex-TEE} may approach zero when the primary inputs extracted from the TEE have a very high ERC or BSR cost, i.e., when rare mineral resources rare biological species, or even whole ecosystems, are consumed or destroyed, even if their exergy contents were small.

Let us consider that EEA is the starting point for calculating the TEE cost and the related resource sustainability indicator. In this case, the exergy equivalent of capital and labor are taken into account through the procedure suggested by Sciubba. Otherwise, the approach suggested by Rocco and Colombo may be followed. In the latter case, the exergy equivalent of capital is implicitly taken into account and could be calculated a posteriori.

If solar energy, as well as other renewable and partially RES, were properly taken into account, the total cost obtained in this way should be the sum of $(C_{ex-RES} + C_{ex-PRS} + C_{ex-NRS})$ in Equation (5). The exergy cost of the BSR (C_{ex-BSR}) of the biological products used as input in the production process must be evaluated following the frame shown in Figure 5, by using some additional information from the fields of agriculture and forest cultivation.

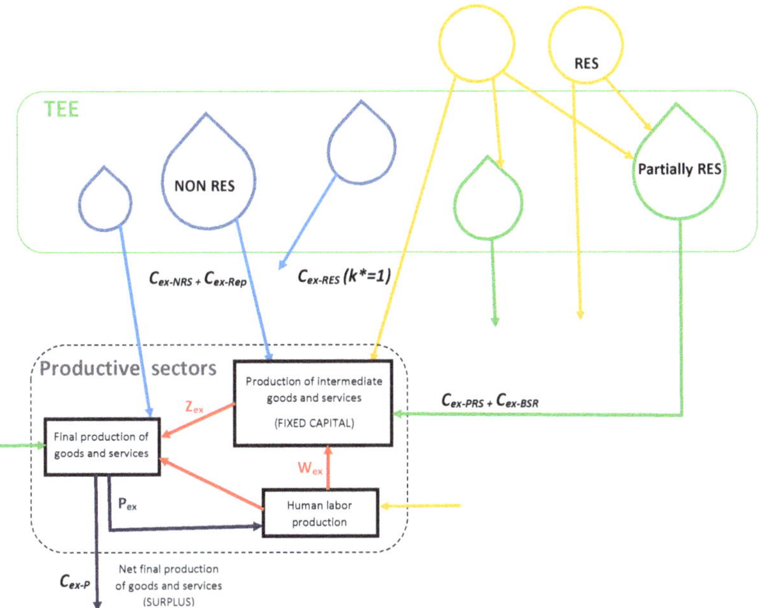

Figure 7. Illustrative sketch of the procedure for the calculation of the TEE cost and the sustainability index.

The C_{ex-Rep} of all mineral resources consumed can be found in the papers by Valero [42,43].

EEA calculate the effect of polluting emissions based on the exergy cost of remediation (Figure 6). In this paper, it is suggested that the actual exergy depletion of the TEE should be calculated, because of its direct and indirect effects. To proceed in this way, the exergy cost of remediation should be eliminated from the total accounting of the exergy cost of the product if the remediation technologies are not actually put in operation. Then, an inventory should be compiled of the depletions in the TEE resulting from the polluting emissions of the production process at hand. The results of an LCA of a similar process taken from the literature may be used as a first attempt. Finally, the depletion of each reservoir should be estimated in terms of its exergy cost, including the exergy cost associated to temperature variations in the zero-exergy matrix. Notice that the depletion of mineral reservoirs must be accounted-for at the cost $C_{ex-NRS} + C_{ex-Rep}$, while the depletion of the reservoirs of biological products must be accounted-for at the cost $C_{ex-RES} + C_{ex-BSR}$. In this way, the effect of polluting emissions will affect all terms in Equations (5) and (6). At this point, an evaluation of each of the five terms on the right-hand side of Equation (5) would be obtained, and the indicator in Equation (6) could be calculated.

4. Conclusions

Exergy cost accounting introduces the sum of direct and indirect exergy consumption as a measure of the resources required to obtain a product. The necessary definition of a proper ultimate boundary of the exergy cost accounting may be carried out by introducing the TEE consistently to assess the sustainability of the production of goods and services. Then, the exergy replacement cost of mineral resources proposed by Valero and Valero, may be introduced as a meaningful improvement to actual exergy cost accounting methodologies, consistently with the frame of the TEE.

Likewise, the proposal by Rocco and Colombo for a definition of the exergy values of labour and capital (directly derived from the input/output algebra by Leontief) was shown to be also consistent with the framework of the TEE and can therefore be used to evaluate the exergy equivalent of capital and human work. In this way, production factors, such as capital investment, human labor and environmental resources, can be unified into a common exergetic description.

To properly take into account the interaction of the production system with biological processes, the bioresource stock replacement cost was here introduced, taking advantage of the idea of partially renewable resources (the living biomass) contained inside the TEE. If using exergy cost accounting to assess the sustainability of a specific product or service, it is important to notice that the TEE framework allows us to assess the impact of polluting emissions based on the actual exergy stock depletion throughout the TEE. Virtuous plants, which effectively include emission neutralization systems, may have a specific exergy cost lower than that of polluting plants, therefore justifying, from an exergy cost accounting point of view, the adoption of devices that strongly reduce polluting emissions.

In addition, an exergy-based sustainability indicator easily arises as the ratio of the exergy cost, calculated neglecting the exergy replacement cost of non-RES and the bioresource stock replacement cost, and the total bioresource stock replacement exergy Thermoeconomic Environment cost, calculated taking all terms into account. This new exergy-based sustainability indicator is expected to be well-suited to expressing the resource sustainability of goods and services. It is equal to one in the ideal case, where all direct and indirect consumption is of RES, giving a clear view of how far the process at hand is from the ideal case, and enabling the calculation of the margin available for possible improvements.

Finally, this method allows us to highlight the advantage of recycling and the usage of non-rare mineral resources, because they both reduce the exergy replacement cost of non-RES and the bioresource stock replacement cost of partial RES. In the same way, the disadvantage of consuming rare mineral or biological resources is properly drawn to our attention, even when their chemical exergy content is small.

Author Contributions: Conceptualization, M.R.; methodology, M.R., S.K. and M.C.; validation, S.K. and M.C.; formal analysis M.R.; investigation, M.R., S.K. and M.C.; resources, M.R., writing—original draft preparation, M.R., S.K. and M.C.; writing—review and editing, M.R. and M.C.; visualization, M.C.; supervision, M.R.; funding acquisition, M.R.; All authors have read and agreed to the published version of the manuscript.

Funding: This research received no external funding.

Institutional Review Board Statement: Not applicable.

Informed Consent Statement: Not applicable.

Data Availability Statement: Not applicable.

Conflicts of Interest: The authors declare no conflict of interest.

Nomenclature

α	natural growth rate of the bioresource in a reservoir
β	extraction rate of the bioresource from a reservoir

ρ	fraction of the bioresource required for the replacement (Figure 5)
k	growth rate in the extended system required for the replacement (Figure 5)
C_{ex-P}	TEEC of the product P
C_{ex-RES}	TEEC of the product P, taking into account only RES
C_{ex-PRS}	TEEC of the product P, taking into account only partially RES
C_{ex-NRS}	TEEC of the product P, taking into account only non-RES
C_{ex-BSR}	TEEC of the product P, taking into account only the exergy BSR cost of partially RES
C_{ex-Bep}	TEEC of the product P, taking into account only the ERC of mineral non-RES
$k*$	unit exergy cost of a flow
i_{ex-TEE}	exergy-based resource sustainability indicator
M	amount of bioresource in a reservoir
M_0	bioresource stock at instant t = 0
P_{ex}	exergy of the product
dR/dt	flow of bioresource required for stock replacement
t	time
Z_{ex}	exergy of the fixed capital

Acronyms

BSR	bioresource stock replacement
CExC	cumulative exergy consumption
EEA	extended exergy accounting
ERC	exergy replacement cost
RES	renewable energy source
TEE	thermoeconomic environment
TEEC	thermoeconomic environment cost

References

1. Reini, M.; Casisi, M. Exergy Analysis with Variable Ambient Conditions, ECOS 2016. In Proceedings of the 29th International Conference on Efficiency, Cost, Optimization, Simulation and Environmental Impact of Energy Systems, Portoros, Slovenia, 19–23 July 2016.
2. Bejan, A.; Tsatsaronis, G.; Moran, M. *Thermal Design and Optimization*; John Wiley & Sons: New York, NY, USA, 1996.
3. Bejan, A. *Advanced Engineering Thermodynamics*, 3rd ed.; John Wiley & Sons Inc.: Hoboken, NJ, USA, 2006.
4. Reini, M.; Casisi, M. The Gouy-Stodola Theorem and the derivation of exergy revised. *Energy* **2020**, *210*, 118486. [CrossRef]
5. Gaggioli, R.A.; Petit, P.J. Use The Second Law First. *Chemtech* **1977**, *7*, 496–506.
6. Rosen, M. A Concise Review of Exergy-Based Economic Methods. In Proceedings of the 3rd IASME/WSEAS Int. Conf. on Energy & Environment, University of Cambridge, Cambridge, UK, 23–25 February 2008.
7. Tiruta-Barna, L.; Benetto, E. A conceptual framework and interpretation of Emergy algebra. *Ecol. Eng.* **2013**, *53*, 290–298. [CrossRef]
8. Gaggioli, R.A. The concept of available energy. *Chem. Eng. Sci.* **1961**, *16*, 87–96. [CrossRef]
9. Dincer, I.; Rosen, M.A. *EXERGY: Energy, Environment and Sustainable Development*, 1st ed.; Elsevier: Burlington, VT, USA, 2007.
10. Gaggioli, R.A.; Paulus, D.M. Available energy—Part II: Gibbs extended. *J. Energy Res. Technol.* **2002**, *124*, 110–115. [CrossRef]
11. Gaggioli, R.A.; Richardson, D.H.; Bowman, A.J. Available energy—Part I: Gibbs revisited. *J. Energy Resour. Technol.–Trans. ASME* **2002**, *124*, 105–109. [CrossRef]
12. Sciubba, E.; Wall, G. A brief commented history of exergy from the beginnings to 2004. *Int. J. Thermodyn.* **2010**, *10*, 1–26.
13. Gaggioli, R.; Reini, M. Panel I: Connecting 2nd Law Analysis with Economics, Ecology and Energy Policy. *Entropy* **2014**, *16*, 3903–3938. [CrossRef]
14. Reini, M.; Daniotti, S. Energy/Exergy Based Cost Accounting in Large Ecological-Technological Energy Systems. In Proceedings of the 12th Joint European Thermodynamics Conference, Brescia, Italy, 1–5 July 2013.
15. Szargut, J. *Exergy Method: Technical and Ecological Applications*; Gateshead (GB), WIT Press: Southampton, UK, 2005.
16. Leontief, W.W. *The Structure of the American Economy, 1919–1939*, 2nd ed.; Oxford University Press: New York, NY, USA, 1951.
17. Valero, A.C.; Valero, A.D. *Thanatia—The Destiny of the Earth's Mineral Resources—A Thermodynamic Cradle-to-Cradle Assessment*; World Scientific Publishing Company: Toh Tuck Link, Singapore, 2014; p. 670.
18. Szargut, J.; Morris, D.R.; Steward, F.R. *Exergy Analysis of Thermal, Chemical, and Metallurgical Processes*; Hemisphere: London, UK, 1987.
19. Kharrazi, A.; Kraines, S.; Hoang, L.; Yarime, M. Advancing quantification methods of sustainability: A critical examination Emergy, exergy, ecological footprint, and ecological information-based approaches. *Ecol. Indic.* **2014**, *37*, 81–89. [CrossRef]
20. Odum, H.T. *Environmental Accounting: Emergy and Environmental Decision Making*, 1st ed.; John Wiley & Sons: New York, NY, USA, 1996.

21. Brown, M.T.; Ulgiati, S. Emergy-based indices and ratios to evaluate sustainability: Monitoring economies and technology toward environmentally sound innovation. *Ecol. Eng.* **1997**, *9*, 51–69. [CrossRef]
22. Sciubba, E. Exergy as a Direct Measure of Environmental Impact. *Adv. Energy Syst.* **1999**, *39*, 573–581. [CrossRef]
23. Sciubba, E. Beyond thermoeconomics? The concept of extended exergy accounting and its application to the analysis and design of thermal systems. *Exergy Int. J.* **2001**, *1*, 68–84. [CrossRef]
24. Sciubba, E. Extended exergy accounting applied to energy recovery from waste: The concept of total recycling. *Energy* **2003**, *28*, 1315–1334. [CrossRef]
25. Sciubba, E.; Bastianoni, S.; Tiezzi, E. Exergy and extended exergy accounting of very large complex systems with an application to the province of Siena, Italy. *J. Environ. Manag.* **2008**, *86*, 372–382. [CrossRef]
26. Koroneos, C.J.; Nanaki, E.A.; Xydis, G.A. Sustainability Indicators for the Use of Resources—The Exergy Approach. *Sustainability* **2012**, *4*, 1867–1878. [CrossRef]
27. Rosen, M.A.; Dincer, I.; Kanoglu, M. Role of exergy in increasing efficiency and sustainability and reducing environmental impact. *Energy Policy* **2008**, *36*, 128–137. [CrossRef]
28. Aydin, H.; Turan, O.; Karakoc, T.H.; Midilli, A. Exergetic Sustainability Indicators as a Tool in Commercial Aircraft: A Case Study for a Turbofan Engine. *Int. J. Green Energy* **2015**, *12*, 28–40. [CrossRef]
29. Dewulf, J.; Van Langenhove, H.; Mulder, J.; Van Den Berg, M.M.D.; van der Kooi, H.J.; de Swaan Arons, J. Illustrations towards quantifying the sustainability of technology. *Green Chem.* **2000**, *2*, 108–114. [CrossRef]
30. Dale, V.H.; Beyeler, S.C. Challenges in the development and use of ecological indicators. *Ecol. Indicat.* **2001**, *1*, 3–10. [CrossRef]
31. Paul, B.D. A history of the concept of Sustainable Development: Literature review. Available online: https://citeseerx.ist.psu.edu/viewdoc/download?doi=10.1.1.532.7232&rep=rep1&type=pdf (accessed on 1 October 2013).
32. Sciubba, E. Can an Environmental Indicator valid both at the local and global scales be derived on a thermodynamic basis? *Ecol. Indicat.* **2013**, *29*, 125–137. [CrossRef]
33. Kehdr, S.; Reini, M.; Casisi, M. A critical review of exergy based cost accounting approaches. In Proceedings of the 6th International Conference on Contemporary Problems of Thermal Engineering CPOTE 2020, Krakow, Poland, 21–24 September 2020.
34. Hau, J.L.; Bakshi, B.R. Expanding exergy analysis to account for ecosystem products and service. *Environ. Sci. Technol.* **2004**, *38*, 3768–3777. [CrossRef] [PubMed]
35. Valero, A.; Serra, L.; Uche, J. Fundamentals of Exergy Cost Accounting and Thermoeconomics. *Part I Theory J. Energy Resour. Technol.* **2006**, *128*, 1–8. [CrossRef]
36. Gaudreau, K.; Fraser, R.A.; Murphy, S. The Tenuous Use of Exergy as a Measure of Resource Value or Waste Impact. *Sustainability* **2009**, *1*, 1444–1463. [CrossRef]
37. Sciubba, E. From Engineering Economics to Extended Exergy Accounting: A Possible Path from Monetary to Resource-Based Costing. *J. Ind. Ecol.* **2004**, *8*, 19–40. [CrossRef]
38. Yang, J.; Chen, B. Extended exergy-based sustainability accounting of a household biogas project in rural China. *Energy Policy* **2014**, *68*, 264–272. [CrossRef]
39. Romero, J.C.; Linares, P. Exergy as a global energy sustainability indicator. A review of the state of the art. *Renew. Sustain. Energy Rev.* **2014**, *33*, 427–442. [CrossRef]
40. Kehdr, S.; Casisi, M.; Reini, M. The role of the Thermoeconomic Environment in the exergy based cost accounting of technological and biological systems, submitted to Energy, previously. In Proceedings of the ECOS 2021—The 34th International Conference on Efficiency, Cost, Optimization, Simulation and Environmental Impact of Energy Systems, Taormina, Italy, 28 June–2 July 2021.
41. Giorgi, F.; Coppola, E.; Raffaele, F. A consistent picture of the hydroclimatic response to global warming from multiple indices: Models and observations. *J. Geophys. Res. Atmos.* **2014**, *119*, 11–695. [CrossRef]
42. Szargut, J.; Stanek, W. Thermo-ecological optimization of a solar collector. *Energy* **2007**, *32*, 584–590. [CrossRef]
43. Szargut, J.; Ziębik, A.; Stanek, W. Depletion of the non-renewable natural exergy resources as a measure of the ecological cost. *Energy Convers. Manag.* **2002**, *43*, 1149–1163. [CrossRef]
44. Valero, A.; Valero, A.; Stanek, W. Assessing the exergy degradation of the natural capital: From Szargut's updated reference environment to the new thermoecological-cost methodology. *Energy* **2018**, *163*, 1140–1149. [CrossRef]
45. Valero, A.D.; Agudelo, A.; Valero, A.D. The crepuscular planet. A model for the exhausted atmosphere and hydrosphere. *Energy* **2011**, *36*, 3745–3753. [CrossRef]
46. Valero, A.; Valero, A.; Gómez, J.B. The crepuscular planet. A model for the exhausted continental crust. *Energy* **2011**, *36*, 694–707. [CrossRef]
47. Rocco, M.V.; Colombo, A.; Sciubba, E. Advances in exergy analysis: A novel assessment of the Extended Exergy Accounting method. *Appl. Energy* **2014**, *113*, 1405–1420. [CrossRef]
48. Rocco, M.; Colombo, E. Internalization of human labor in embodied energy analysis: Definition and application of a novel approach based on environmentally extended Input-Output analysis. *Appl. Energy* **2016**, *182*, 590–601. [CrossRef]
49. Lozano, M.; Valero, A. Theory of the exergetic cost. *Energy* **1993**, *18*, 939–960. [CrossRef]

Article

Life Cycle Assessment for Integration of Solid Oxide Fuel Cells into Gas Processing Operations

Khalid Al-Khori *, Sami G. Al-Ghamdi, Samir Boulfrad and Muammer Koç

Division of Sustainable Development, College of Science and Engineering, Hamad Bin Khalifa University, Qatar Foundation, Doha 34110, Qatar; salghamdi@hbku.edu.qa (S.G.A.-G.); sboulfrad@hbku.edu.qa (S.B.); mkoc@hbku.edu.qa (M.K.)
* Correspondence: k.alkhori@gmail.com

Abstract: The oil and gas industry generates a significant amount of harmful greenhouse gases that cause irreversible environmental impact; this fact is exacerbated by the world's utter dependence on fossil fuels as a primary energy source and low-efficiency oil and gas operation plants. Integration of solid oxide fuel cells (SOFCs) into natural gas plants can enhance their operational efficiencies and reduce emissions. However, a systematic analysis of the life cycle impacts of SOFC integration in natural gas operations is necessary to quantitatively and comparatively understand the potential benefits. This study presents a systematic cradle-to-grave life cycle assessment (LCA) based on the ISO 14040 and 14044 standards using a planar anode-supported SOFC with a lifespan of ten years and a functional unit of one MW electricity output. The analysis primarily focused on global warming, acidification, eutrophication, and ozone potentials in addition to human health particulate matter and human toxicity potentials. The total global warming potential (GWP) of a 1 MW SOFC for 10 years in Qatar conditions is found to be 2,415,755 kg CO_2 eq., and the greenhouse gas (GHG) impact is found to be higher during the operation phase than the manufacturing phase, rating 71% and 29%, respectively.

Keywords: emissions; CO_2; GWP; functional unit; natural gas; SOFC

Citation: Al-Khori, K.; Al-Ghamdi, S.G.; Boulfrad, S.; Koç, M. Life Cycle Assessment for Integration of Solid Oxide Fuel Cells into Gas Processing Operations. *Energies* **2021**, *14*, 4668. https://doi.org/10.3390/en14154668

Academic Editor: Attilio Converti

Received: 12 June 2021
Accepted: 16 July 2021
Published: 1 August 2021

Publisher's Note: MDPI stays neutral with regard to jurisdictional claims in published maps and institutional affiliations.

1. Introduction

Energy must provide a broad range of essential societal services even though it comes with significant adverse environmental impacts depending on the energy source and technology used. The continuing development of more sustainable energy resources and deployment of new technologies aims to reduce such negative energy impacts while maximizing its benefits to a better balance between opportunities and energy cost—and ultimately to overcome difficulties related to efficiency.

Fossil fuels are currently the world's primary energy source, generating more than 75% of the total energy demand; this is estimated to remain as such for many years to come [1]. However, fossil fuel combustion is the primary source of GHG emissions that cause global warming [2]. Current economic development depends heavily on exploiting fossil fuels, which will be difficult to sustain indefinitely [3]. Therefore, the energy sector must seek to reduce CO_2 emissions significantly. There is a logical and moral obligation to consider the negative environmental impact of GHG emissions and broaden the industry's focus beyond economic wealth creation. Therefore, to achieve sustainable development in the energy sector, industries should look for alternative processes that reduce CO_2 emissions.

Solid Oxide Fuel Cells (SOFCs) are a new, cleaner technology based on hydrogen and electrochemical reaction to generate electricity. SOFCs are highly efficient at producing minimal emissions [4] and can be used to improve the efficiency of oil and gas plants while reducing CO_2 emissions. Integrating SOFC into the oil and gas industry could be an effective method of efficient energy production and application. More details on SOFC types, usages, and challenges can be found in "Integration of Solid Oxide Fuel Cells into Oil and Gas Operations: Needs, Opportunities, and Challenges" [5].

However, long-term, broader, and multi-dimensional impacts of SOFC integration into natural gas operations must be studied to demonstrate their benefits quantitatively. A Life Cycle Assessment (LCA) can be used to evaluate the environmental impact of SOFC manufacturing and operations, compare different integration scenarios in oil and gas plants, and clarify how different kinds of integration of SOFC in oil and gas can reduce emissions. An LCA consists of several multipurpose steps used to gather and explore all inputs and outputs of a product in addition to its possible environmental impacts. This is done for the entire life cycle, from the raw material collection and manufacturing, usage, maintenance, and disposal or repurposing [6]. The environmental management standards ISO 14040 and ISO 14044 (both issued in 2006) form the basis for the systematic LCA.

This study presents a systematic cradle-to-grave life cycle assessment (LCA) based on the ISO 14040 and 14044 standards using a planar, anode-supported SOFC with a lifespan of 10 years and a functional unit of 1 MW electricity output designed to be integrated into a proper natural gas operation.

This study will have the following features:

- The SOFC in this LCA study is fueled by natural gas and operates in a natural gas plant, while many other similar studies on SOFC LCA are based on a fuel other than pure natural gas or used in a domestic and residential area.
- Utilizing the SOFC in a natural gas plant will eliminate unnecessary flaring of natural gas.
- Take advantage of the presence of natural gas in Qatar at a reasonable cost as fuel to SOFC.
- GWP from the operational phase of SOFC in Qatar is much less than operating SOFC in other countries.
- Availability of data from this LCA study will allow for comparison with LCA results of traditional power generation used in the gas processing plant.
- The ratio of GWP between the manufacturing and operation phases is aligned with the results from other SOFC LCA studies.
- The gas plant can generate its own electricity using SOFC which will result in less environmental impact compared to other traditional power generation.

The following section comparatively summarizes and analyzes the relevant literature on the LCA and its application to energy, oil and gas operations, and SOFC; the next section presents the methodology followed in this study and describes the conditions, data collection, and justifications; the third section presents the results, findings, and discussions in detail, and the conclusions are comparatively presented in the fourth section.

1.1. Background

The LCA has become an appropriate and effective tool for evaluating matters related to resource depletion and environmental degradation [6]. From 1994 to 2014, LCA studies in the energy sector increased by 60% [6]. An LCA involves four stages: goal and scope, life cycle inventory analysis (LCI), life cycle impact assessment (LCIA), and interpretation. The first stage identifies the functional unit and system boundaries and then defines the assumptions, limitations, and allocations (if any) in addition to selecting the LCIA method. The second phase involves conducting an inventory of flows, including all inputs of water, energy, and raw materials and emissions to the air, land, and water. The third phase selects the impact categories, category indicators, and characterization models. The last step evaluates the results' completeness, sensitivity, and consistency while providing conclusions and recommendations.

Most LCA studies of SOFCs follow the cradle-to-grave approach for system boundaries. However, many exclude the manufacturing and disposal stages due to the assumption that most environmental impacts are caused by operation and fuel production [7–10]. The decision to add or remove a specific step is based on the goals set in a particular study.

It is possible to leave out activities that do not affect the overall understanding of the analysis and continue to consider the relevant issues with the LCA [7]. Environmental, health, economic, and political problems have influenced studies analyzing SOFC environmental performance and compared them with traditional power generation in many different fields [7].

Using different methodologies in SOFC LCA studies led to differences in FU and system boundary choices, among others. This, along with the unavailability of up-to-date inventory data, made comparisons and evaluations of these studies' outcomes complex [7]. One common problem is data availability [11], as detailed data for materials used in SOFC production are not released by manufacturers due to concerns regarding confidentiality and market competitiveness [7].

The CML method is commonly used in LCAs; it includes ten impact categories, is flexible, gives accurate results, and is transparent [7]. Buchgeister used three different LCIA methods (Eco-indicator 99, CML 2001, and Impact 2002) for SOFCs fueled by gasified biomass fuel, noting that overall environmental impacts are differed [12]. The study recommended using more than one method to reach a uniform outcome, as a single approach could provide weak signals due to discrepancies in LCIA methods. For example, a midpoint (or problem-oriented) approach usually focuses on actions like emissions relief and resource usage along impact pathways like GWP, AP, and EP. An endpoint (or damage-oriented) approach focuses on the final impacts linked to outcomes (like human health) along impact pathways like AoP [7]. The IPCC methodology mainly focuses on GHGs such as CO_2, CH_4, and N_2O.

CO_2 emissions from SOFC production are negligible compared to those generated during operation. In contrast, the harmful emissions produced by other gases (e.g., SO_2, CO, NO_x, and SO_x) are negligible during the process but are generated during production—though it only affects the area around the factory [8]. The latter contributes to acidification in local land and water systems and the development of volatile compounds that influence ozone levels and human health [11]. A SOFC operation contributes most life cycle emissions due to the fuels used should not be an excuse to eliminate the manufacturing phase from its LCA, as the supply chain may vary by location, energy mix, and other factors [7].

Nigel and Brunel demonstrated a 75–93% reduction of CH_4, NMHCs, PM, and CO through SOFCs rather than traditional systems [13]. Jakob and Hirshberg [14] showed that SOFC use reduced GWP by 50% and required up to 20% less energy than traditional systems like gas boilers. Herron proved that SOFCs in three American cities produced almost no harmful gases like NO_x, $PM_{2.5}$, PM 10, and SO_x, unlike three types of traditional systems (natural gas combined with cycle plants, coal-fired plants, and nuclear plants) and achieved superior performance [9].

The SOFC disposal phase has been ignored by many studies, especially those before 2011, mainly due to unknown strategies for SOFC end-of-life practices and a lack of relevant data. Mehmeti et al. considered the inclusion of SOFC disposal in LCA studies to be optional, recommending conducting a sensitivity analysis to account for the LCA's numerous uncertainties and so that this process could detect significant factors affecting the overall performance [7].

1.2. Literature Review

The electrical mix plays a significant role in determining emissions results for CO_2 and GWP, as the operational phase of SOFCs accounts for approximately 75–97% of life cycle environmental influences [15]. As fuel production and supply contribute to around 97% of the impact on ODP, ADP, and PED, the selection of fuel types has a significant effect on the results of the SOFC LCA. During SOFC production using the UK mix grid, 60% of CO_2 release occurs during sintering of the SOFC cell, and the remainder is related to fuel processor and DC/AC power converter production as part of the BoP [16]. A recycling rate of 75% will reduce GWP by 8–11% [16].

Lee et al. argued that the manufacturing and end-of-life phases have little impact on the environmental burden of a SOFC system (2.1–9.5% and <0.6%, respectively; [10]. Within the manufacturing stage, stack production contributes approximately 72% of the total environmental impact for planar SOFC and 28% for BoP, mainly due to the utilization of stainless steel and chromium alloys that require more energy during production. Strazza et al. noted that the level of environmental impact during manufacturing is based on the quantity of steel and the type of energy mix used; the worst case is coal-based energy, demonstrating that the ecological implications of SOFC manufacturing depend on the production location [17]. That study also observed that, when using the midpoint LCA approach, natural gas is recommended for lower AP and POCP, but that for ADP, EP, GWP, ODP, and PED, biogas shows better results. Although H_2 has better results than GWP and ODP, it is higher in POCP, EP, and AP. However, leaks from gas plants or pipelines can contribute significantly to overall emissions since CH_4 contributes to GWP by a mass ratio of 25 CH_4 to 1 CO_2 [18].

Sadhukhan emphasized that GWP, AP, and POCP are the most essential categories for evaluating different technologies (based on the Monte Carlo analysis). The SOFC LCA for these categories produces lower results than internal combustion engines micro gas turbines for distributed power generation [19]. Lin et al. stated that GWP is the only category impacted by a SOFC operation, during which CO_2 emissions represent 80% of total life cycle emissions [20]. Sadhukhan also argued that using SOFCs could eliminate $PM_{2.5}$ and N_2O emissions, which contribute more to GWP than CO_2 given the mass ratio of 298:1 between N_2O and CO_2; N_2O also impacts EP and POCP [19]. Strazza et al. provided further details on environmental impact per category, stating that ADP, ODP, GWP, and PED accounted for as little as 2% of the total life cycle environmental impact during the manufacturing and end-of-life stages. However, this rose to approximately 10–32% for AP, EP, and POCP [17].

Nease and Adams used the ReCiPe 2008 method to evaluate the life cycle impact, showing that the effects of the manufacturing phase affected climate change mainly for the fossil fuel and metal depletion categories, equaling approximately 9% for one year [21,22]. Nease and Adams claimed that, even with the uncertainties, SOFCs were better than NGCCs and were better than supercritical pulverized coal by 45.8% in life cycle influence [22]. Baratto and Diwekar also showed that using SOFCs instead of diesel engines in trucks decreased PM by 82.08%, HC by 92.65%, NO_x by 99.1%, CO by 97.77%, and CO_2 by 64.32% [8]. Reenaas et al. analyzed several fuel types for combined SOFC-GTs in Norway. They confirmed that, mainly due to less transportation, domestic LNG had 60% less GWP, 85% less POCP, and 90% less AP when compared with imported LNG and sulfur-free diesel [23]. Besides, the use of SOFC-GT in marine applications instead of diesel engines contributed to reductions of 35–93% in GWP, POCP, and AP [23]. Lin et al. conducted a fuel-type comparison for SOFCs between APUs and diesel engines in trucks, finding that a SOFC fueled by biodiesel was the most environmentally friendly system [20]. This required 14.5 times less fuel than diesel engines and emitted five times fewer GHGs, while no re-design of the fuel system was necessary since the fuels had similar physical and chemical compositions [20].

Concerning disposal, Cánovas et al. argued that a 70% recycling rate would result in a 7.5% reduction in life cycle impact, primarily due to lower carcinogen emissions [15]. Strazza et al. stated that nickel was the most recyclable portion of SOFCs while ceramics will end up in landfills [17]. However, Nease and Adams showed that decommissioning large SOFCs can require large amounts of fossil fuel and produce more emissions [22].

Reenaas also confirmed that overall the LCA was not sensitive to changes in the durability of SOFC-GTs. Increasing SOFC life to 10 years from 5 years would only increase the GWP, AP, and POCP by 3% at most but would reduce SOFC efficiency by 20% and lead to 6% increases in GWP and AP and 33% in POCP [23].

1.3. Objectives

The general purpose of this LCA study is to understand the impact of SOFCs on environmental values and provide a basis for comparisons of different types of integration approaches in gas processing plants. The results are intended to serve as a reference for future researchers interested in integrating SOFCs in a gas plant and providing manufacturers and decision-makers with valuable data. As shown in Figure 1, the overall SOFC system includes the different phases of SOFC, including the raw material up to the manufacturing and operational phase. However, the disposal phase is not part of the LCA study since its life cycle environmental impact is less than 3% of the total impact, as based on the literature review.

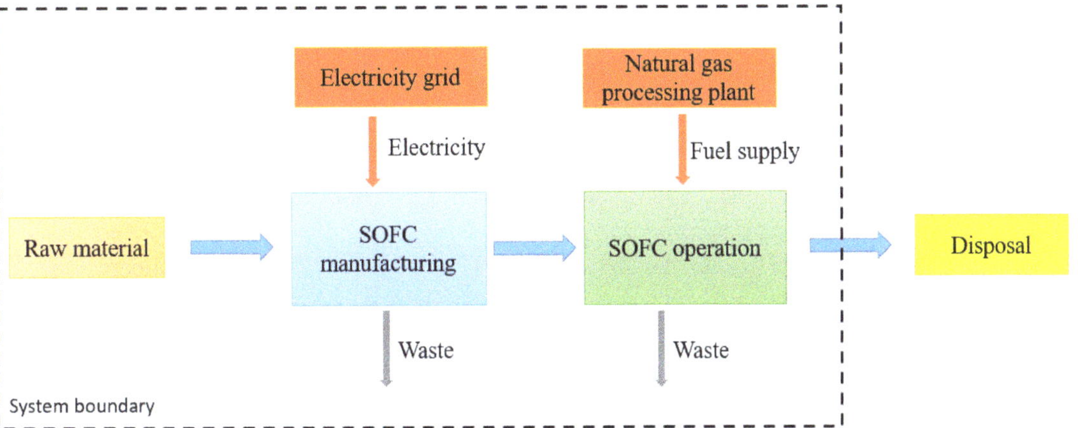

Figure 1. The lifespan of SOFC from raw materials to disposal.

The aim of conducting an LCA for SOFCs is to identify the materials and processes involved in SOFC manufacturing and their environmental impact and evaluate the ecological effect of SOFC operation. The resulting information can improve decision-making about resource depletion and environmental degradation [7].

2. Method

SETAC developed an LCA code of practice, and it encouraged the ISO to create a standardized set of steps for the LCA process. The methodological framework defined in the ISO 14040 standard includes four main phases:

- Goal and scope definition: the study's aim, breadth, and depth are outlined, setting the functional unit and system boundaries.
- LCI: data collection is performed, including calculation and allocation.
- LCIA: potential environmental effects related to the inventory analysis results are evaluated.
- Interpretation: the LCIA results are analyzed and summarized concerning the goal and scope.

Interpretation is the last stage of the systemic LCA procedure in which the results of the inventory exploration and impact assessments are evaluated for completeness, sensitivity, and consistency. This process also clarifies uncertainties and assumptions regarding improvements to environmental performance while defining further limitations and informing recommendations.

2.1. Goal and Scope

This first phase describes the study's purpose, scope, allocation procedures, and assumptions or limitations (if any). The system boundaries are defined in this stage, along with the functional unit and the LCIA method.

Defining the FU is not always straightforward, mainly when the product produces several useful outputs, but its selection should reflect the actual condition related to the product and market needs [7]. This helps normalize all inputs (e.g., materials, energy resources, and outputs like heat) and allows for comparative analyses. The FU for SOFCs can be presented in terms of stack power capacity (kW) or total energy output (power plus heat, kWh) if heat is considered a beneficial energy outcome. The unit scale (kWh vs. MWh) will not impact the final results since all emissions outcomes are linear to the chosen FU [24]. Proper unit selection ensures that the LCA results are accurate and improves the outcome and interpretation stages [7].

In this study, the FU was defined as 1 MW of net electricity generated by the SOFC system (and after that utilized by the gas plant) during its service lifespan of 10 years. The study's scope was to evaluate the potential environmental impacts of integrating and operating a 10 MW SOFC system fueled by natural gas in a gas plant. It was assumed that the rest of the plant (offshore, pipeline, and onshore infrastructure) were not part of the LCA, though the natural gas input as fuel was considered.

The system boundary for the LCA included SOFC manufacturing and operation. End-of-life was not quantitatively defined because insufficient data were available to quantify the disposal or recycling of SOFC materials properly; it can, however, be qualitatively assessed for completeness [6]. Details of the system boundary and the inputs and outputs of each process are given for the manufacturing phase in Figure 2 and the operational phase in Figure 3. Transportation of raw material and natural gas was not considered in this LCA. One crucial assumption is related to the material; since the Gabi software did not have the Yttria-stabilized-zirconia in its database, another type of ceramic, Alumina, was chosen to replace the YSZ in the LCA study.

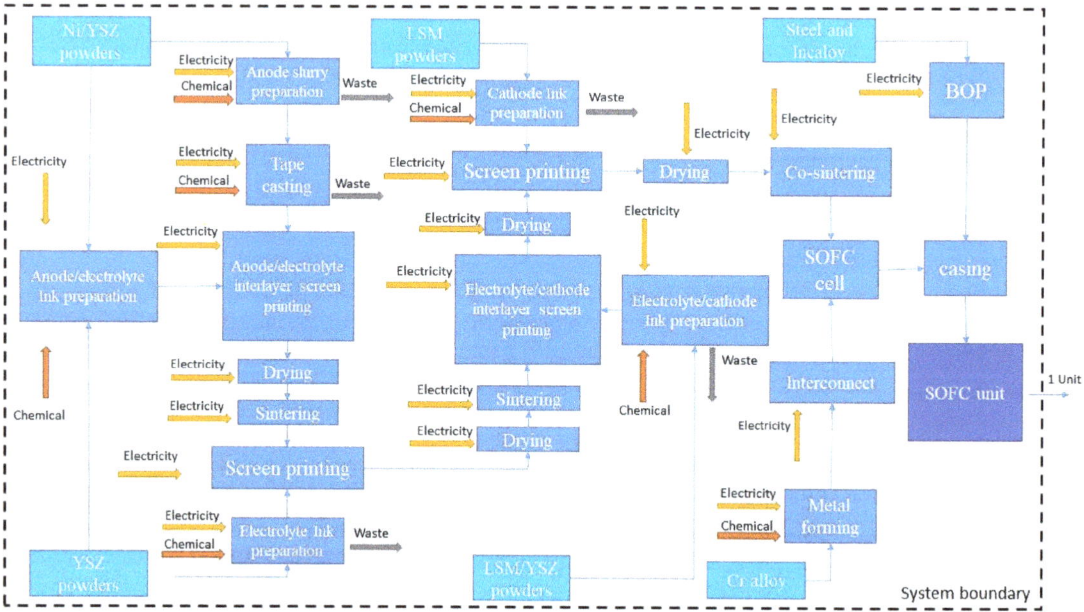

Figure 2. System boundary for SOFC LCA manufacturing phase.

The detail drawing of the SOFC manufacturing steps as developed in GaBi software is provided in Figure S1.

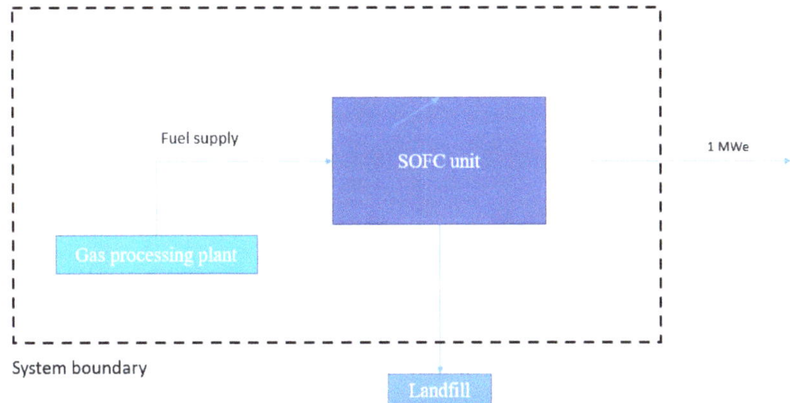

Figure 3. System boundary for SOFC LCA operational phase and the functional unit.

2.2. Life-Cycle Inventory

The LCI collects data from the studied system or product to produce the functional unit. This stage delineates materials, water, and used energy along with waste discharges to air, land, or water (e.g., emissions, wastewater, and solid waste) for all life cycle phases from raw material to operation. Two data types should be considered for SOFC LCIs [8]: foreground and background data. Foreground data are the outcome parameters from SOFC manufacturing and operation, while background data are related to materials and energy used for SOFC production and operation and delivering the FU. Assumptions may be necessary when data are not available or obsolete [1], resulting in uncertainties in the LCI. This is commonly the case for SOFCs, for which material details, energy inputs, and waste data are unavailable due to manufacturer confidentiality.

This study identified three input types (materials, chemicals, and energy) and collected data from previous research [13,24–31]; these were re-calculated to suit the current FU. Table 1 lists the material inventory and quantity required to generate one FU (1 MW).

Table 1. SOFC cell material inventory list.

Material Description	Material Weight (Kg/MW)
Anode (Ni 70% wt)	1116.00
Anode (Alumina 30% wt)	332.00
Electrolyte (Alumina)	39.00
Cathode (LSM)	78.00
Anode/electrolyte interlayer (NiO 50% vol)	20.00
Anode/electrolyte interlayer (Alumina 50% vol)	20.00
Electrolyte/cathode interlayer (LSM 50% vol)	20.00
Electrolyte/cathode interlayer (Alumina 50% vol)	20.00

Table 2 lists the chemicals required for different processes during SOFC cell manufacturing. Most binders and solvents are the same for several functions but use different quantities for each method. Table 3 lists the energy requirements for various processes during SOFC manufacturing. Materials related to the BoP are also part of the manufacturing phase; these are listed in Table 4 with their material type, weight, and power consumption. In addition to inputs, these processes produced primarily waste consisting of CO_2 and evaporated solvent. Table 5 shows the amount and quantity of these outputs for all functions during the manufacturing phase.

Table 2. Chemical inventory list for SOFC manufacturing.

Process Type	Chemical Description	Material Weight (Kg/MW)
Anode Slurry preparation	Plasticizer (Sanitizer)	132.00
	Butvar-76 (binder)	131.70
	n-Butyl acetate (solvent)	394.80
Tape casting	Carbone black (pore former)	87.60
Electrolyte ink preparation	Butvar-76 (binder)	3.40
	n-Butyl acetate (solvent)	10.15
Anode/electrolyte interlayer ink	Methocel A4M (binder)	22.32
	2-Butoxyethanol (solvent)	12.60
Electrolyte/cathode interlayer ink	Methocel A4M (binder)	22.32
	2-Butoxyethanol (solvent)	12.60
Cathode ink preparation	Methocel A4M (binder)	44.40
	2-Butoxyethanol (solvent)	25.32

Table 3. Energy consumption by the process.

Process Description	Energy Input (MJ/MW)
Anode slurry preparation	40
Anode tape casting	30
Anode/electrolyte interlayer ink	70
Anode/electrolyte interlayer screen printing	60
Drying	1710
Sintering	10,530
Electrolyte ink preparation	140
Screen printing	130
Drying	1710
Electrolyte/cathode interlayer ink	70
Electrolyte/cathode interlayer screen printing	60
Drying	1710
Cathode ink preparation	150
Screen printing	130
Drying	1710
Co-Sintering	8600
Metal forming (for interconnect)	430

Table 4. List of BoP inventory.

Description	Material Type	Material Weight (Kg/MW)	Energy Input (MJ/MW)
Air blower	Steel	10,000.00	235,200
Fuel blower	Steel	10,000.00	2,355,200
Air heat exchanger	Incoloy/Steel	2000.00	49,400
Fuel heat exchanger	Incoloy/Steel	2000.00	49,400
Heater for startup	Steel	5000.00	270,600
Casing	Steel	10,000.00	235,200

Table 5. Waste output during the SOFC manufacture phase.

Waste Output (Type)	Quantity (kg/MW)
CO_2 (air emissions)	432
n-Butyl acetate (evaporated solvent)	444
2-Butoxyethanol (evaporated solvent)	55

The detail results of the SOFC LCI data analysis as generated by Gabi software are provided in Table S1.

2.3. Life Cycle Impact Assessment

The LCIA considers the possible footprint concerning LCI flows using either the problem-oriented or damage-oriented approach within a cause-effect structure. Here, impact categories and impact indicators are selected, and characterization models are specified. LCIAs are mainly used to recognize and assess the degree and importance of the potential environmental effects of the product [1].

In this study, the LCIA indicators related to the Global Warming Potential (GWP), Acidification Potential (AP), Eutrophication Potential (EP), Ozone Depletion Potential (ODP), Human Health Particulate Matter Potential (HHPM), and Human-Toxicity Potential (HTP). Table 6 lists all indicators evaluated and their scale boundary.

Table 6. LCIA indicators and related impact categories.

Indicator	Impact Category	Scale	Characterization Factor
CO_2			
CH_4	GWP	Global	CO_2 equivalent
N_2O			
SO_x	AP	Regional	SO_2 equivalent
NO_x		Local	
NO	EP	Local	N equivalent
NO_2			
CFCs	ODP	Global	CFC 11 equivalent
HCFCs			
PM_{10}	HHPM	Regional	$PM_{2.5}$ equivalent
$PM_{2.5}$		Local	
LC_{50}	HTP	Regional	CTUh
		Local	

Impact indicators are typically characterized using the following equation:

$$\text{Inventory Data} \times \text{Characterization Factor} = \text{Impact Indicators}$$

For GWP, all greenhouse gases are expressed in CO_2 equivalents by multiplying the relevant LCI results by a CO_2 characterization factor and then combining the resulting impact indicators to determine an overall indicator of GWP. The characterization will put these different quantities of chemicals on an equal scale to provide the impact each one has on global warming.

3. Results

The total LCA results for all impact categories for SOFC manufacturing and operation are detailed in Table 7. The ratios of total emissions between the manufacturing phase and operation phase are shown in Figure 4. The emissions related to GWP, AP, and HHPM occur more readily during the operation phase, while they are more prevalent for EP, ODP, and HTP during the manufacturing phase.

Table 7. Total LCA results of impact category and per phase.

SOFC Phase	GWP	AP	EP	ODP	HHPM	HTP
Manufacturing	703,755	2000	77	4.38×10^{-8}	103	3.51×10^{-4}
Operation	1,712,000	3848	63	1.46×10^{-8}	223	5.94×10^{-5}
Total	2,415,755	5848	141	5.84×10^{-8}	326.33	4.10×10^{-4}

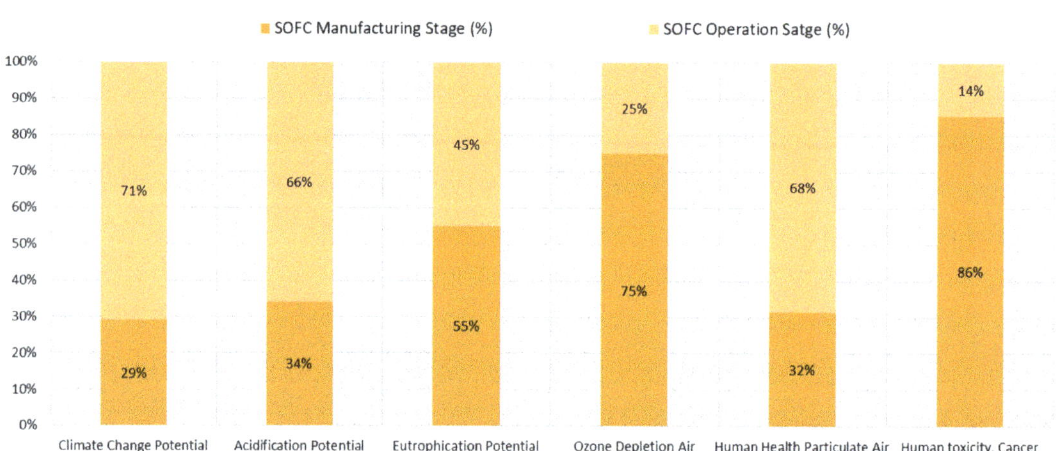

Figure 4. The ratio of emissions between the manufacturing and operation phases of the SOFC.

Figure 5 shows the detailed results of the six selected impact categories during the manufacturing phase of the SOFC. Total GWP for all production stages during the manufacturing phase is equal to 703,755 kg CO_2 eq. The production of the BoP accounts for 81.41% of the total GWP and the fuel blower process for 74.55%. Therefore, the fuel blower is responsible for 60.7% of total GHG emissions produced during the 1 MW SOFC. For AP, the total emissions are equal to 2000 kg SO_2 eq. Out of 24 stages of SOFC manufacturing, two phases of the process account for 70.68%: the fuel blower and slurry preparation, each comprising 40.94% and 29.74%, respectively. The total result for EP is equal to 77 kg N eq. The fuel blower appears to be the stage that most impacts this category at 62.29%. The ODP, the total is 4.38×10^{-8} kg CFC 11 eq. with 92.15% accounted for the slurry preparation stage. In HHPM, the total result is equal to 103 kg $PM_{2.5}$ eq. The slurry preparation and fuel blower account for 66.21%, where 39.58% is for slurry preparation and 26.63% for the fuel blower production stage. And finally, the HTP results are equal to 3.51×10^{-4}, where fuel blower and slurry preparations account for 35.60% and 18.39%, respectively.

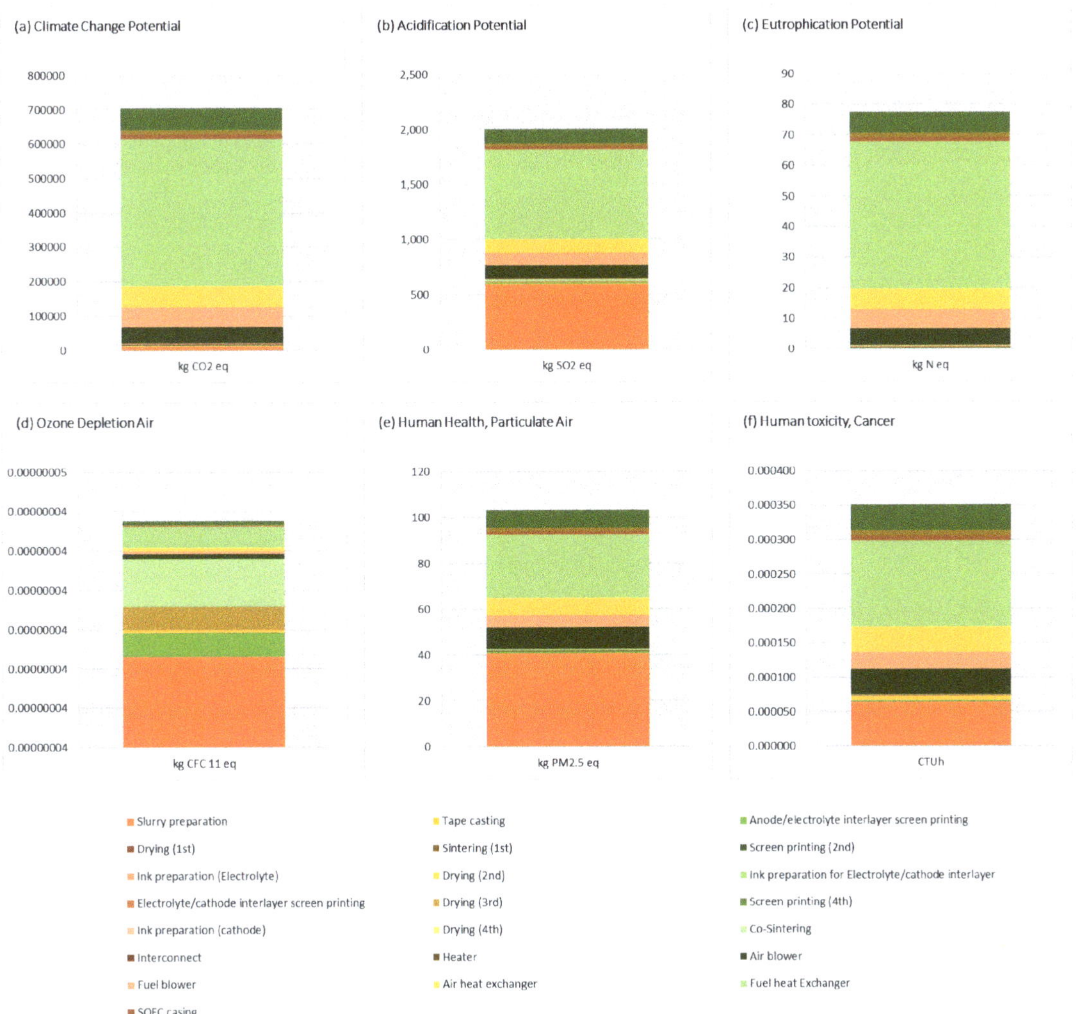

Figure 5. LCA results for six impact categories for the process during the SOFC manufacturing phase.

3.1. Global Warming Potential (GWP)

The total climate change emissions for the life span of 10 years or 80,000 h of 1 MW SOFC is 2,415,755 Kg CO_2 eq., where 71% is emitted during the operation phase due to the usage of natural gas as fuel for the SOFC. Generating 1 MW of electricity will require 7 MM Btu of natural gas per hour and an operation spanning 80,000 h—this will emit 1.712 MM kg CO_2 eq.

The manufacturing phase accounts for the remaining emissions, 703,755 kg CO_2 eq., a quantity generated mainly from services and goods used to produce a 1 MW SOFC system. Figure 6, which shows the 24 different processes in the manufacturing phase, proves that most GWP is generated from only eight processes—mainly because of electricity use.

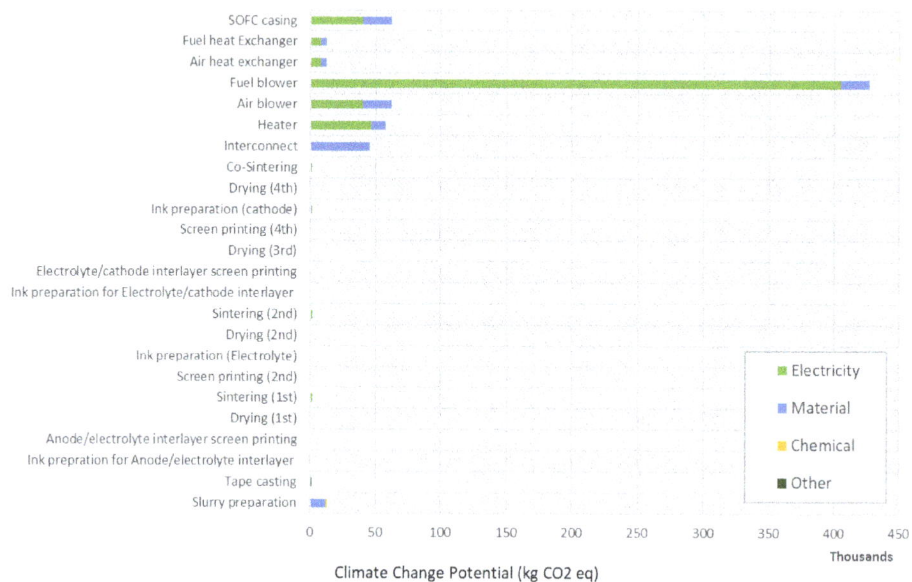

Figure 6. Climate Change Potential (kg CO_2 eq.) for all manufacturing stages.

Most processes consuming electricity, and thus the most influential factor in GWP, is the fuel blower process. The material is the most significant source of GWP during the interconnect process. Additionally, Figure 7 shows the primary three services or goods used during the manufacturing phase, and they are scaled for easy comparison.

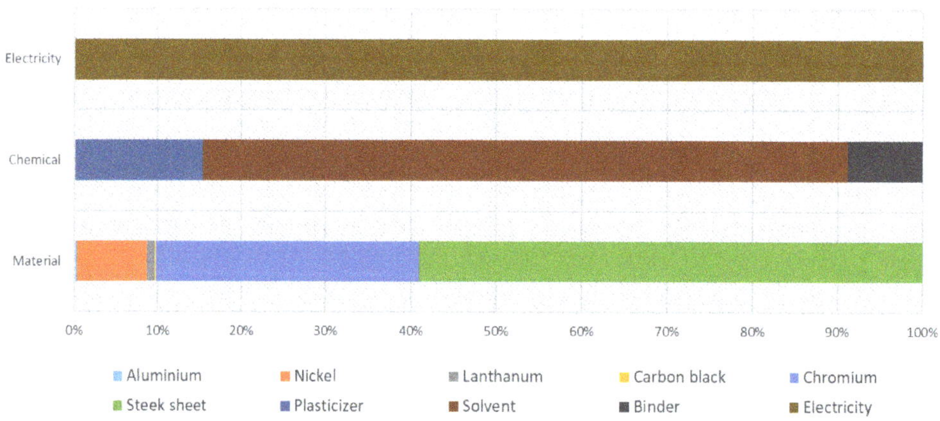

Figure 7. GWP output for a different type of service or goods.

The chemical most impactful on GWP is the solvent, with more than 70% of GWP generated due to using chemicals in the manufacturing phase, and the least is the binder, with 15%. The material that most contributes to GWP is the steel sheet, with almost 60%, followed by the Chromium, with 30%. The remaining 10% of material usage is mainly nickel.

The SOFC consists of four main parts, where each one is manufactured as a standalone piece before being assembled. Figure 8 shows these four main parts and their contributions to the GWP.

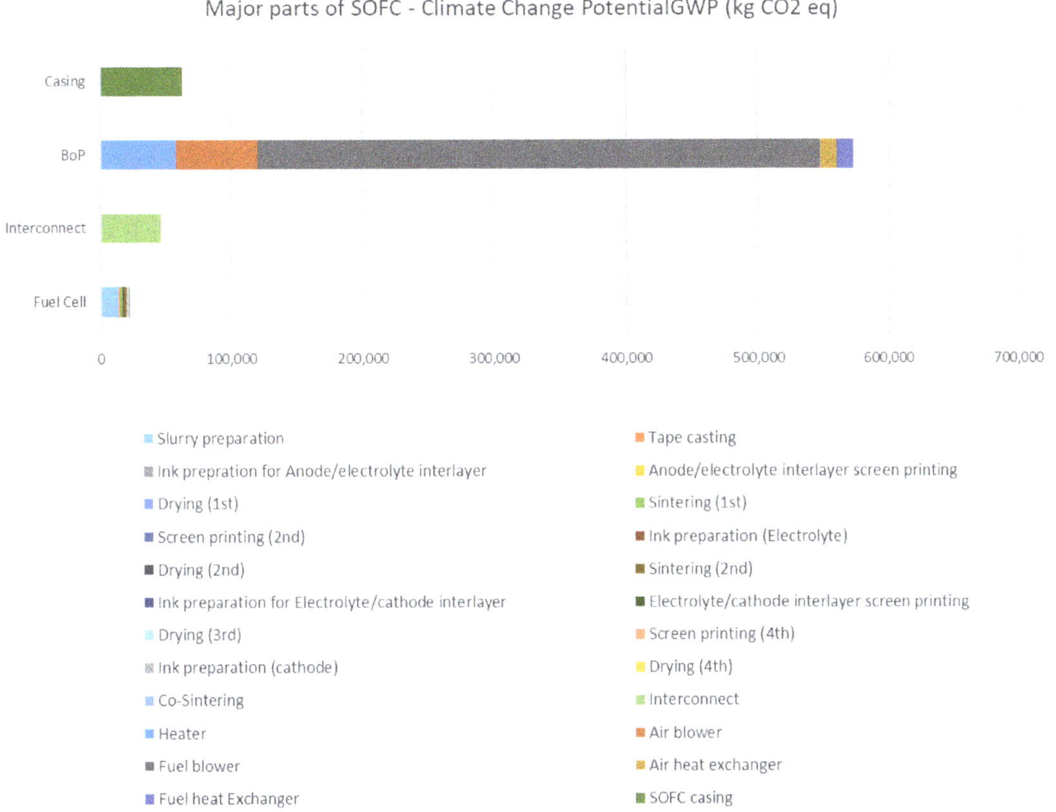

Figure 8. GWP for significant parts of SOFC manufacturing.

The BoP is the component of SOFC that generates the highest GWP at 572,920 kg CO_2 eq, which is 81% of the total GWP of the manufacturing phase, and the fuel blower is the aspect of the BoP that most contributes to the GWP at 427,100 kg CO_2 eq. This represents 60% of the emissions from the manufacturing phase. Following BoP is the casing, then interconnect, and the last is the fuel cell itself with 22,498 kg CO_2 eq., which represents less than 1% of the total GWP of the SOFC.

Another potential analysis is the primary type of process for the fuel cell. Figure 9 shows these processes and contributions of each one of them concerning the GWP. The ink preparation is the type of process that contributes the most to the GWP with 16,649 kg CO_2 eq. It accounts for 75% of GWP's fuel cell manufacturing. Slurry preparation for the anode is 82% of all GWP generated from the ink preparation process of fuel cell manufacturing. The type of process in fuel cell manufacturing that provides the second-largest contribution of GWP is the sintering process, followed by drying—each accounting for 22% and 5%, respectively.

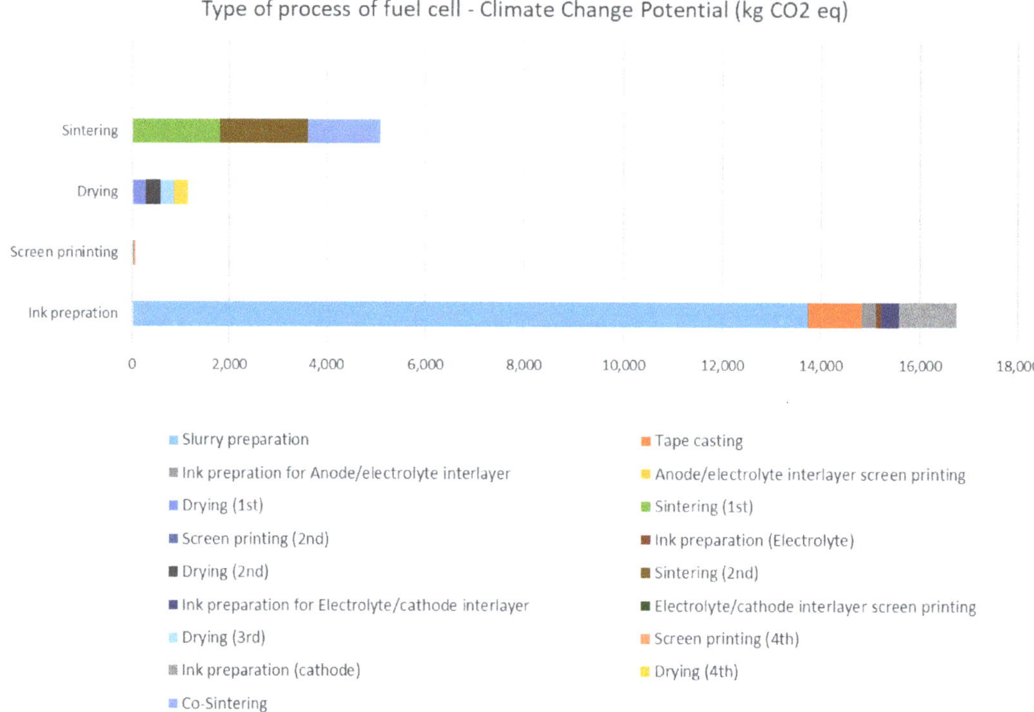

Figure 9. Type of process of the fuel cell manufacturing.

The average hourly GHG of the 1 MW SOFC is approximately 30 kg CO_2 eq. /MWh. By contrast, traditional power generation, like Natural Gas Combined Cycle (NGCC), is between 417 kg CO_2 eq. /MWh and 557 kg CO_2 eq./MWh [32,33]. There is a difference in the LCA life span of each technology. The SOFC life span lasts ten years, while the NGCC life span lasts approximately 30 years. By replacing the SOFC each year, the total GHG emissions will be around 90 kg CO_2 eq. /MWh, which remains below the GHG emissions of the NGCC. Thus, SOFC technology produces 80% fewer GHGs than traditional power generations.

3.2. Sensitivity Analysis

The process contributing most substantially to the GWP in the manufacturing phase is the fuel blower, which accounts for 60% of emissions generated during the manufacturing phase. The fuel blower is mainly used to increase the fuel pressure, in this case, methane, to meet the SOFC operating pressure of approximately 7 bar. However, suppose the SOFC is being used in a gas processing plant. In that case, the fuel gas compressor is already available, and the 7-bar pressure exists in the plant, leaving no need for a separate or dedicated compressor for the SOFC fuel. Eliminating the need for the fuel blower from BoP assembly of the SOFC unit will reduce the overall GWP of SOFC by 17%, and the total emissions will be below 2 MM kg CO_2 eq. The percentage GWP during the manufacturing phase dropped from 29% to 14%, while the operation GWP remains the same at 1.712 MM kg CO_2 eq.

Geography and location play a role in the SOFC manufacturing process—indeed, the LCA of the SOFC found that the impact on total climate change is based on the electricity mix used in each location. In general, the effect is minimal and not particularly significant in the manufacturing phase. However, there is a substantial difference in GWP from one country to another during the operation phase. This is mainly caused by natural gas, used as a fuel for SOFC during the operation phase to generate the 1 MW power. Figure 10 shows the total GWP of 1 MW SOFC in four different countries.

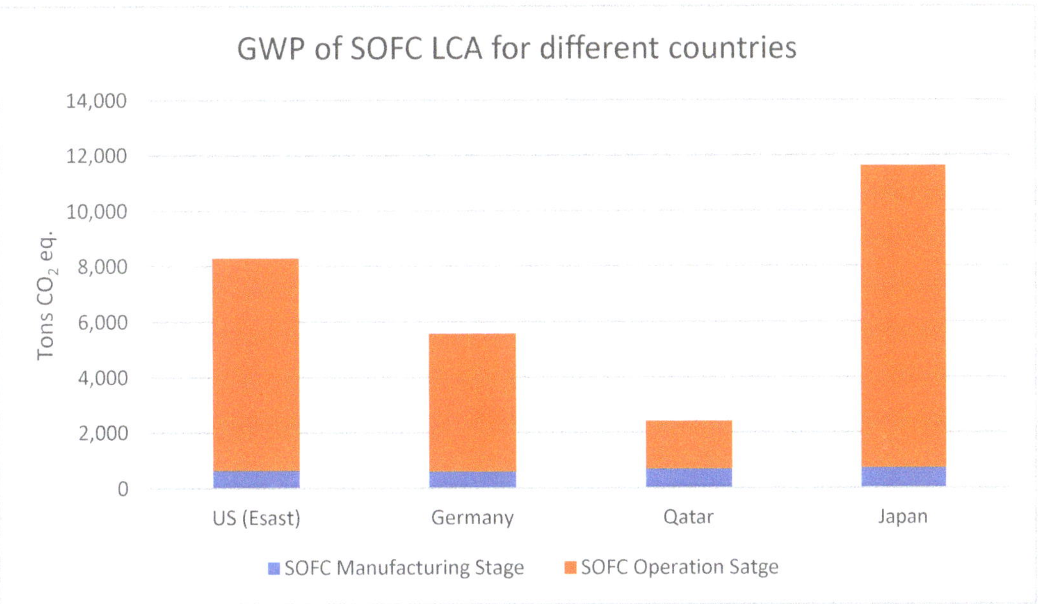

Figure 10. Total GWP of 1 MW SOFC in four different countries.

Because of natural gas resources, Qatar has the least GWP compared to other countries like Germany, the US, and Japan. The natural gas in Germany is mainly imported via pipelines from Russia, so the transportation of such resources is added to its total GWP. Similarly, in the US, the pipeline network of natural gas spread across the country. Japan, which has an enormous GWP impact, gets its natural gas requirement from ships and overseas tankers.

4. Conclusions

This study's primary objectives were to understand better the impact of SOFC integration on the natural gas processing plant in terms of environmental values and provide a basis for comparing different types of integration approaches in gas processing plants.

The operational phase of the SOFC has the most significant impact on global warming potential (GWP), acidification potential (AP), and human health particulate matter (HHPM). In contrast, the effect is higher during the manufacturing phase for eutrophication potential (EP), ozone depletion (ODP), and human-toxicity potential (HTP).

In summary, 1 MW SOFC used in gas processing plants for 80,000 total running hours (10 years) will have the following impact category:

- The total GWP is 2,415,755 kg CO$_2$ eq. with 29% during the manufacturing phase.
- Total AP is 5848 kg SO$_2$ eq. with 34% during manufacturing.
- Total EP is 141 kg N eq. with 55% during manufacturing.
- Total Ozone Depletion Air is 5.84×10^{-8} kg CFC 11 eq. with 75% during manufacturing.
- Total Human Health Particulate Air is 326 kg PM$_{2.5}$ eq. with 32% during manufacturing.
- Total Human Toxicity, Cancer is 4.10×10^{-4} CTUh with 86% during manufacturing.

The study results are supported by similar studies where the ratio between the manufacturing phases is almost 30 to 70 in the operation phase for GWP. The GWP during the manufacturing stage for the Qatar case is almost like manufacturing cases in other countries like the US, Germany, and Japan. It is a little higher than the US and Germany but less than Japan, which could be due to the availability and transportation of raw materials. However, for the operation phase, the difference is huge between Qatar and the other three countries. This is mainly due to fuel transportation which in the case of Qatar it is the lowest environmental impact. In addition, there is potential to reduce the total emissions produced by the SOFC if specific processes with the highest impact on climate change potential can be eliminated. For example, pressurized methane is already available in typical natural gas processing plants. Thus, an advantage of using SOFC in gas plants and particularly in Qatar by eliminating the requirement of having an additional fuel blower as part of SOFC assembly will save approximately 17% of the total GHG of SOFC LCA.

The hourly GHG released into the atmosphere for each 1 MW of electricity generated using SOFC is approximately 30 kg of CO_2 eq. Comparing CO_2 eq. to the emissions emitted from traditional power generation like Natural Gas Combined Cycle (NGCC), the difference is more than 80%. And this proves that fossil-fuel power generation's impact on the environment depends on the technology used. As fossil fuel remains the primary energy source for at least a few decades to come, such a study of LCA can determine the best technology that has less impact on the environment and has lower GWP.

Qatar is a small country with the largest natural gas resources and exports, which gives it a high dependence on natural gas for its revenues and a high GHG emissions per capita. Therefore, SOFC integration can lead to significant country-wide reductions in emissions improvements to efficiency.

The findings of this study are expected to serve as a reference for future researchers interested in the integration of SOFCs into natural gas processing plants and provide decision-makers with reliable quantitative analysis and data.

Supplementary Materials: The following are available online at https://www.mdpi.com/article/10.3390/en14154668/s1, Figure S1: SOFC manufacturing steps with quantities, Table S1: SOFC LCI data analysis.

Author Contributions: Conceptualization, K.A.-K.; Data curation, K.A.-K.; Formal analysis, K.A.-K.; Methodology, S.G.A.-G.; Supervision, M.K.; Validation, S.B.; Visualization, K.A.-K.; Writing—original draft, K.A.-K.; Writing—review & editing, S.G.A.-G. All authors have read and agreed to the published version of the manuscript.

Funding: This research received no external funding.

Informed Consent Statement: Not applicable.

Acknowledgments: The authors would like to acknowledge the support of Hamad Bin Khalifa University, Qatar.

Conflicts of Interest: The authors declare no conflict of interest.

Nomenclature

ADP	Abiotic depletion potential
AOP	Area of production
AP	Acidification potential
APU	Auxiliary power unit
BoP	Balance of plants
CH_4	Methane
CO	Carbon monoxide
CO_2	Carbone dioxide
CTUh	Comparative Toxic Unit for human
EP	Eutrophication potential
FU	Functional unit
GHG	Greenhouse gases
GT	Gas turbine
GTL	Gas-to-liquid
GWP	Global warming potential
H_2	Hydrogen
H_2O	Water
H_2S	Hydrogen sulfide
HHPM	Human Health Particulate Matter Potential
HTP	Human-Toxicity Potential
IPCC	Intergovernmental Panel on Climate Change
ISO	International Standard Organization
Kw	Kilowatt
kWh	Kilowatt-hour
LC_{50}	Lethal concentration required to kill 50% of the population
LCA	Life cycle assessment
LCI	Life cycle impact
LCIA	Life cycle impact assessment
LSM	Lanthanum strontium manganite
MGT	Micro gas turbine
MW	Megawatt
N_2	Nitrogen
N_2O	Nitrous oxide
NGCC	Natural gas combined cycle
Ni	Nickel
NiO	Nickel oxide
NMHCs	Nonmethane hydrocarbons
NO_x	Nitrogen oxides
ODP	Ozone depletion potential
PEP	Product environmental profile
PM	Particulate matter
POCP	Photochemical ozone creation potential
Pt	Platinum
PV	Photovoltaic
SETAC	Society for Environmental Toxicology and Chemistry
SO_2	Sulfur dioxide
SOFC	Solid oxide fuel cell
SO_x	Sulfur oxides
VOC	Volatile organic compounds
YSZ	Yttria-stabilized-zirconia

References

1. Yu, K.M.K.; Curcic, I.; Gabriel, J.; Tsang, S.C.E. Recent Advances in CO_2 Capture and Utilization. *ChemSusChem* **2008**, *1*, 893–899. [CrossRef]
2. Weldu, Y.W.; Assefa, G. Evaluating the environmental sustainability of biomass-based energy strategy: Using an impact matrix framework. *Environ. Impact Assess. Rev.* **2016**, *60*, 75–82. [CrossRef]

3.   Langlois, E.V.; Campbell, K.; Prieur-Richard, A.H.; Karesh, W.B.; Daszak, P. Towards a better integration of global health and biodiversity in the new sustainable development goals beyond Rio+20. *Ecohealth* **2012**, *9*, 381–385. [CrossRef]

4.   Jewell, J. Ready for nuclear energy?: An assessment of capacities and motivations for launching new national nuclear power programs. *Energy Policy* **2011**, *39*, 1041–1055. [CrossRef]

5.   Al-Khori, K.; Bicer, Y.; Koc, M. Integration of Solid Oxide Fuel Cells into Oil and Gas Operations: Needs, Opportunities, and Challenges. *J. Clean. Prod.* **2020**, *245*, 118924. [CrossRef]

6.   Grubert, E. Implicit prioritization in life cycle assessment: Text mining and detecting metapatterns in the literature. *Int. J. Life Cycle Assess.* **2016**, *22*, 148–158. [CrossRef]

7.   Mehmeti, A.; Mcphail, S.J.; Pumiglia, D.; Carlini, M. Life cycle sustainability of solid oxide fuel cells: From methodological aspects to system implications. *J. Power Sources* **2016**, *325*, 772–785. [CrossRef]

8.   Baratto, F.; Diwekar, U.M. Life cycle assessment of fuel cell-based APUs. *J. Power Source* **2010**, *139*, 188–196. [CrossRef]

9.   Herron, S. *Life Cycle Assessment of Residential Solid Oxide Fuel Cells Life*; Arizona State University: Tempe, AZ, USA, 2012; Available online: https://repository.asu.edu/attachments/82943/content/ASU_SSEBE_CESEM_2012_CPR_001.pdf (accessed on 7 July 2021).

10.   Lee, Y.D.; Ahn, K.Y.; Morosuk, T.; Tsatsaronis, G. Environmental impact assessment of a solid-oxide fuel-cell-based combined-heat-and-power-generation system. *Energy* **2015**, *79*, 455–466. [CrossRef]

11.   Guine, J.B. Selection of impact categories and classification of LCI results to impact categories. In *Life Cycle Impact Assessment*; Springer: Berlin/Heidelberg, Germany, 2015. [CrossRef]

12.   Buchgeister, J. *Comparison of Sophisticated Life Cycle Impact Assessment Methods for Assessing Environmental Impacts in a LCA Study of Electricity Production*; Firenze University Press: Firenze, Italy, 2015.

13.   Hart, H.; Brandon, N.; Shemilt, J. The environmental impact of solid oxide fuel cell manufacturing. *Fuel Cells Bull.* **1999**, *2*, 4–7. [CrossRef]

14.   Jakob, M.; Hirschberg, S. *Total Greenhouse Gas Emissions and Costs of Alternative Swiss Energy Supply Strategies*. 2001. Available online: https://www.researchgate.net/profile/Martin-Jakob-2/publication/253789416_Total_greenhouse_gas_emissions_and_costs_of_alternative_Swiss_energy_supply_strategies/links/00b7d529e481e28e68000000/Total-greenhouse-gas-emissions-and-costs-of-alternative-Swiss-energy-supply-strategies.pdf (accessed on 17 July 2021).

15.   Cánovas, A.; Zah, R.; Gassó, S. Comparative Life-Cycle Assessment of Residential Heating Systems, Focused on Solid Oxide Fuel Cells. *Sustain. Energy Build.* **2013**, 659–668. [CrossRef]

16.   Staffell, I.; Ingram, A.; Kendall, K. Energy and carbon payback times for solid oxide fuel cell based domestic CHP. *Int. J. Hydrogen Energy* **2012**, *37*, 2509–2523. [CrossRef]

17.   Strazza, C.; Del Borghi, A.; Costamagna, P.; Traverso, A.; Santin, M. Comparative LCA of methanol-fuelled SOFCs as auxiliary power systems on-board ships. *Appl. Energy* **2010**, *87*, 1670–1678. [CrossRef]

18.   National Renewable Energy Laboratory; Mann, M.K.; Whitaker, M.; Driver, T.; Mueller, M.; Franco, G.; Spiegel, L.; Hope, L.; Oglesby, R. *Life Cycle Assessment of Existing and Emerging Distributed Generation Technologies in California*; California Energy Commission: Sacramento, CA, USA, 2011. Available online: https://citeseerx.ist.psu.edu/viewdoc/download?doi=10.1.1.414.9181&rep=rep1&type=pdf (accessed on 17 July 2021).

19.   Sadhukhan, J. Distributed and micro-generation from biogas and agricultural application of sewage sludge: Comparative environmental performance analysis using life cycle approaches. *Appl. Energy* **2014**, *122*, 196–206. [CrossRef]

20.   Lin, J.; Babbitt, C.W.; Trabold, T.A. Life cycle assessment integrated with thermodynamic analysis of bio-fuel options for solid oxide fuel cells. *Bioresour. Technol.* **2013**, *128*, 495–504. [CrossRef]

21.   Nease, J.; Adams, T.A. Comparative life cycle analyses of bulk-scale coal-fueled solid oxide fuel cell power plants. *Appl. Energy* **2015**, *150*, 161–175. [CrossRef]

22.   Nease, J.; Ii, T.A.A. Life Cycle Analyses of Bulk-Scale Solid Oxide Fuel Cell Power Plants and Comparisons to the Natural Gas Combined Cycle. *Can. J. Chem. Eng.* **2015**, *93*, 1349–1363. [CrossRef]

23.   Reenaas, M. Solide Oxide Fuel Cell Combined with Gas Turbine Versus Diesel Engine as Auxiliary Power Producing Unit. Master's Thesis, Faculty of Information Technology Mathematics and Electrical Engineering, Norwegian University of Science and Technology, Oslo, Norway, 2005.

24.   Rebitzer, G.; Ekvall, T.; Frischknecht, R.; Hunkeler, D.; Norris, G.; Rydberg, T. Life cycle assessment Part 1: Framework, goal and scope definition, inventory analysis, and applications. *Environ. Int.* **2004**, *30*, 701–720. [CrossRef] [PubMed]

25.   Wincewicz, K.C.; Cooper, J.S. Taxonomies of SOFC material and manufacturing alternatives. *J. Power Sources* **2005**, *140*, 280–296. [CrossRef]

26.   Smith, L.; Ibn-mohammed, T.; Yang, F.; Reaney, I.M.; Sinclair, D.C.; Koh, S.C.L. Comparative environmental profile assessments of commercial and novel material structures for solid oxide fuel cells. *Appl. Energy* **2019**, *235*, 1300–1313. [CrossRef]

27.   Richards, M. *Solid Oxide Fuel Cell Manufacturing Overview*; 2011. Available online: https://www.energy.gov/sites/prod/files/2014/03/f12/mfg2011_iia_richards.pdf (accessed on 17 July 2021).

28.   Gandiglio, M.; Sario, F.D.; Lanzini, A.; Bobba, S.; Santarelli, M.; Blengini, G.A. Life Cycle Assessment of a Biogas-Fed Solid Oxide Fuel Cell (SOFC) Integrated in a Wastewater Treatment Plant. *Energies* **2019**, *12*, 1611. [CrossRef]

29. Scataglini, R.; Mayyas, A.; Wei, M.; Chan, S.H.; Lipman, T.; Gosselin, D.; D'Alessio, A.; Breunig, H.; Colella, W.G.; James, B.D. A Total Cost of Ownership Model for Solid Oxide Fuel Cells in Combined Heat and Power and Power-Only Applications. In *Solid Oxide Fuel CellManufacturing Overview, Proceedings of the Hydrogen and Fuel Cell Technologies Manufacturing R&D Workshop, Washington, DC, USA, 11–12 August 2011*; Springer: Washington, DC, USA, 2015.
30. Birnbaum, K.U.; Zapp, P. *Solid Oxide Fuel Cells, Sustainability Aspects*; Springer: New York, NY, USA, 2013. [CrossRef]
31. Carlson, E.J. *Solid Oxide Fuel Cell Manufacturing Cost Model: Simulating Relationships between Performance, Manufacturing, and Cost of Production*. 2004, pp. 1–73. Available online: https://www.researchgate.net/publication/236512323_SOLID_OXIDE_FUEL_CELL_MANUFACTURING_COST_MODEL_SIMULATING_RELATIONSHIPS_BETWEEN_PERFORMANCE_MANUFACTURING_AND_COST_OF_PRODUCTION (accessed on 17 July 2021).
32. Yin, L.; Liao, Y.; Zheng, K.; Liu, J. Comparative analysis of gas and coal-fired power generation in ultra-low emission condition using life cycle assessment (LCA). *IOP Conf. Ser. Mater. Sci. Eng.* **2017**, *199*, 012054. [CrossRef]
33. Meng, F.; Dillingham, G. Life Cycle Analysis of Natural Gas-Fired Distributed Combined Heat and Power versus Centralized Power Plant. *Energy Fuels* **2018**, *32*, 11731–11741. [CrossRef]

Article

Optimal Sizing of Hybrid Wind-Solar Power Systems to Suppress Output Fluctuation

Abdullah Al-Shereiqi [1], Amer Al-Hinai [1,*], Mohammed Albadi [1] and Rashid Al-Abri [1,2]

[1] Department of Electrical and Computer Engineering, Sultan Qaboos University, P.O. Box 33, Al-Khodh, Muscat 123, Oman; s53299@student.squ.edu.om (A.A.-S.); mbadi@squ.edu.om (M.A.); arashid@squ.edu.om (R.A.-A.)

[2] Sustainable Energy Research Center, Sultan Qaboos University, P.O. Box 33, Al-Khodh, Muscat 123, Oman

* Correspondence: hinai@squ.edu.om

Abstract: Harnessing wind energy is one of the fastest-growing areas in the energy industry. However, wind power still faces challenges, such as output intermittency due to its nature and output reduction as a result of the wake effect. Moreover, the current practice uses the available renewable energy resources as a fuel-saver simply to reduce fossil-fuel consumption. This is related mainly to the inherently variable and non-dispatchable nature of renewable energy resources, which poses a threat to power system reliability and requires utilities to maintain power-balancing reserves to match the supply from renewable energy resources with the real-time demand levels. Thus, further efforts are needed to mitigate the risk that comes with integrating renewable resources into the electricity grid. Hence, an integrated strategy is being created to determine the optimal size of the hybrid wind-solar photovoltaic power systems (HWSPS) using heuristic optimization with a numerical iterative algorithm such that the output fluctuation is minimized. The research focuses on sizing the HWSPS to reduce the impact of renewable energy resource intermittency and generate the maximum output power to the grid at a constant level periodically based on the availability of the renewable energy resources. The process of determining HWSPS capacity is divided into two major steps. A genetic algorithm is used in the initial stage to identify the optimum wind farm. A numerical iterative algorithm is used in the second stage to determine the optimal combination of photovoltaic plant and battery sizes in the search space, based on the reference wind power generated by the moving average, Savitzky–Golay, Gaussian and locally weighted linear regression techniques. The proposed approach has been tested on an existing wind power project site in the southern part of the Sultanate of Oman using a real weather data. The considered land area dimensions are 2 × 2 km. The integrated tool resulted in 39 MW of wind farm, 5.305 MW of PV system, and 0.5219 MWh of BESS. Accordingly, the estimated cost of energy based on the HWSPS is 0.0165 EUR/kWh.

Keywords: optimal layout; wake effect; fluctuation; wind farm; ramping rate

Citation: Al-Shereiqi, A.; Al-Hinai, A.; Albadi, M.; Al-Abri, R. Optimal Sizing of Hybrid Wind-Solar Power Systems to Suppress Output Fluctuation. *Energies* **2021**, *14*, 5377. https://doi.org/10.3390/en14175377

Academic Editors: Sergio Ulgiati and Adrian Ilinca

Received: 23 July 2021
Accepted: 24 August 2021
Published: 30 August 2021

1. Introduction

In most countries, the demand for electricity is growing rapidly. One of the challenges in the electricity sector is to meet this demand while supplying customers with reliable and stable power simultaneously. Conventional power resources are being supplemented with renewable resources. Most of the power suppliers depend on fossil fuels as the primary energy source due to their ready availability and lower cost compared to other resources. However, the increase in demand, along with increased oil and gas production costs, drive the use of other energy resources [1]. In addition, political integration via a common energy policy or climate-change mitigation is one of the motives to use renewable energy sources. According to the International Renewable Energy Agency (IRENA) report, most of the investment in renewable resources is in wind and solar resources; in 2018, around 80% of the investment in renewable energy was in these two resources.

The main challenges of utilizing renewable resources are the high capital cost and the fluctuations of wind and solar power output. However, the recent development of renewable energy technologies shows a declining trend in cost. There have also been advancements in the integration of renewable resources into the existing conventional power resources [2]. This requires the mitigation of the vulnerabilities imposed on the grid through the intermittent nature of these resources. Variability and ramp events in power output are the key challenges for system operators due to their impact on the system in both the long and short term [3]. These impacts include, but are not limited to, system balancing, reserve management, scheduling, and the commitment of generation units.

Previous research has examined several sizing approaches to find the ideal size of hybrid plants that include wind, solar, and battery storage. Most prior research focused on improving the scale of hybrid wind-solar photovoltaic power systems (HWSPS) by lowering costs while ensuring the necessary degree of power supply dependability. Smoothing techniques can be divided into two groups. The first group uses energy storage systems such as flywheels, batteries, and capacitors. For instance, a 51 MW wind farm system can be stabilized using a 34 MW battery [4]. The second group is based on power curtailment strategies such as pitch, inertia, and DC link voltage controls [5]. According to the literature, the battery-energy storage system (BESS) is widely used with renewable resources to resolve the fluctuation issue [6]. The choice of smoothing sources depends on the difference between the actual signal and the smoothed signal. The smoothed signal is defined in terms of the reference wind power. The reference signal is the key to determining the supportive resources needed to satisfy the system reliability. Different smoothing techniques are used to generate the reference signal, including moving average [7], wavelet decomposition [8], Gaussian [9], Savitzky–Golay [10], and low pass filter [11] techniques.

Different studies have implemented the smoothing techniques using the BESS. For example, the authors of [12] used the BESS to meet the reference signal of the wind power generated by the moving average (MAV) technique. Kim et al. [13] used wavelet decomposition to obtain the smoothed signal of wind power, where hybrid storage combining an ultra-capacitor and BESS is used as the smoothing source. A simple MAV technique and low pass filter are used in [14] to control the BESS in order to stabilize the PV output power. However, in all these studies, fixed wind farm (WF) and PV plant sizes are used. Most of the studies smooth the fluctuation using the minimum BESS size. As an example, the optimization in [15,16] was conducted to minimize the BESS size while reducing wind power intermittency. It is important to investigate different methods of achieving stable output power using the optimal source sizes.

The investigation performed in [17] shows that the Gaussian technique is more effective than the MAV for smoothing the wind and solar generation to an acceptable ramping rate level. The authors used solar and wind plants as the primary sources and used the BESS for smoothing. In [18], the Savitzky–Golay (SG), MAV, and Gaussian techniques are used with the BESS to mitigate PV output fluctuation. The results show a smoother output power with the SG than with the other algorithms. The authors of [19] used a wavelet decomposition method while using an ultra-capacitor with the BESS to prolong the BESS's lifetime.

According to the literature, MAV is widely used to smooth noisy signals [20]. The BESS is operated to make up the difference between the actual signal and the MAV signal. However, the MAV approach depends on past time series data, which are different from the current value of the fluctuating variable. The problem of the MAV and most other techniques is the memory effect feature, meaning that the approach depends on data from the past [21]. This means that the BESS is operated excessively, which shortens its lifespan. Many factors affect the size of the needed memory in the used smoothing technique. For instance, the window size and the number of the samples in the window with the type of data determine the required memory. The aforementioned factors in the memory could also cause over-smoothing, which ultimately increases the required BESS capacity [22]. The optimal sizing of the HWSPS while utilizing the MAV technique has already been investigated in our previous study [23]. To avoid the aforementioned factors used for the MAV technique, our research [23] investigates other smoothing techniques.

Hence, in addition to the MAV method, this study considers the locally weighted linear regression, Gaussian, and SG techniques. No previous studies have investigated the use of locally weighted linear regression or SG for smoothing the wind power fluctuations. The developed tool is suitable for a site with ample solar irradiation and copious wind resources to take advantage of the complimentary nature of both the PV and wind resources.

The wake impact is not taken into account in any of the HWSPS sizing studies. This study is unique in that it integrates a model of the wake effect into the size of the HWSPS in order to decrease wind power losses due to turbine layout. Furthermore, instead of using the load demand profile to create an HWSPS, the MAV, locally weighted linear regression, Gaussian, and SG filters/techniques are utilized to provide a smooth reference power within the operational ramping rates. The integrated tool is designed to use the selected site effectively for a hybrid renewable power plant in a grid-connected mode. This means this approach attempts to utilize the site effectively to generate maximum output power to the grid at a constant level periodically based on the availability of the renewable energy resources. The PV plant and BESS are sized to provide the reference power generated. As a result, an integrated strategy is created that combines a genetic algorithm with a numerical iterative technique to determine the appropriate size of the HWSPS.

The reminder of this paper is organized as follow. Section 2 describes the sizing methodology to mitigate the wind output power fluctuation. Section 3 presents the smoothing techniques to obtain a smooth signal. Section 4 shows the data needed to run the proposed case study. The results and discussions are highlighted in Section 5. The importance of the wake effect is described in Section 6, while Section 7 comprises the effect of the contribution factor on sizes of the PV plant and BESS. Finally, the conclusions based on the results are discussed in Section 8.

2. Sizing Methodology

Wind farm sizing is a complex optimization problem that cannot be solved with traditional optimization methods. Therefore, GA is used to solve the wind farm layout optimization problem. The stochastic and intermittent nature of wind speed contributes to the intermittency of wind power. Therefore, many studies focus on estimating wind speed. However, the wake effect that occurs among the wind turbines is also a critical factor that must be considered when designing wind farms. Jensen's wake effect model [24] is used in the wind farm layout optimization (WFLO) problem. Jensen's model is one of the recommended models with a strong performance [25,26]. Hence, this model is the one used for further analysis in this study, and its mathematical model is described in detail in [27].

To obtain the optimal wind farm size, the optimization problem is modeled in Equations (1)–(4).

$$Min(COE) = Min\left[\frac{N\left(\frac{2}{3} + \frac{1}{3}e^{(-0.00174\,N^2)}\right)}{\sum_{\tau=0}^{360^\circ}\sum_{i=1}^{N}[P_i[(x,y),v_i,\theta_\tau)]\,P_r(v_i,\theta_\tau)]}\right] \tag{1}$$

subject to

$$0 \le x_k \le l \quad \& \quad 0 \le y_k \le w \quad \forall k \in [i,n], \ l,w = 2000 \tag{2}$$

$$D_{in} = \sqrt{[x_i - x_n]^2 + [y_i - y_n]^2} \ge (5 \times D_{r_0}) \quad x_i,y_i,x_n,y_n \in S, \ \forall i,n = 1,...., N, \ i \ne n \tag{3}$$

$$N > 0 \tag{4}$$

where

$$v_i = v_0\left[1 - \sqrt{\sum_{i=1}^{N}(1 - v/v_0)^2}\right] \tag{5}$$

The objective function given in Equation (1) is to maximize the output power and minimize the costs. In other words, it is mainly to minimize the cost of energy. The major goal of this objective function is to find the optimal wind turbine layout by introducing Jensen's wake effect model. The numerator of Equation (1) considers economy of scale that is directly depending on the number of the wind turbines (N). Moreover, the denominator of the objective function represents the total power produced. The power output of each wind turbine depends on its location within the farm (x, y), and the probability of occurrence (P_r) of a scenario of a combination of wind speed (v) and direction angle (θ). Jensen's wake model given by Equation (5) is applied if there is wake effect among the turbines; otherwise, free stream wind speed v_0 is used. The optimization problem constraints included the upper and lower limits of the wind farm terrain as defined in Equation (2). This is defined by the width (w) and length (l) of the selected site to limit the locations of the turbines (x, y) within the wind farm site. In addition, the spacing constraint of five times rotor diameters (D_{r_0}) between any two wind turbines using the Euclidean distance formula is given in Equation (3). The other inequality constraint is the minimum number of wind turbines (N), as presented in Equation (4). The genetic algorithm is used to solve the optimization problem (i.e., 3000 generations, 600 populations size). When the improvement in the fitness value falls below a certain threshold for a number of consecutive steps, or when the maximum number of iterations is achieved, the optimization process is terminated. According to the available literature, the objective function and the parameters of Grady et al.'s [28] study are widely used as a benchmark. Therefore, the objective function in Grady et al.'s study was used to cross-check the developed wake model and the formulated optimization problem [27].

The major goal of this research is to size a wind farm, which serves as the foundation for calculating the sizes of the PV plant and BESS. The optimum wind farm size is determined using a genetic algorithm [29]. The methodology is determined by the size of the chosen site, the position of the turbines, and the wind speeds and directions. After determining the wind farm size, the PV plant size and BESS capacity are identified as smoothing sources. Different smoothing techniques are used as described in the following section to generate reference power. A numerical iterative algorithm (NIA) is used to determine the PV plant size and the BESS capacity following the methodology and the evaluation criteria explained in our previous study [23] by deploying the contribution factor. In the proposed NIA, each PV module's output power was estimated based on the historical data of solar irradiance and temperature. Next, the size of the PV system was calculated by involving a contribution factor (S) with a value between 0 and 1, in steps of 0.02. A search space range was established by the contribution factor. When S = 0, no PV power is required; when S = 1, the PV plant's power generated equals $P_{ref}(t)$. The BESS is sized depending on the cumulative net energy after obtaining the PV system size. The BESS is used as a secondary source for smoothing, reducing capacity and lowering system costs.

The integrated approach's main goal is to size the HWSPS such that it is both cost-effective and dependable. The planned HWSPS is evaluated using cost of energy (COE) as a major performance indicator. The COE formula takes into account all of the components' capital costs, operating and maintenance expenses, replacement cost, and salvage cost. The loss of the power supply probability (LPSP) is utilized for techno-economic assessment and comparison in this study. The LPSP is defined as the likelihood that the supply would be unable to meet demand, and its value varies from zero to one. The flow chart in Figure 1 shows the phases of the suggested strategy for scaling the HWSPS. The procedure of defining the HWSPS capacity is divided into two major steps. In the first step, an ideal wind farm is determined using the evolutionary algorithm, subject to site dimensions and turbine spacing, while Jensen's wake effect model is used to reduce power losses caused by wind turbines layouts. Based on the reference wind power obtained by the MAV, SG, Gaussian, and LWLR methods, a numerical iterative algorithm is used in the second stage to determine the best combination of PV plant and BESS in the determined search space.

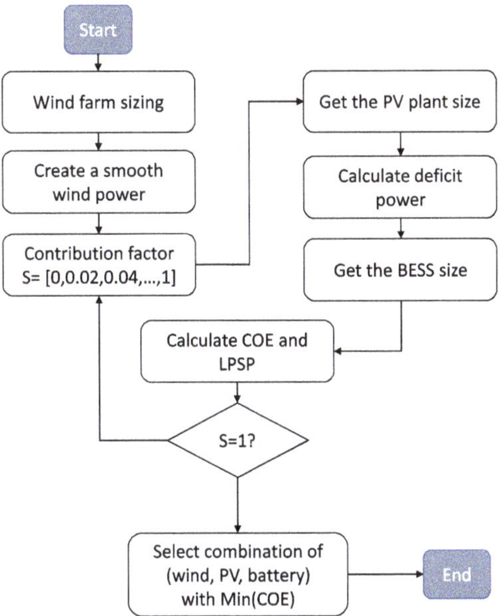

Figure 1. Process flow for sizing the HWSPS.

3. Smoothing Techniques

To obtain a smoothed reference output power approximating the load demand, many smoothing techniques are utilized. An integrated approach has been developed, which uses moving average, locally weighted linear regression, Gaussian, and Savitzky–Golay techniques. The proposed strategy is different from previous studies in that it does not involve a load demand profile. The sizing approach was designed to utilize the selected site effectively for a grid-connected system. The focus of this research is to maximize the output power from a hybrid renewable power plant at a constant level based on the availability of renewable energy resources. It is assumed that the load demand variations are absorbed by the grid.

3.1. Moving Average

In this investigation, the moving average smoothing technique (MAV) is used to obtain a smoothed reference output power P_{ref}, which represents the load demand. The P_{ref} value is the reference for the optimal HWSPS plant. The smoothing wind window of MAV [30,31] must be manipulated carefully to reach the desired reference power. The mathematical interpretation of the k-period MAV is presented by Equation (6):

$$P_{ref}(t) = \frac{P_{wind}(t) + P_{wind}(t-1) + P_{wind}(t-2) + \ldots + P_{wind}(t-k)}{k} \tag{6}$$

3.2. Locally Weighted Linear Regression

Linear regression is a technique for determining the linear connections between input and output. For non-linear relationships between the input and the output, locally weighted linear regression (LWLR) [32] is used. Unlike normal linear regression, LWLR does not use fixed parameters (β); thus, it is a non-parametric algorithm used to smooth noisy signals. It is a memory-based method and uses training data that are local to the point of interest. The mathematical model of the LWLR is given by Equation (7).

$$\Re(\beta) = \sum_{i=1}^{m} \mathcal{W}^{(i)} \left[P_{wind}^{(i)} - \left(\beta_0 + \beta_1\, t^{(i)} \right) \right]^2 \tag{7}$$

As with the moving average, specifying the span length is critical for the LWLR. The span is defined by the fraction (i.e., 0.057%) of the data points closest to the target point t_0. The data points within the span determine the smoothed value. The points outside the span have a zero weight. Thus, the smoothing process is local, since it uses only the local points in each span. Quadratic and linear models can be used in the regression. In this study, linear models are used. The tricube function [33] is used to calculate the weights, as given by Equation (8):

$$\mathcal{W}^{(i)} = \left[1 - \left| \frac{t_0 - t_i}{d(t)} \right|^3 \right]^3 \tag{8}$$

The fitted model is obtained for the target point t_0, and the same process is repeated for all the data points. The mathematical representations and calculations for the LWLR are summarized in the following points:

-   Define the span length;
-   Obtain the regression weights for each data point in the span;
-   Solve the LWLR problem to obtain β_0 and β_1 by taking the first derivative for minima, as shown in Equations (9)–(11):

$$\frac{d\mathfrak{R}(\beta)}{d\beta_0} = -2 \sum_{i=1}^{m} \mathcal{W}^{(i)} \left[P_{wind}^{(i)} - \left(\beta_0 + \beta_1 \, t^{(i)} \right) \right] \tag{9}$$

$$\frac{d\mathfrak{R}(\beta)}{d\beta_1} = -2 \sum_{i=1}^{m} \mathcal{W}^{(i)} \left[P_{wind}^{(i)} - \left(\beta_0 + \beta_1 \, t^{(i)} \right) \right] t^{(i)} \tag{10}$$

$$\begin{bmatrix} \beta_0 \\ \beta_1 \end{bmatrix} = \begin{bmatrix} \sum \mathcal{W}^{(i)} & \sum \mathcal{W}^{(i)} t^{(i)} \\ \sum \mathcal{W}^{(i)} t^{(i)} & \sum \mathcal{W}^{(i)} t^{(i)} t^{(i)} \end{bmatrix}^{-1} \begin{bmatrix} \sum \mathcal{W}^{(i)} P_{wind}^{(i)} \\ \sum \mathcal{W}^{(i)} P_{wind}^{(i)} t^{(i)} \end{bmatrix} \tag{11}$$

-   Finally, to obtain the corresponding P_{ref}-value for the target point (t_0), substitute in the line equation for β_0 and β_1.

3.3. Gaussian Distribution

The Gaussian distribution is also known by other names, such as normal distribution. The Gaussian function is widely used as a smoothing operator for noisy data points. In addition, it is used in defining the probability distribution (histogram) of data. The Gaussian function follows the bell-shaped curve. It is a function of non-zero value and is symmetrical about the mean $(t = \mu)$.

The probability distribution is described mathematically using the expression in Equation (12) [34]:

$$p(t) = \frac{1}{\sigma \sqrt{2\pi}} e^{-\frac{1}{2} \left(\frac{t-\mu}{\sigma} \right)^2} \tag{12}$$

where μ and σ represent the mean and standard deviation, respectively. The term $\frac{1}{\sigma \sqrt{2\pi}}$ is a constant that acts as a normalizer. The exponential term decays quickly as t diverges from μ. The rate of decay is influenced by the value of σ. The curve of the Gaussian function is divided into three segments, defined as follows:

-   Segment (I): $(\mu - \sigma) \leq t \leq (\mu + \sigma)$;
-   Segment (II): $(\mu - 2\sigma) \leq t \leq (\mu + 2\sigma)$;
-   Segment (III): $(\mu - 3\sigma) \leq t \leq (\mu + 3\sigma)$.

These three segments combined have the highest probability value (99.72%), with respective weights of 68.26% (segment (I)), 27.18% (segment (II)), and 4.28% (segment (III)).

3.4. Savitzky–Golay

Instead of smoothing by averaging the data and making an aggressive change in the original signal, Savitzky–Golay [35] is a simple smoothing algorithm that follows the pattern of the original signal. SG is performed by fitting a least square in each window. In general, it is a moving polynomial fit with constant weighting coefficients. These coefficients are called convolution integers. These sets of integers are selected based on the window size and the polynomial degree. A reference table for 25 window sizes is presented in [36] and contains different values of convolution integers for quadratic polynomials. Alternatively, convolution integers could be defined based on approximation by polynomials. In our case, a window size of 31 points, i.e., $(m = 31)$ with a quadratic polynomial ($\{=2$) is used. Thus, the smoothed signal for a set of data points (t_i, P_{wind_i}) with the length n is calculated using the formula in Equation (13). The convolution integers are calculated using the Vandermonde matrix (M_{vand}) as given by Equation (14):

$$P_{ref_j} = \sum_{i=-\frac{m-1}{2}}^{\frac{m-1}{2}} Q_i \, P_{wind_{j+i}} \qquad \frac{m+1}{2} \le j \le n - \frac{m-1}{2} \tag{13}$$

$$Q = \left(M_{vand}{}^T \, M_{vand}\right)^{-1} M_{vand}{}^T \tag{14}$$

where

$$M_{vand} = \begin{bmatrix} 1 & b_1 & b_1^2 & \cdots & b_1^f \\ 1 & b_2 & b_2^2 & \cdots & b_2^f \\ 1 & b_3 & b_3^2 & \cdots & b_3^f \\ 1 & \vdots & \vdots & \ddots & \vdots \\ 1 & b_i & b_i^2 & \cdots & b_i^f \end{bmatrix}, b = [\frac{1-m}{2}, \ldots, 0, \ldots, \frac{m-1}{2}]$$

The power fluctuation $\Delta P_{ref}(t)$ is measured at each smoothing technique's time instant, taking the difference between the subsequent output powers, as shown in Equation (15):

$$\Delta P_{ref}(t) = |P_{ref}(t+1) - P_{ref}(t)| \tag{15}$$

4. Case Study

A case study has been provided to show the applicability of the suggested technique. The weather data from Thumrait, Dhofar Governorate, Oman, were utilized. This study makes use of wind speed and direction data collected throughout a year at a resolution of 10 min intervals. This histogram shown in Figure 2 will be utilized in the optimization of the wind farm layout. After determining the farm size, the instant wind power $P_{wind}(t)$ is calculated using real 10-minute wind profiles, whose wind magnitude is shown in Figure 3. In addition, the PV output power is determined by utilizing the global horizontal irradiance of the mentioned site [23]. Table 1 lists the key technical and economic characteristics utilized in this analysis for the PV module, wind turbines, and BESS. For this study, a square land area is considered with dimensions of 2 × 2 km. The wind farm site is assumed flat with a surface roughness of 0.3 m.

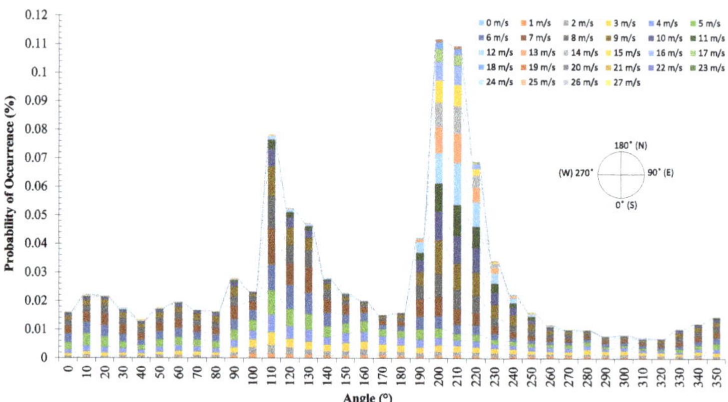

Figure 2. Histogram of wind speeds and directions for a year.

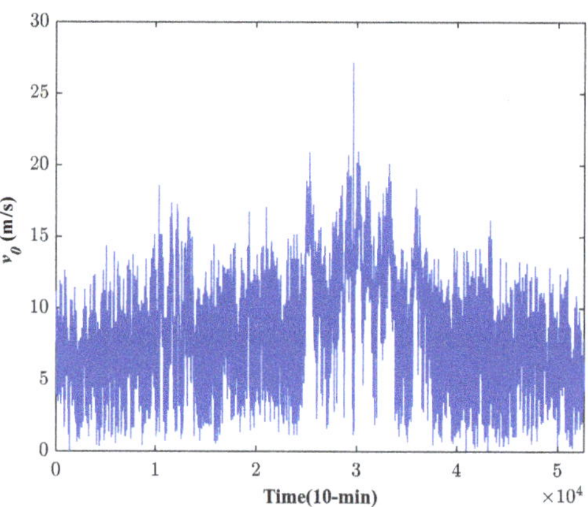

Figure 3. Wind speedprofile for a year.

Table 1. Specifications of the used wind turbine, PV module, BESS, and inverter [23,37–39].

Wind Turbine		PV	
Rated power	3 MW	Model	Polycrystalline
Hub height	84 m	Maximum power at STC (P_{rpv})	275 W
Rotor diameter	82 m	Temperature coefficient of (P_{rpv})	$-0.47\%/C°$
Capital cost	1784 EUR/kW	Capital cost	598.62 EUR/kW
O&M cost	3% capital cost/year	O&M cost	1% capital cost/year
Lifetime	20 years	Lifetime	20 years
BESS		**Inverter**	
Nominal capacity	1000 Ah	Rated power	115 kW
Nominal voltage	2 V	efficiency (η_{inv})	90%
Capital cost	213 EUR/kWh	Capital cost	117.26 EUR/kW
Replacement cost	213 EUR/kWh	Replacement cost	117.26 EUR/kW
O&M cost	9.8 EUR/kWh/year	O&M cost	0.92 EUR/ kW /year
Lifetime	5 years	Lifetime	20 years

5. Results and Discussion

Running the simulation with the data given produces results with 13 as the optimal number of turbines. They are distributed mainly along the WF's boundaries to decrease the wake impact as shown in Figure 4.

To demonstrate the wake effect's impact on wind power generation in more detail, Figure 5 depicts the power generation of the wind farm with and without the wake effect using the optimum configuration. The wake effect model produces more precise and realistic results for the power generation of a wind farm [40]. The overall velocity loss owing to the turbine wake effect is approximately 2.72% of the available wind speed. This equates to a 5.31% decrease in the wind farm's generating power. The distinction between the two situations is obvious, and utilizing the wake effect model enables planners to estimate more realistic wind farm power.

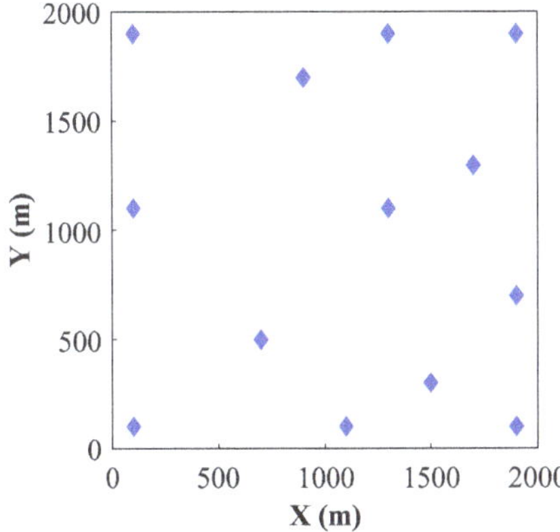

Figure 4. Optimal wind farm layout.

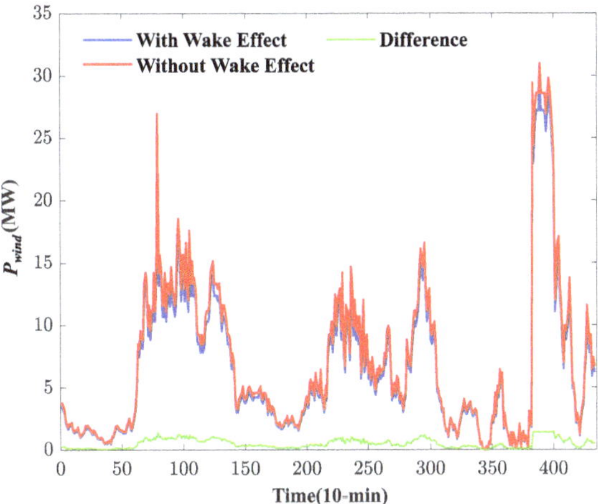

Figure 5. Power generated by the wind farm for three days.

The suggested strategy then proceeds to create a smooth wind output power (P_{ref}) utilizing the LWLR, MAV, Gaussian, and SG approaches to decrease variations in the wind farm's output power with the help of a PV plant and BESS. The primary goal of smoothing wind generation is to minimize the ramping rate, as seen in Figure 6. The ramping rates are displayed before and after smoothing. Before smoothing, the maximum ramping is 32.33 MW. After smoothing, the maximum ramping rates are 2.14 (with LWLR), 3.85

(with MAV), 2.40 (with Gaussian), and 3.20 MW (with SG). This means that the maximum ramping in $P_{ref_{LWLR}}$, $P_{ref_{MAV}}$, $P_{ref_{Gaussian}}$, and $P_{ref_{SG}}$ correspond to 5.49%, 9.87%, 6.15%, and 8.21% of the wind farm's capacity, respectively. These results are adequate to move on with this case study.

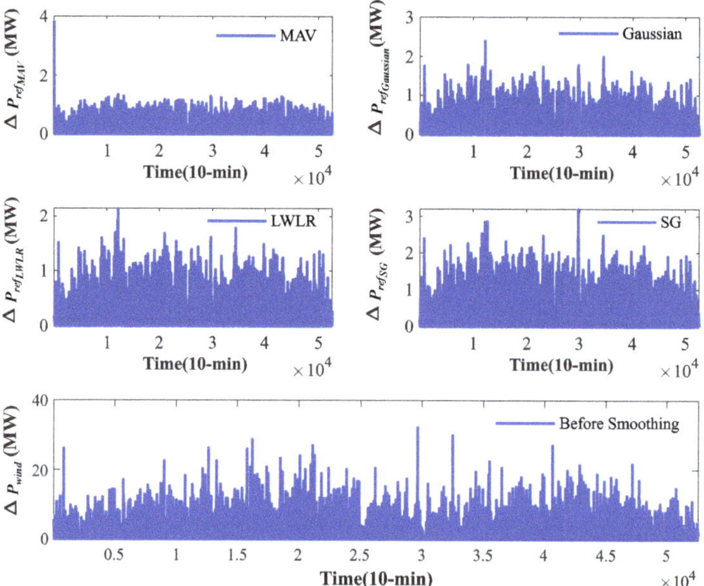

Figure 6. The yearly ramping rate of the wind farm.

Each smoothing technique was used with 30 data points, and the generated P_{ref} for each technique is shown in Figure 7. It is obvious that P_{ref} is smoother than P_{wind}. The purple-bounded regions in Figure 8 present how much P_{wind} falls short of P_{ref}. The gap between P_{wind} and $P_{ref_{MAV}}$ is greater than that from other smoothing techniques. This means more resources are needed to compensate the generations. As a result, a PV plant and BESS with 50 contribution factors (S) were installed to alleviate the shortfall, as stated in Section 2.

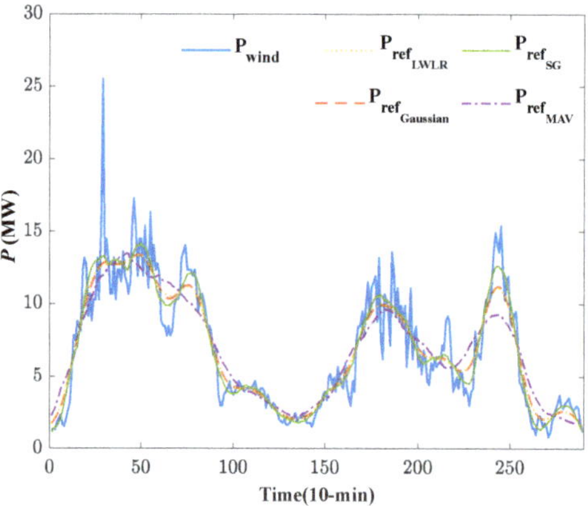

Figure 7. Actual and smoothed wind power samples.

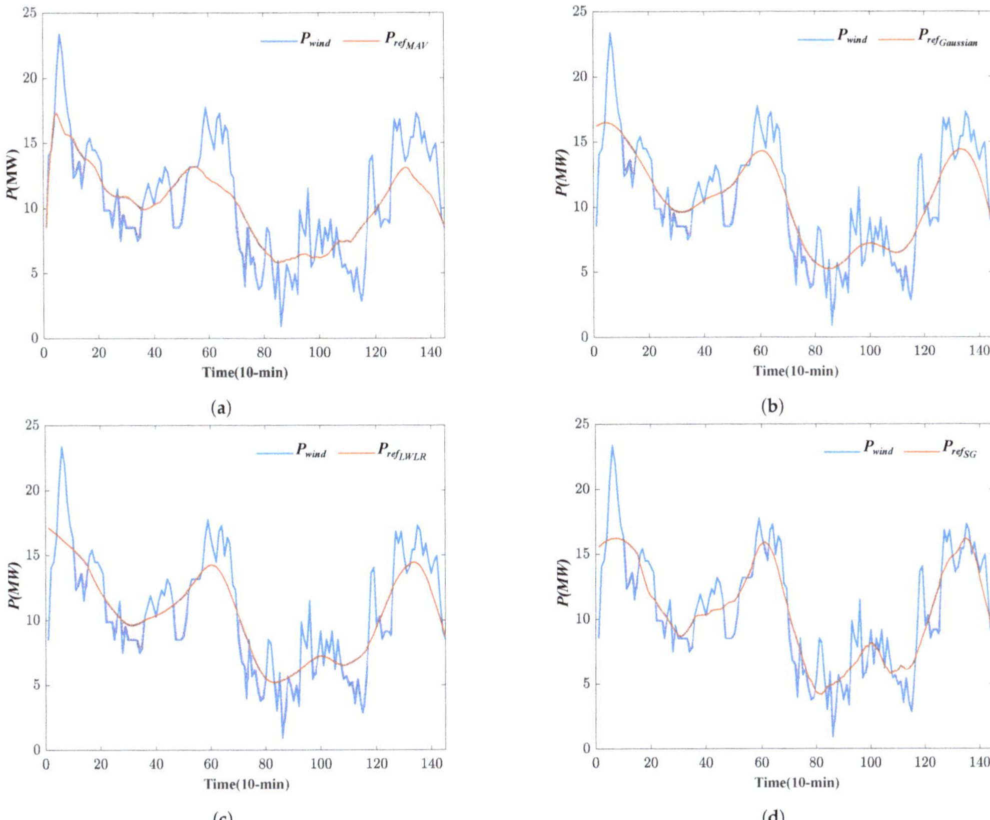

Figure 8. One day sample for shortages of P_{wind} to fulfill the P_{ref}: (**a**) MAV; (**b**) Gaussian; (**c**) LWLR; (**d**) SG.

Running the suggested method in NIA produced several PV plant and BESS setups with COE and LPSP, as illustrated in Figure 9. As the sizing of BESS is meant to minimize oversizing by considering just the minimal yearly negative accumulative net energy, the BESS size dropped with each increase in S for each smoothing approach, and finally the size was fixed. The minimum CE values for the whole year are 0.72609 (for LWLR), 0.6680 (for MAV), 0.57942 (for Gaussian), and 0.41753 MWh (for SG), as shown in Figure 10.

The capacity of the BESS began at 18.74 MWh for S = 0, and then reached the minimum value of 0.8350 MWh for S = 0.04 (with the MAV). In addition, the minimum COE was 0.0229 EUR/kWh, with 10.716 MW for the PV plant sizes. The simulation results also attest to the effect of LWLR, Gaussian, and SG in smoothing the wind power to overcome the negative impact of the memory effect in MAV. Compared to MAV, the optimal sizes of the BESS and PV for LWLR, Gaussian, and SG were attained while S = 0.02. Thus, the optimal COE was 0.0237 EUR/kWh (for LWLR), 0.0203 EUR/kWh (for Gaussian), and 0.0165 EUR/kWh (for SG). However, the COE for LWLR was the highest, due to the size of the BESS, which reached 0.9076 MWh.

On the other hand, the BESS for the SG was 0.2024 MWh lower than the BESS size for the Gaussian. The PV size was 5.329 (for LWLR), 5.329 (for Gaussian), and 5.305 MW (for SG). These techniques yielded similar PV size values, in contrast with MAV, which yielded a value of 10.716 MW. Table 2 shows an overview of the best outcomes.

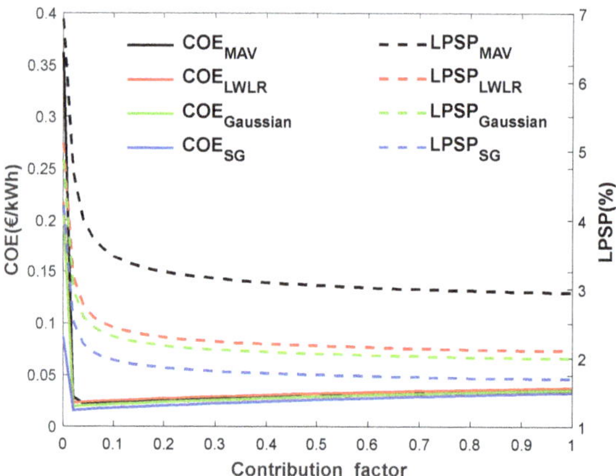

Figure 9. COE and LPSP for various contributing factors.

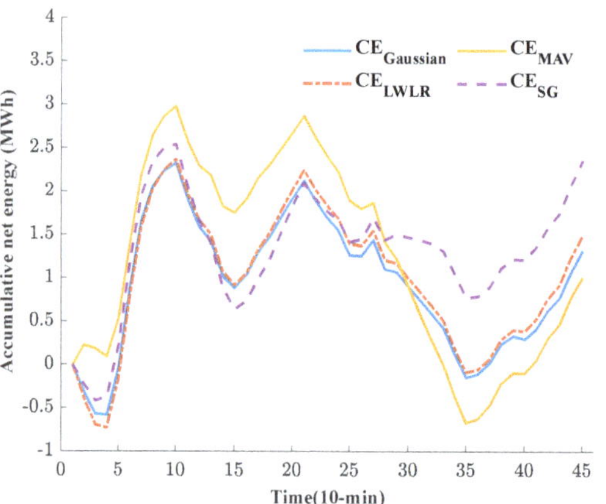

Figure 10. Minimum accumulative net energy.

Table 2. Optimal HWSPS configurations based on the suggested method, taking into account various smoothing approaches.

	MAV	**LWLR**	**Gaussian**	**Savitzky–Golay**
Wind farm (MW)	39	39	39	39
$C_{battery}$(MWh)	0.8350	0.9076	0.7242	0.5219
PV (MW)	10.716	5.329	5.329	5.305
COE (EUR/kWh)	0.0229	0.0237	0.0203	0.0165
LPSP (%)	4	3.17	2.99	2.52
S	0.04	0.02	0.02	0.02

As shown in Figure 11, the values still fall short of the reference power, but the whole LPSP values for the entire year are 4% for MAV, 3.17% for LWLR, 2.99% for Gaussian, and 2.52% for SG. The largest loss comes at night when PV generation is not available. The size of the PV plant grows linearly as the contribution factor S increases, resulting in larger fluctuations and more damped output.

With an increase in a contribution factor, the power deficit reduces, but the COE rises. Simultaneously, the BESS SOC varies substantially between SOC_{min} and SOC_{max}. (charged and discharged condition). During the day, there is enough $P_{pv\,MAV}$ to compensate for the wind generation shortfall. A portion of the surplus energy is used to power the BESS. Due to the tiny size of the PV plant, the BESS is usually employed if the LWLR, Gaussian, or SG are used. The BESS is the most commonly used with LWLR, followed by Gaussian and then SG. Due to the huge size of the PV plant, the MAV uses the BESS the least. MAV, on the other hand, has a greater COE and a lower LPSP than SG.

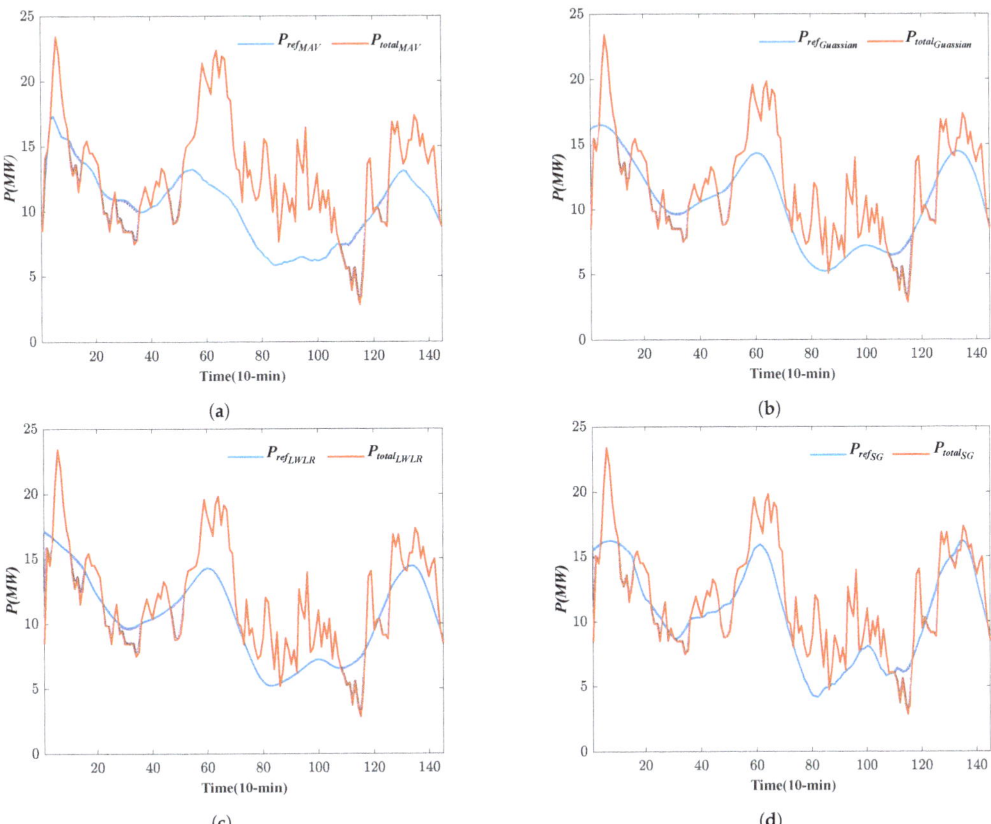

Figure 11. Sample of total output power and smoothedwind power: (**a**) MAV; (**b**) Gaussian; (**c**) LWLR; (**d**) SG.

6. Sizing without the Wake Effect

A simulation was run to determine the best HWSPS without taking into account the wake effect. For each smoothing approach, the simulation was run using the optimum contribution factor. With the MAV set to S = 0.04, this resulted in a HWSPS with a PV plant of 11.372 MW and a BESS of 0.8984 MWh. The PV plant and BESS sizes rose by 6.12% and 7.59%, respectively. The optimum PV and BESS sizes for the SG for S = 0.02 were 5.63 MW and 0.5319 MWh, respectively. Running the simulation with S = 0.02 and the Gaussian model resulted in a HWSPS with a 5.66 MW PV plant and a 0.7418 MWh BESS. Finally, when compared to the wake findings, the sizes of PV and BESS rose by 6.12% and 2.91%, respectively, using the LWLR method. This illustrates the overestimation that happens when estimating the HWSPS size without taking the wake effect into account.

7. The Influence of Contribution Factor on PV Plant and BESS Sizes

The wind farm's cost alone is less than the optimal cost of the proposed HWPS. However, the LPSP of a HWSPS with MAV is 4%, compared to 6.91% for a stand-alone wind farm, implying that the HWSPS is roughly 2.91% more reliable. These findings demonstrate that a wind farm alone is not a dependable solution.

Increasing the size of the PV plant and the BESS improves the LPSP but raises the COE. Furthermore, the surplus electricity grows, resulting in a decrease in revenue. Thus, the availability of other resources on a wind farm benefits the electrical system, but only up to a certain degree of resource penetration.

8. Summary

In this paper, an approach was developed to mitigate the wind output power fluctuation. It focuses on scaling the HWSPS to decrease the impact of renewable energy resource intermittency and provide maximum output power to the grid at a consistent level on a timely manner based on renewable energy resource availability. The PV and BESS are sized dependent on the production of the wind farm. Unlike earlier research, this technique does not consider the load profile when sizing the HWSPS. The appropriate size of the HWSPS is determined using the smoothed wind power signal as a reference. The smoothed signal is generated using the MAV, LWLR, SG, and Gaussian smoothing techniques.

Furthermore, the GA is employed in the initial step of the method, with Jensen's wake effect model being applied to provide more precise and realistic results. The research focuses on sizing the HWSPS in order to decrease production variations and enhance dependability of the wind farm. The NIA is used to size the PV plant and the BESS, which is a trade-off between system cost and reliability.

The suggested method was shown using real GHI data from a wind power plant site in Oman with a multi-speed and multi-directional wind profile. The wake impact on turbines was shown, and the power outputs of the wind farm with and without the wake effect model were compared. The optimal HWSPS had a wind to PV ratio of 3.64:1 with MAV and around 7.32:1 with other smoothing techniques. The corresponding BESS capacity represented 1.68% of the HWSPS's rating for MAV, 2.05% for LWLR, 1.18% for SG, and 1.63% for Gaussian.

The results also show that the SG smoothing technique is more suitable for this task than MAV, Gaussian, or LWLR techniques. It was also found that the window size plays a vital role in the smoothing of noisy output, but this smoothness has a negative impact on the cost of energy. The memory effect feature of the presented smoothing techniques led to a delay between the smoothed signal and the actual signal. In general, this implies an increase in the smoothing source capacities.

In addition, an evaluation of the influence of the wake effect and the contribution factor on the sizing of the HWSPS was conducted. This evaluation revealed the importance of the wake effect to avoid overestimating the HWSPS size. As a result, the suggested method is efficient in conducting a feasibility study for the size of a HWSPS in order to achieve a cost-effective and dependable system.

Author Contributions: Conceptualization, A.A.-S.; methodology, A.A.-S.; investigation, M.A.; resources, A.A.-H.; writing—original draft preparation, A.A.-S.; writing—review and editing, A.A.-H. and R.A.-A.; A.A.-S.; software, A.A.-S.; validation, A.A.-H., M.A. and R.A.-A.; supervision, A.A.-H. All authors have read and agreed to the published version of the manuscript.

Funding: This research is funded by HMTF awarded to Sultan Qaboos University research team (SR/ENG/ECED/17/01).

Institutional Review Board Statement: Not applicable.

Informed Consent Statement: Not applicable.

Acknowledgments: The authors would like to express their gratitude to Sultan Qaboos University and Oman Rural Areas Electricity Company (Tanweer) for their assistance with this research.

Conflicts of Interest: The authors declare no conflict of interest.

References

1. Al-Maamary, H.M.; Kazem, H.A.; Chaichan, M.T. The impact of oil price fluctuations on common renewable energies in GCC countries. *Renew. Sustain. Energy Rev.* **2017**, *75*, 989–1007. [CrossRef]
2. Appavou, F.; Brown, A.; Epp, B.; Leidreiter, A.; Lins, C.; Murdock, H.; Musolino, E.; Petrichenko, K.; Farrell, T.; Krader, T.; et al. Renewables 2017 global status report. In *Renewable Energy Policy Network for the 21st Century*; REN21: Paris, France, 2017.
3. Cretì, A.; Fontini, F. *Economics of Electricity: Markets, Competition and Rules*; Cambridge University Press: Cambridge, UK, 2019.
4. Kawakami, N.; Iijima, Y.; Fukuhara, M.; Bando, M.; Sakanaka, Y.; Ogawa, K.; Matsuda, T. Development and field experiences of stabilization system using 34MW NAS batteries for a 51MW wind farm. In Proceedings of the 2010 IEEE International Symposium on Industrial Electronics, Bari, Italy, 4–7 July 2010; pp. 2371–2376.
5. Ochoa, D.; Martinez, S. Frequency dependent strategy for mitigating wind power fluctuations of a doubly-fed induction generator wind turbine based on virtual inertia control and blade pitch angle regulation. *Renew. Energy* **2018**, *128*, 108–124. [CrossRef]
6. Jabir, M.; Azil Illias, H.; Raza, S.; Mokhlis, H. Intermittent smoothing approaches for wind power output: A review. *Energies* **2017**, *10*, 1572. [CrossRef]
7. Abdalla, A.A.; Khalid, M. Smoothing Methodologies for Photovoltaic Power Fluctuations. In Proceedings of the 2019 8th International Conference on Renewable Energy Research and Applications (ICRERA), Brasov, Romania, 3–6 November 2019; pp. 342–346.
8. Guo, T.; Liu, Y.; Zhao, J.; Zhu, Y.; Liu, J. A dynamic wavelet-based robust wind power smoothing approach using hybrid energy storage system. *Int. J. Electr. Power Energy Syst.* **2020**, *116*, 105579. [CrossRef]
9. Desta, A.; Courbin, P.; Sciandra, V.; George, L. Gaussian-based smoothing of wind and solar power productions using batteries. *Int. J. Mech. Eng. Robot. Res.* **2017**, *6*, 154–159. [CrossRef]
10. Schafer, R.W. What is a Savitzky-Golay filter?[lecture notes]. *IEEE Signal Process. Mag.* **2011**, *28*, 111–117. [CrossRef]
11. Stroe, D.I.; Zaharof, A.; Iov, F. Power and energy management with battery storage for a hybrid residential PV-wind system—A case study for Denmark. *Energy Procedia* **2018**, *155*, 464–477. [CrossRef]
12. Shimizukawa, J.; Iba, K.; Hida, Y.; Yokoyama, R. Mitigation of intermittency of wind power generation using battery energy storage system. In Proceedings of the 45th International Universities Power Engineering Conference UPEC2010, Cardiff, UK, 31 August–3 September 2010; pp. 1–4.
13. Kim, T.; Moon, H.j.; Kwon, D.h.; Moon, S.i. A smoothing method for wind power fluctuation using hybrid energy storage. In Proceedings of the 2015 IEEE Power and Energy Conference at Illinois (PECI), Champaign, IL, USA, 20–21 February 2015; pp. 1–6.
14. Johnson, J.; Schenkman, B.; Ellis, A.; Quiroz, J.; Lenox, C. Initial operating experience of the 1.2-MW La Ola photovoltaic system. In Proceedings of the 2012 IEEE 38th Photovoltaic Specialists Conference (PVSC) PART 2, Austin, TX, USA, 3–8 June 2012; pp. 1–6.
15. Khalid, M.; Savkin, A. Minimization and control of battery energy storage for wind power smoothing: Aggregated, distributed and semi-distributed storage. *Renew. Energy* **2014**, *64*, 105–112. [CrossRef]
16. Luo, F.; Meng, K.; Dong, Z.Y.; Zheng, Y.; Chen, Y.; Wong, K.P. Coordinated operational planning for wind farm with battery energy storage system. *IEEE Trans. Sustain. Energy* **2015**, *6*, 253–262. [CrossRef]
17. Addisu, A.; George, L.; Courbin, P.; Sciandra, V. Smoothing of renewable energy generation using Gaussian-based method with power constraints. *Energy Procedia* **2017**, *134*, 171–180. [CrossRef]
18. Atif, A.; Khalid, M. Saviztky–Golay filtering for solar power smoothing and ramp rate reduction based on controlled battery energy storage. *IEEE Access* **2020**, *8*, 33806–33817. [CrossRef]
19. Jiang, Q.; Hong, H. Wavelet-based capacity configuration and coordinated control of hybrid energy storage system for smoothing out wind power fluctuations. *IEEE Trans. Power Syst.* **2012**, *28*, 1363–1372. [CrossRef]
20. Cheng, F.; Willard, S.; Hawkins, J.; Arellano, B.; Lavrova, O.; Mammoli, A. Applying battery energy storage to enhance the benefits of photovoltaics. In Proceedings of the 2012 IEEE Energytech, Cleveland, OH, USA, 29–31 May 2012; pp. 1–5.
21. Alam, M.; Muttaqi, K.; Sutanto, D. A novel approach for ramp-rate control of solar PV using energy storage to mitigate output fluctuations caused by cloud passing. *IEEE Trans. Energy Convers.* **2014**, *29*, 507–518.
22. Sukumar, S.; Mokhlis, H.; Mekhilef, S.; Karimi, M.; Raza, S. Ramp-rate control approach based on dynamic smoothing parameter to mitigate solar PV output fluctuations. *Int. J. Electr. Power Energy Syst.* **2018**, *96*, 296–305. [CrossRef]
23. Shereiqi, A.A.; Al-Hinai, A.; Albadi, M.; Al-Abri, R. Optimal Sizing of a Hybrid Wind-Photovoltaic-Battery Plant to Mitigate Output Fluctuations in a Grid-Connected System. *Energies* **2020**, *13*, 3015. [CrossRef]
24. Jensen, N.O. *A Note on Wind Generator Interaction*; Risø National Laboratory: Roskilde, Denmark, 1983
25. Archer, C.L.; Vasel-Be-Hagh, A.; Yan, C.; Wu, S.; Pan, Y.; Brodie, J.F.; Maguire, A.E. Review and evaluation of wake loss models for wind energy applications. *Appl. Energy* **2018**, *226*, 1187–1207. [CrossRef]
26. Gao, X.; Li, Y.; Zhao, F.; Sun, H. Comparisons of the accuracy of different wake models in wind farm layout optimization. *Energy Explor. Exploit.* **2020**, *38*, 1725–1741. [CrossRef]
27. Al Shereiqi, A.; Mohandes, B.; Al-Hinai, A.; Bakhtvar, M.; Al-Abri, R.; El Moursi, M.; Albadi, M. Co-optimisation of wind farm micro-siting and cabling layouts. *IET Renew. Power Gener.* **2021**, *15*, 1848–1860. [CrossRef]
28. Grady, S.; Hussaini, M.; Abdullah, M.M. Placement of wind turbines using genetic algorithms. *Renew. Energy* **2005**, *30*, 259–270. [CrossRef]
29. Gao, X.; Yang, H.; Lin, L.; Koo, P. Wind turbine layout optimization using multi-population genetic algorithm and a case study in Hong Kong offshore. *J. Wind Eng. Ind. Aerodyn.* **2015**, *139*, 89–99. [CrossRef]

30. Lam, R.K.; Yeh, H.G. PV ramp limiting controls with adaptive smoothing filter through a battery energy storage system. In Proceedings of the 2014 IEEE Green Energy and Systems Conference (IGESC), Long Beach, CA, USA, 24 November 2014; pp. 55–60.
31. Lee, H.J.; Choi, J.Y.; Park, G.S.; Oh, K.S.; Won, D.J. Renewable integration algorithm to compensate PV power using battery energy storage system. In Proceedings of the 2017 6th International Youth Conference on Energy (IYCE), Budapest, Hungary, 21–24 June 2017; pp. 1–6.
32. Atkeson, C.G.; Moore, A.W.; Schaal, S. Locally weighted learning. *Lazy Learn.* **1997**, *11*, 11–73. [CrossRef]
33. Zhang, J.; Chung, C.; Han, Y. Online damping ratio prediction using locally weighted linear regression. *IEEE Trans. Power Syst.* **2015**, *31*, 1954–1962. [CrossRef]
34. Lacouture, Y.; Cousineau, D. How to use MATLAB to fit the ex-Gaussian and other probability functions to a distribution of response times. *Tutor. Quant. Methods Psychol.* **2008**, *4*, 35–45. [CrossRef]
35. Dai, W.; Selesnick, I.; Rizzo, J.R.; Rucker, J.; Hudson, T. A nonlinear generalization of the Savitzky-Golay filter and the quantitative analysis of saccades. *J. Vis.* **2017**, *17*, 10. [CrossRef]
36. Savitzky, A.; Golay, M.J. Smoothing and differentiation of data by simplified least squares procedures. *Anal. Chem.* **1964**, *36*, 1627–1639. [CrossRef]
37. Malheiro, A.; Castro, P.M.; Lima, R.M.; Estanqueiro, A. Integrated sizing and scheduling of wind/PV/diesel/battery isolated systems. *Renew. Energy* **2015**, *83*, 646–657. [CrossRef]
38. Singh, S.; Singh, M.; Kaushik, S.C. Feasibility study of an islanded microgrid in rural area consisting of PV, wind, biomass and battery energy storage system. *Energy Convers. Manag.* **2016**, *128*, 178–190. [CrossRef]
39. Rcon, E. ENERCON Product Overview. Available online: https://wind-turbine.com/download/101655/enercon_produkt_en_06_2015.pdf2015 (accessed on 10 February 2021).
40. Tao, S.; Kuenzel, S.; Xu, Q.; Chen, Z. Optimal Micro-Siting of Wind Turbines in an Offshore Wind Farm Using Frandsen–Gaussian Wake Model. *IEEE Trans. Power Syst.* **2019**, *34*, 4944–4954. [CrossRef]

Article

Understanding the Sustainability of the Energy–Water–Land Flow Nexus in Transnational Trade of the Belt and Road Countries

Gengyuan Liu [1,2,*], Asim Nawab [1], Fanxin Meng [1], Aamir Mehmood Shah [1], Xiaoya Deng [3], Yan Hao [1,2], Biagio F. Giannetti [1,4], Feni Agostinho [1,4], Cecília M. V. B. Almeida [1,4] and Marco Casazza [5]

1   State Key Joint Laboratory of Environment Simulation and Pollution Control, School of Environment, Beijing Normal University, Beijing 100875, China; asimuop839@yahoo.com (A.N.); fanxin.meng@bnu.edu.cn (F.M.); aamirmehmood55@hotmail.com (A.M.S.); haoyan@bnu.edu.cn (Y.H.); biafgian@unip.br (B.F.G.); feni@unip.br (F.A.); cmvbag@unip.br (C.M.V.B.A.)
2   Beijing Engineering Research Center for Watershed Environmental Restoration & Integrated Ecological Regulation, Beijing 100875, China
3   State Key Laboratory of Simulation and Regulation of Water Cycle in River Basin, Department of Water Resources, China Institute of Water Resources and Hydropower Research, Beijing 100038, China; dengxy@iwhr.com
4   Post-Graduation Program in Production Engineering, Paulista University, São Paulo 04026-002, Brazil
5   Department of Science and Technology, University of Naples 'Parthenope', Centro Direzionale, Isola C4, 80143 Naples, Italy; marco.casazza001@studenti.uniparthenope.it
*   Correspondence: liugengyuan@bnu.edu.cn

Citation: Liu, G.; Nawab, A.; Meng, F.; Shah, A.M.; Deng, X.; Hao, Y.; Giannetti, B.F.; Agostinho, F.; Almeida, C.M.V.B.; Casazza, M. Understanding the Sustainability of the Energy–Water–Land Flow Nexus in Transnational Trade of the Belt and Road Countries. *Energies* **2021**, *14*, 6311. https://doi.org/10.3390/en14196311

Academic Editor: Helena M. Ramos

Received: 26 July 2021
Accepted: 2 September 2021
Published: 2 October 2021

Publisher's Note: MDPI stays neutral with regard to jurisdictional claims in published maps and institutional affiliations.

Abstract: Increasing economic and population growth has put immense pressure on energy, water and land resources to satisfy national and supra-national demand. Through trade, a large proportion of such a demand is fulfilled. With trade as one of its key priorities, the China Belt and Road Initiative is a long-term transcontinental investment program. The initiative gained significant attention due to greater opportunities for economic development, large population and different levels of resource availability. The nexus approach has appeared as a new viewpoint in discussions on balancing the competing sectoral demands. However, following years of work, constraints exist in the scope and focus of studies. The newly developed multi-regional input–output (MRIO) models covering the world's economy and its use of resources permit a comprehensive analysis of resource usage by production and consumption at different levels, and bring more knowledge about resource nexus problems. Using the MRIO model, this work simultaneously tracks energy, water and land use flows and investigates the transnational resource nexus. A nexus strength indicator is proposed which depends on ternary diagrams to grade countries based on their combined resources' use and sectoral weighting. Equal sectoral weighting is assigned. The analysis presented a sectorally balanced nexus approach. Findings support existing work by recognizing energy, water and land as the robust transnational connections, from both production and consumption points of view. Resource nexus issues differ from country to country owing to inequalities in industrial set-up, preferences in economic policy and resource endowments. The paper outlines how key resource nexus problems can be identified and prioritized in view of alternative and often opposing interests.

Keywords: multi-regional input–output; nexus; trade; Belt and Road

1. Introduction

In today's globalized world, increasing economic and population growth present distinctive challenges when it comes to safeguarding enough energy, water and land resources to satisfy national and supra-national demand. Trade and imports fulfill an increasing share of that demand [1]. Resources therefore need to be managed more sustainably [2]. The independent treatment of water, energy and land systems may lead to the formulation and implementation of ineffective policies and actions. Efficient methods that take into account the interdependencies of resource use are thus required.

With trade as one of its key priorities, the China Belt and Road Initiative is a long-term transcontinental investment program. In 2013, China unveiled its plans to build a Silk Road Economic Belt and a 21st Century Maritime Silk Road (known as the "Belt and Road"—indicated as BRI hereafter), which immediately attracted worldwide attention [3,4]. It is predicted that this scheme will improve resource movements and trade efficiency by connecting more than 65 countries, which represent around 62% of the global population, about 35% of the global trade and over 31% of the world's GDP [5]. Studies have reported that trade and economic expansion may contribute to environmental degradation [6,7]. Since numerous countries along the BRI are not as developed as China, doubts exist that China's international trade and investment may lead to transferring resource exploration and environmental pressures to less-developed regions. The transnational trade networks can have serious consequences on demand for energy, water and land resources. More importantly, resources are interlinked, and any single-sector interventions may cause unintended side-effects in other sectors. At present, some studies associated with the BRI have been conducted linked to virtual water [8], energy efficiency [9], trade impacts [10] or carbon emissions [11]. Mostly, studies concentrated on single resource categories in supply chains and did not perform integrated assessments. To avoid the assumption of unsustainable development patterns and to promote fruitful models of sustainable cooperation, policies based on integrated research are required to allow more balanced use of natural resources.

In recent times, the nexus approach has become an especially important perspective among researchers. The Bonn 2011 nexus conference promoted the idea of a nexus, where the overall issues concerning economic development were understood from the viewpoint of the water–energy–food nexus. Nexus thinking prevents the negative consequences of a single resource development policy and improves resource use efficiency. However, hardly any consensus has emerged on the nexus meanings, with varying interpretations in different disciplines, in diverse situations and by different scholars [12,13]. The absence of a defined framework renders it difficult to determine what produces an efficient assessment of nexus, and presents major problems when formulating nexus-oriented plans. That is to say, it has been challenging to decide how to implement the nexus and formulate workable solutions [14].

The nexus research can be carried out from different viewpoints, depending on the sector in question. From the water point of view, energy and food are generated (as output) and water is an input resource. From the energy point of view, food is produced, but water may either be input, as in the case of hydroelectricity, or sometimes output, when energy is utilized in the treatment of water. If the food point of view is adopted, resource inputs are energy and water [15]. In any situation, the viewpoint considered will influence the policy design. This is attributed to different sector preferences as well as the data and knowledge. Considering the existing approaches to the resource nexus, the two-sector nexus concept, as water–energy [16], is common, and the three-sector nexus, as water–energy–food [17], is the most commonly recognized nexus concept in research and policy-making groups. The three-sector nexus often disregards the position of the land component. In accordance with other researchers [18,19], land involvement in a resource nexus approach can be regarded as vital due to its important ecological functions. Land plays a key role in nutrient recycling, production of food and water, supply of energy and provides resources for livelihood and development. It is very challenging to take an integrated view of these interrelated matters, given that nexus problems occur in different ways in different regions, with different resources and technology applications, governance and development priorities. Thus, for sustainable management of resources, an effective method should be one that can measure resource flows and interdependencies.

Input–output analysis (IOA) in conjunction with newly developed worldwide multi-regional input–output (MRIO) databases [20,21], with their extensive worldwide coverage of industrial interlinkages and usage of resources, may provide novel insights into the resource nexus. Such databases explain inter-industry links inside state economies and across foreign trade. Additionally, they are built with greater sector-based information and ecological stress depiction [22]. The databases enable analysis of resource nexus problems for all industries and different resources and to look deeper at their economic factors from the viewpoints of production as well as consumption. The ground-breaking work on the relationship between nexus structure and IOA mainly used case studies to tackle the nexus between energy and water resources. Of which, Marsh [23] proposed numerous input–output procedures, such as linkage, multiplier and dependence analysis, for dealing with various aspects of nexus problems. In recent times, and in the view of growing interregional and foreign trade significance, nexus scholars utilized MRIO and ecological network analysis (ENA) to investigate structural features and sectoral relations of economic networks [1,24,25].

In this manuscript, we establish a quantitative indicator for transnational resource nexus analysis based on the MRIO model. This indicator is used to compare and grade resource nexus issues resulting from transnational economic activities involving production as well as consumption. The indicator named as nexus strength in this study attempts to classify key resource nexuses on the basis of combined absolute resources' use. In other words, which resource nexuses of a country add more towards the transnational use of natural resources? This article aims to contribute primarily in two ways to the present understanding and management of the nexus problems. Firstly, the MRIO application enables the analysis of potentially ignored nexuses and related synergies and co-benefits. Secondly, a measurable indicator will help users to recognize the most complex nexuses, possibly assisting better sectoral and spatial scale analyses. The analysis presents the findings of an application in Belt and Road countries. The rest of the paper is structured as follows: Section 2 discusses the method, nexus strength indicator, ternary diagram and data, Section 3 presents the key results, and then the results are discussed in Section 4, and finally, Section 5 concludes the paper.

2. Method

2.1. Multi-Regional Input–Output Modeling (MRIO)

Up to now, MRIO models are among the most frequently used methods to study the economic and resource interdependence between different regions [26]. The input–output analysis (IOA) is based on data contained in IO tables. Each entry in the i-th row and j-th column demonstrates the flow from the i-th sector to the j-th sector. The IOA, composed by N linear equations, describes the production of a set of N economic sectors, as denoted in Equation (1):

$$x_i = \sum_{j=1}^{N} z_{ij} + y_i \tag{1}$$

where, N stands for the number of sectors in an economy, x_i represents the total output of the i-th sector, y_i is the final demand of sector i, while z_{ij} is the monetary flow from the i-th sector to the j-th sector. The MRIO model extends the standard IOA matrix to a bigger economy, which involves each sector in each country or region having a separate row and column. The MRIO denotes all of the input–output interactions of the defined economy.

The key input–output balance can be written in matrix form as follows:

$$\begin{pmatrix} x^1 \\ x^2 \\ x^3 \\ \vdots \\ x^m \end{pmatrix} = \begin{pmatrix} A^{11} & A^{12} & A^{13} & \cdots & A^{1m} \\ A^{21} & A^{22} & A^{23} & \cdots & A^{2m} \\ A^{31} & A^{32} & A^{33} & \cdots & A^{3m} \\ \vdots & \vdots & \vdots & \ddots & \vdots \\ A^{m1} & A^{m2} & A^{m3} & \cdots & A^{mm} \end{pmatrix} \begin{pmatrix} x^1 \\ x^2 \\ x^3 \\ \vdots \\ x^m \end{pmatrix} + \begin{pmatrix} \sum_s y^{1s} \\ \sum_s y^{2s} \\ \sum_s y^{3s} \\ \vdots \\ \sum_s y^{ms} \end{pmatrix} \tag{2}$$

where, coefficient matrix A depicts the intermediate input matrix across sectors and regions. Vector x shows the total output of each economic sector in each region.

The mathematical structure of embodied environmental impacts with respect to energy, water and land use can be expressed as:

$$E = R_e(I - A)^{-1}Y \tag{3}$$

$$W = R_w(I - A)^{-1}Y \tag{4}$$

$$L = R_l(I - A)^{-1}Y \tag{5}$$

where, E, W and L represent the embodied energy, water and land matrix induced by the final demand of the whole economic system. R_e, R_w and R_l are the diagonal matrices, representing the pressure coefficient of energy, water and land consumption. The diagonal elements are the direct energy consumption coefficient (R_{ei}), direct water consumption coefficient (R_{wi}) and direct land consumption coefficient (R_{li}). $L = (I - A)^{-1}$ is the Leontief inverse matrix, that captures both direct and indirect inputs. Y is a diagonal matrix, whereas the diagonal element Y_j shows the final demand of products and services in the sector j.

2.2. Nexus Strength Indicator

The nexus approach has constituted the focus of numerous research activities, but there is a lack of agreement on suitable methods to tackle the multidimensionality of the nexus [27]. Scholars have debated that current nexus frameworks mostly remain as partially preferring one sector over others [16,28]. More efforts are required to streamline nexus methods and concepts for policy-makers to make them widely available and usable [17]. So far, different approaches have analyzed the complex interactions between water, energy and food [27,29–32], yet methods vary significantly in their goals, scope and perspective. Some studies applied several performance indicators to assess nexuses among resources, mostly from consumption and intensity perspectives. For instance, they included the energy consumption rate of water [33] or the energy return spent on water [34]. A few program-based indexes were also implemented that concentrated on the weight and reliance of the social economic structure [35,36]. However, no current quantitative measures are easily acceptable to compare resource nexuses concerning numerous resources and countries at the same time. We tackle such problems in this article through ternary diagrams and sectoral weighting. Equal sectoral weighting is assigned via the average method (1/3 each). The ternary diagrams approach is very advantageous and relevant for resource nexus analysis provided the meaning of each line in the diagram is carefully understood.

Ternary diagrams used as graphic tools deal with the multiple resource issues. The nexus ternary diagram has three resources: energy (E), water (W) and land (L). An equilateral triangle represented these resources, where each corner of the triangle represents one of the resources, E, W or L, and each side represents a binary resource system. The location of a point within the internal area of the triangle promptly provides a series of information. Lines that cut the point position represent aggregated use of a given resource. The size of each point/circle inside the triangle shows the combined usage of the three resources (from 0 to 1). Ternary diagrams ensure to present all possible resource use combinations (for nexuses). The combination of any points on the ternary plot can be decided by reading from 0, along the basal line at the bottom of the diagram, to 1 (or 100 percent) at the apex of the triangle. The result of the ternary diagram is labeled as the nexus strength of a specific country. Following this approach, we can evaluate which country has a high nexus strength.

Mathematically, the summation of nexus strength for each country/sector can be expressed as:

$$Nexus\ strength = p_w d_{w,i} + p_e d_{e,i} + p_l d_{l,i};\ with\ i \in I\ ;\ I = \{1,\ldots,n\} \tag{6}$$

$$d_{w,i} = \frac{g_{w,i}}{g_w}\ ;\ g_w = \max(\{g_{w,i}\})_{i \in I} \tag{7}$$

$$d_{e,i} = \frac{g_{e,i}}{g_e} \; ; \; g_e = \max(\{g_{e,i}\})_{i \in I} \tag{8}$$

$$d_{l,i} = \frac{g_{l,i}}{g_l} \; ; \; g_l = \max(\{g_{l,i}\})_{i \in I} \tag{9}$$

$$\sum_{R}^{n}(p_n) = 1 \; ; R = \{w, e, l\} \tag{10}$$

where, w, e and l stand for water, energy and land respectively, i stands for each industry and I represents the set of all industries. Taking water as an example, $g_{w,i}$ represents the water consumption of industry i, g_w represents the largest industrial water consumption among all industries, $d_{w,i}$ represents the deviation between the largest industrial water consumption and water consumption of industry i, p is a weight that determines the relative importance of a given resource and p_w represents the weight of a water resource.

2.3. Reading Ternary Diagrams

For convenience in reading, it is necessary to understand certain ways and rules related to the use of ternary diagrams. Widely used in physical sciences, phase diagrams express equilibrium states in which two or more phases of matter exist together in solutions or in pure substances. Initially, Gibbs proposed the phase rule for multi-component analysis of a system [37], which in the literature is known by different names, such as ternary graph, triangle plot, Gibbs triangle or de Finetti diagram. For reading, the points below should be considered (follow Figure 1):

(1) In Figure 1a, each vertex of the equilateral triangle denotes 1, or 100% of one element, and 0% of the remaining two elements. Point 'x' within a triangle represents a three-resource system. The three lines (EW, WL, LE) connecting the vertexes represent the combinations of E, W and L, and they represent a binary system. When moving along the edge of the diagram so as to symbolize the concentrations in a binary system, it is not important whether we advance in a clockwise or anticlockwise direction, as long as we are constant. For instance, take side EW: if we go in the direction of W, it denotes a binary system of E and W, having increasing concentrations of W and correspondingly decreasing concentrations of E, likewise for WL and LE.

(2) The ternary diagram may be ruled with lines parallel to the sides, and the composition at different points can then be read directly (Figure 1b). For instance, to find the pattern of E, W and L at position 'a' in the triangle, the triangle side EL opposite to vertex W signifies a binary system comprised of E and L, in which the concentration of W is zero. The lines drawn parallel to side EL show increasing W from 0% to 100%, and the line that cuts 'a' is equal to 15% of W and 85% of $E + L$. Likewise, along the line EW, $L = 0$. The lines parallel to EW illustrate increasing the concentration of L from 0% to 100%. The line parallel to EW that cuts 'a' is equal to 20% of L. Hence, E can be calculated as $100 - (W + L) = 100 - (15 + 20) = 65\%$. Other examples shown in Figure 1b are: point $b = 30\%E + 20\%W + 50\%L$ and point $c = 0\%E + 60\%W + 40\%L$.

(3) Any line which is parallel to any side of the triangle represents the ternary systems in which the proportion of any one component is constant (in Figure 1c, example 'gh'). In this particular situation, E is constant and the composition of W and L is changing.

Though our ternary diagram approach is simple in nature, it is versatile to be extended in a variety of ways related to research on the resource nexus. These extensions can be incorporated through weighing's and objectives. The suggested nexus strength provides a simple depiction of the important resource nexuses in the economic system. Nonetheless, the operational value of this indicator will rest on the particular local ecological, economic and political situation.

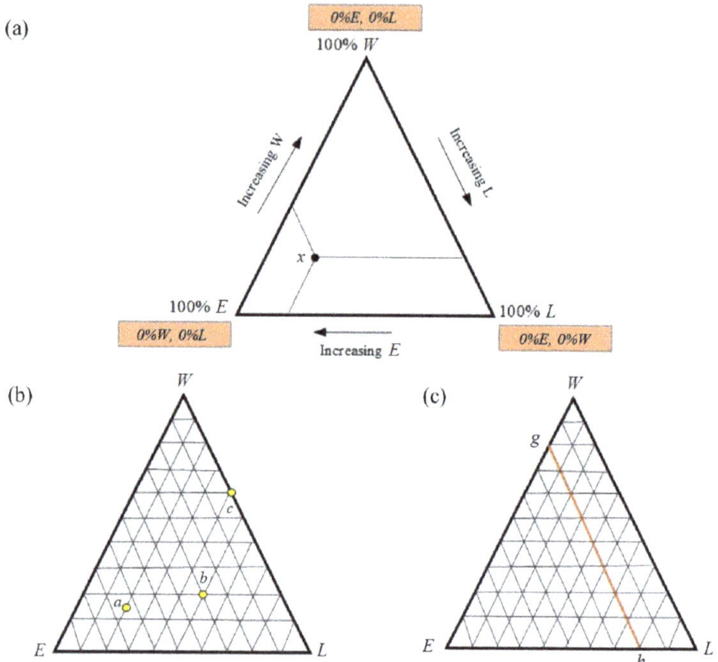

Figure 1. Representation of ternary diagram. (**a**) three-resource system, (**b**) the composition at different points, (**c**) line parallel to any side of the triangle.

2.4. Data Source

The current study utilizes energy, water and land use data of World Input Output Database (WIOD) satellite accounts. The database encompasses 27 EU countries and 13 other important countries worldwide, plus an aggregated region named as Rest of the World (RoW), with 35 sectors per region [21]. The study considers primary energy usage (referred to as energy flow), blue water and green water use, except gray water (water flow), and land usage, i.e., arable area, permanent crops, pastures and forest area (land flow). The BRI is a global open cooperation initiative, welcoming the participation of countries. Therefore, there are no specific boundaries. From WIOD, only those countries were considered that fall along the BRI. Thus, we were able to analyze only 15 Belt and Road countries in the current study. The research year is 2010 considering the availability of environmental accounts. The specific research scope and the names of the associated regions are presented in Table 1. Information on sectors' aggregation can be found in the Appendix A (see Table A1).

Table 1. Selected countries along the Belt and Road.

Section	Specific Regions	Total
East Asia	China, South Korea	2
South East Asia	Indonesia	1
North Asia	Russia	1
South Asia	India	1
Central and Eastern Europe	Bulgaria, Czech Republic, Estonia, Hungary, Lithuania, Latvia, Poland, Romania, Slovakia, Slovenia	10

3. Results

3.1. Interwoven Trade Relations among Economies

This section presents the current intertwined trade relations of energy, water and land use between the BRI countries (excluding the rest of the world). In the first three figures, the fifteen regions are represented around a circle. The trade volume of each region is represented by the corresponding arc length around the circle, while chords, representing different economy couplings, represent the overall bilateral trade relations.

Figure 2 demonstrates the relations between these fifteen economies in transnational trade flows of energy. Among these flows, the largest one was related to South Korean exports to China. In particular, about 7.48×10^4 Kiloton of coal equivalent (Ktce) of embodied energy was exported from South Korea, of which 72% was exported to China. Meanwhile, China also exported a considerable amount of embodied energy to South Korea and India.

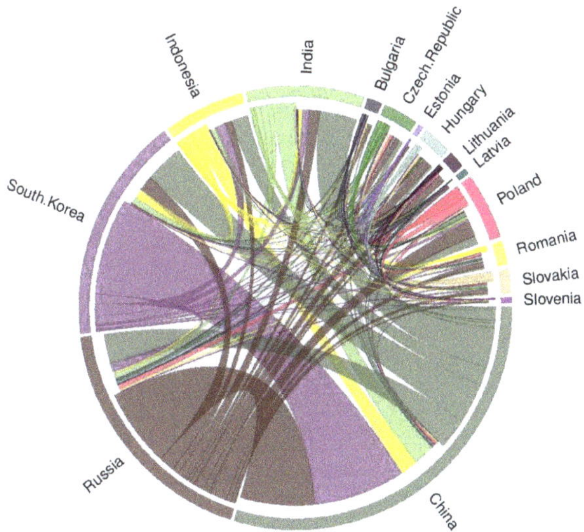

Figure 2. Interlinked relations of energy use between the fifteen economies in trade (excluding RoW).

As the second largest economy in the world, China needs huge amounts of energy imported from foreign countries. China imported about 1.26×10^5 Ktce of embodied energy in total, making it the largest importer of embodied energy among the fifteen countries. China largest energy flow associated with its imports occurred in its trade with Russia, 4.38×10^4 Ktce, accounting for 34% of its total imports. According to the analysis, China was the leading receiver of embodied energy from South Korea and Russia. This demonstrates that South Korea and Russia were the most significant trade partners of China for embodied energy. Notable export–import pairs supporting large energy flows were South Korea–China, Russia–China, China–India and Indonesia–China.

The transnational trade flows of water between selected countries are presented in Figure 3. The biggest flow was associated with the Indian exports to China. Around 1.83×10^4 Million ton (Mt) of embodied water was exported from India, of which 60% went to China. In addition, China was a prominent receiver of embodied water from Indonesia and Russia. As illustrated, among the countries, China also served as a supplier. China exported about 1.79×10^4 Mt of embodied water, of which 25% went to Russia and 24% to South Korea, respectively. Major export–import pairs supporting large water flows were India–China, Indonesia–China, Russia–China and China–South Korea.

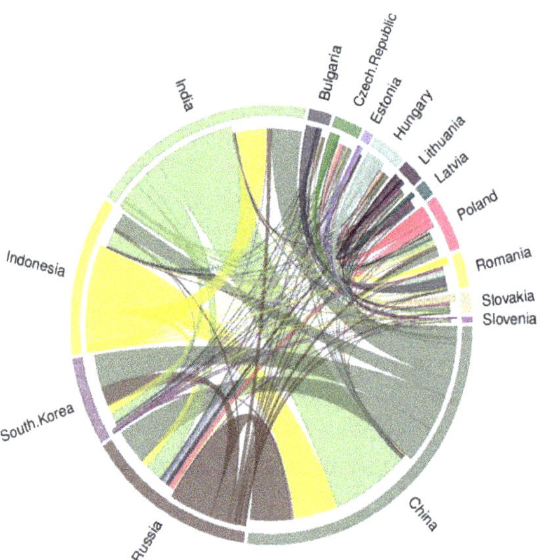

Figure 3. Interlinked relations of water use between the fifteen economies in trade (excluding RoW).

The transnational trade links of embodied land between these economies are portrayed in Figure 4. It can be seen that the largest flow was related to Chinese exports to Russia. About 1.47×10^4 Kilo hectare (Kha) of embodied land was exported from China, of which 59% went to Russia and 13% to South Korea. However, China also imported a substantial amount of embodied land from Russia and India (59% and 21%, respectively), where South Korea was the prominent receiver of embodied land from Russia. Major export–import pairs supporting large land flows were China–Russia, Russia–South Korea, China–South Korea and India–China. As can be understood, regions such as China, India, Russia and South Korea, etc., serve as hubs in transnational trade, that play key roles in both global exporting and importing markets.

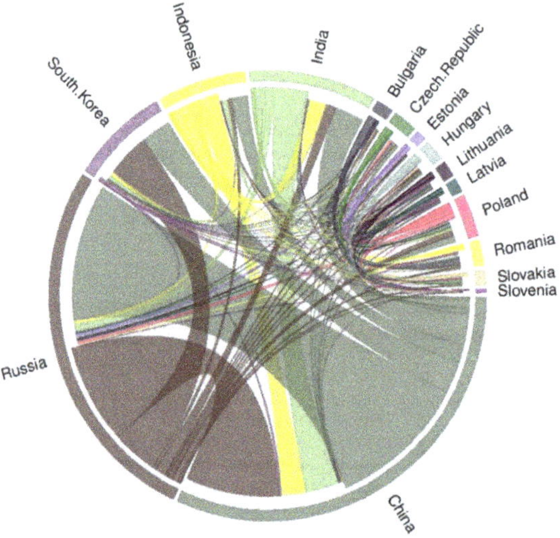

Figure 4. Interlinked relations of land use between the fifteen economies in trade (excluding RoW).

3.2. Self-Sufficiency by Source and Sink

This section presents the energy, water and land use self-sufficiency rates by sources and sinks, assessed through the indicators defined in an earlier study for arable land use [38]. Regions within the world economy extract different resources (energy, water and land) from the local environment and offer these resources for their own or foreign regions' final use. Thus, for each region, the self-sufficiency rate by source can be defined as the ratio of a resource (i.e., energy, water or land) exploited locally for its own final use to the total available resource (i.e., energy, water or land) exploited locally. For each region, this rate evaluates the contribution of local energy, water or land resources to its final consumption. Correspondingly, for a sink region in the supply chain, numerous resources are needed to satisfy its final demand. Along with the local environment, energy, water and land resources are also imported from overseas partners. The self-sufficiency rate by sink of a region can therefore be defined as the ratio of a resource (i.e., energy, water or land) exploited locally for its own final use to the region's resource (i.e., energy, water or land) use represented by resource use embodied in the goods used as its final consumption.

With respect to energy, China and India displayed the maximum energy self-sufficiency rate by source (see Table 2), being respectively 81.56% and 81.18%. This indicates that, from the supply side, most of the energy resources extracted from the local environment were used for domestic final consumption. Russia was a major energy source, having a rate of 63.16%. Thus, it served as a region that mostly provided energy resources to foreign countries. As the largest sink regions, the energy self-sufficiency rates for China and India were high, representing 83.66% and 82.96% of embodied energy in China and India's final use. This energy was self-provided. Conversely, for South Korea, being among the largest sink regions, the rate was 60.53%, illustrating that major energy resources embodied in South Korea's final use were imported from abroad. For some European regions, such as Latvia and Slovenia, their energy use self-sufficiency rates by source were respectively 64.92% and 57.78%, while those by sink were respectively 33.44% and 31.50%. Countries as a beneficiary of foreign resources would suffer the biggest impact, if these countries ran into supply problems.

Table 2. Self-sufficiency rate of selected countries along the BRI route by source and sink.

Region	Self-Sufficiency Rate by Source			Self-Sufficiency Rate by Sink		
	Energy	Water	Land	Energy	Water	Land
China	81.57%	86.93%	89.26%	83.66%	79.92%	79.35%
Russia	63.17%	91.36%	93.09%	88.74%	79.88%	86.76%
South Korea	51.41%	84.84%	82.07%	60.54%	24.81%	14.98%
Indonesia	61.01%	87.06%	87.06%	57.90%	93.74%	81.96%
India	81.19%	90.69%	90.60%	82.97%	96.19%	89.07%
Bulgaria	48.46%	46.79%	46.62%	61.07%	75.03%	65.83%
Czech Republic	52.43%	55.58%	63.83%	56.28%	46.05%	49.14%
Estonia	48.67%	38.48%	38.74%	47.69%	59.34%	55.78%
Hungary	52.26%	53.34%	53.17%	44.92%	75.38%	60.14%
Lithuania	46.29%	51.73%	73.40%	45.06%	74.92%	78.24%
Latvia	64.92%	63.44%	60.76%	33.44%	69.13%	73.43%
Poland	61.86%	69.77%	77.46%	60.61%	67.28%	64.81%
Romania	72.84%	83.83%	83.55%	64.56%	83.05%	81.26%
Slovakia	46.56%	63.72%	61.17%	49.48%	46.21%	28.61%
Slovenia	57.78%	67.03%	74.79%	31.50%	36.50%	37.00%

Regarding water, India and China, being the largest sources, had the maximum water self-sufficiency rates (90.69% and 86.92%), showing that the vast majority of water resources extracted in the two regions were mostly used to satisfy their own final requirements. As a sink region, the rate was much higher for India, with 96.18% of embodied water in

India's final use being self-provided. For China, being among the largest sink regions, its self-sufficiency rate by sink was 79.92%, showing that more than 20% of its water use was dependent on resources from foreign areas. An interesting situation is noted for South Korea, whose water use self-sufficiency rate by source and that by sink were respectively 84.84% and 24.80%. As witnessed, the welfare with respect to domestic water resources was almost preserved within this country, while more than 75% of the water use originated from abroad.

With regard to land, for China and Russia, being the largest exploiters of land resources, their land use self-sufficiency rates by source were respectively 89.26% and 93.08%, while those by sink were 79.34% and 86.75%. For India, its land self-sufficiency rates by source and by sink were 90.59% and 89.06%. In contrast, South Korea showed some interesting features, since its land use self-sufficiency rates by source and that by sink were 82.06% and 14.97%. This indicates that countries' land use mainly depended on resources from foreign areas. The increasing resource needs of South Korea's economy, as a country with limited resources, may influence this phenomenon.

3.3. Sectoral Contribution

Figure 5 shows the sectoral contributions to international trade of the top five net importers and exporters of energy, water and land resources' use in order to help understand the trade structure and resource balance in these regions.

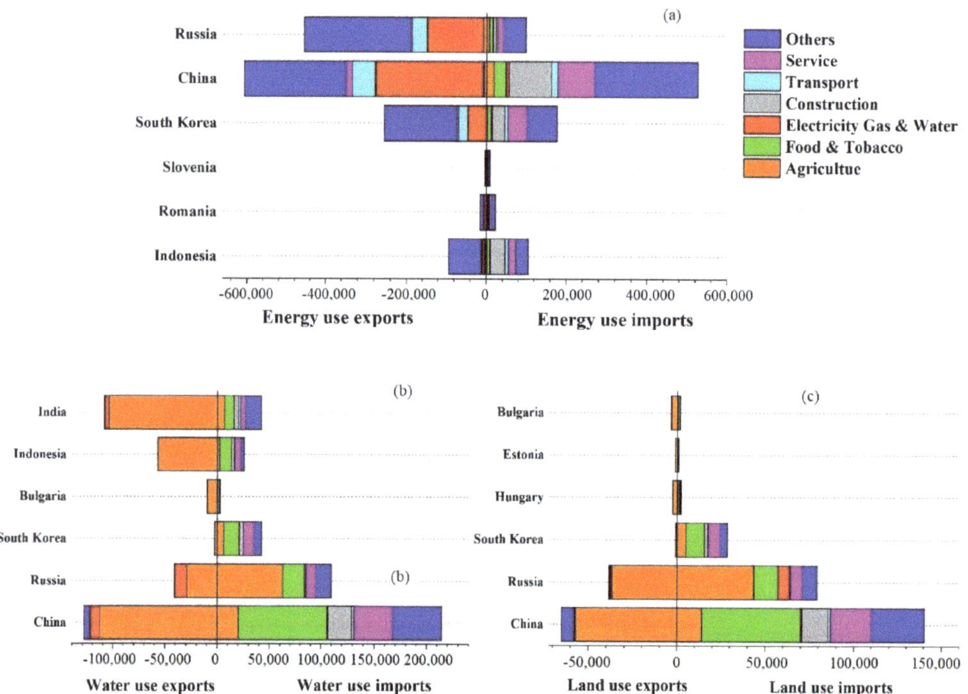

Figure 5. Sectoral contributions to trade of the top three net importers and exporters of energy–water–land use. (**a**) Energy, (**b**) water, (**c**) land.

Regarding energy, Indonesia, Romania and Slovenia appeared among the top three net importers in trade of energy use, while Russia, China and South Korea proved to be the three leading net exporters, as shown in Figure 5a. For Indonesia, the Construction sector shared the largest proportions (33.74%) of energy use embodied in Indonesia imports, followed by the others sector (30.10%), etc. The situation was similar for Romania and Slovenia, where the others sector remained the largest contributor to their imports of energy use, followed by the Service sector. As prominent net exporters of energy use in trade, Russia was dominated by the others sector (58.49%), whereas, for China, the Electricity Gas and Water sector (44.07%) dominated. Meanwhile, for South Korea, the others sector largely contributed to embodied energy exports. China had larger resource imports and exports, showing its significant role as a world trading center, with massive embodied resources flowing in and out.

Regarding water, China, Russia and South Korea were the top three net importers in trade of water use. Meanwhile, India, Indonesia and Bulgaria were the three leading net exporters, as shown in Figure 5b. For China, the Food and Tobacco sector shared the major proportion (39.66%) of water use embodied in China's imports, followed by the others sector (22.34%). For Russia, the Agriculture sector dominated, while, for South Korea, the Food and Tobacco industry remained the largest contributor to their water use imports. Meanwhile, the water use exports for India, Indonesia and Bulgaria were mostly related to the Agriculture sector, revealing their status as a resource-intensive economic structure.

With respect to land, China, Russia and South Korea were the top three net importers in trade of land use. Bulgaria, Estonia and Hungary were the three leading net exporters, as shown in Figure 5c. For China, the Food and Tobacco industry shared the biggest proportion (39.84%) of land use embodied in China's imports, proving China's intensive requirements for food products from foreign areas. For Russia, the Agriculture sector (55.04%), and for South Korea the Food and Tobacco industry (36.52%), remained the largest contributors to land use imports. With regard to land use exports, the Agriculture industry played a dominant role.

3.4. Nexus Strength by Country

The nexus strength by country can be seen in Figures 6 and 7. Findings of the research have been evaluated using equal weights, considering each resource as equally important. Thus, nexus strength only relates to the total use of resources. The findings are in line with the production-based view (territorial, i.e., represents resource usage inside national borders) and the consumption-based view (caused by final demand). For all country-level values, the same scaling factor is used, and thus they are comparable with one another. The nexus strength is somewhat consistent with domestic output levels, as large economies displayed high nexus strength, as shown by the point size within the nexus ternary diagram. Figure 6 illustrates the status of all countries from a production perspective, and activities mostly appear at the middle of the ternary plot. Particularly in China, the energy–water–land nexus seems to be strong, given the wide combined use of resources revealed by its point size in the plot. The lines that cut its position in the plot provide a series of information that can be used to compare its resource use composition with other countries. China used about 35% energy, 30% water and 35% land. This can be associated mainly with the role of its Others, Agriculture and Food industries. Nevertheless, the amount and accessibility of resources causes variations in the strength and makeup of these industries' related nexuses. Two other influential economies after China, i.e., Russia and India, also tend to have a strong connection between energy and water–land. However, they are somewhat at the margins of the plot, indicating a large use of a single resource compared to their use of the other two resources. As can be seen from lines that cut its point location, Russia used large portions of land, around 54%, while its energy and water usage was only about 22% and 24%, respectively. India used large portions of water, around 59%, while its energy and land usage was only about 18% and 23%, respectively. The agriculture industry, for example, in India, has many key drivers, including the existence of energy/gas reserves, domestic policy and technology. Figure 7 shows all countries' status from a consumption perspective. China maintained its large combined resource usage, i.e., about 34% energy, 33% water and 34% land. As for the two perspectives, there is hardly any big change noted. It may be because China is the world's second-largest economy, driving global production and consumption. Again, Russia is driven by large proportions of land, around 53%, while energy and water usage is only about 17% and 30%, respectively. India is largely water-driven, around 60%, while energy and land usage is only about 19% and 21%, respectively.

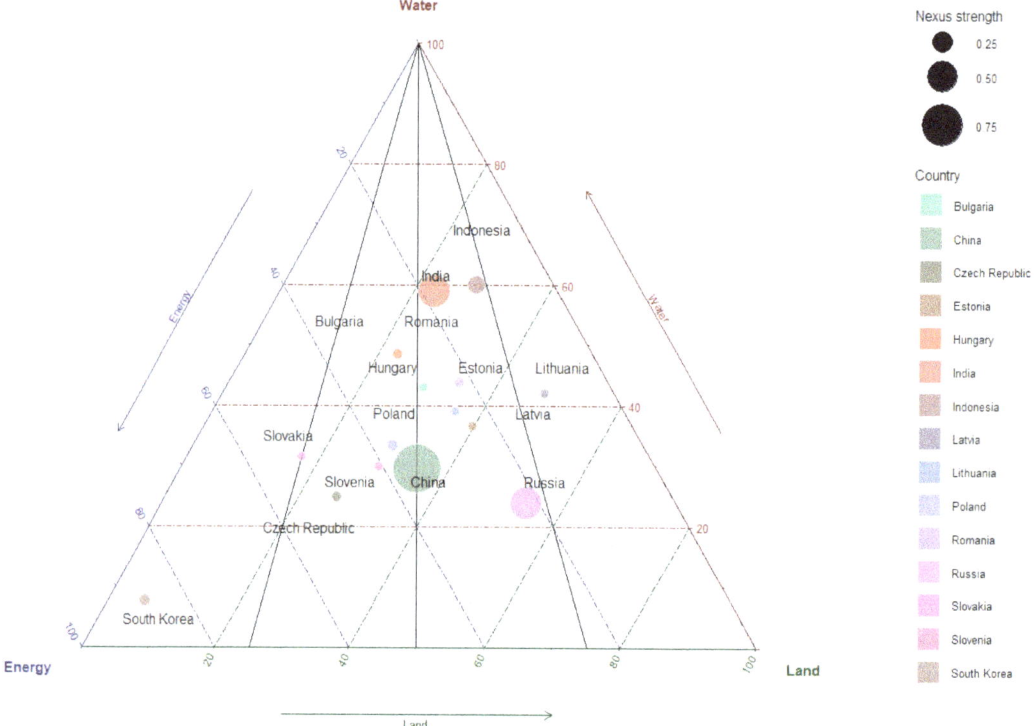

Figure 6. Nexus strength results by country based on the production viewpoint. Lines that cut the point position represent aggregated usage of a given resource. Size of the point/or circle inside the triangle shows the combined contribution of the three resources. Location of a point within the triangle provides a series of key information that can be used to compare its resource use combination with other countries.

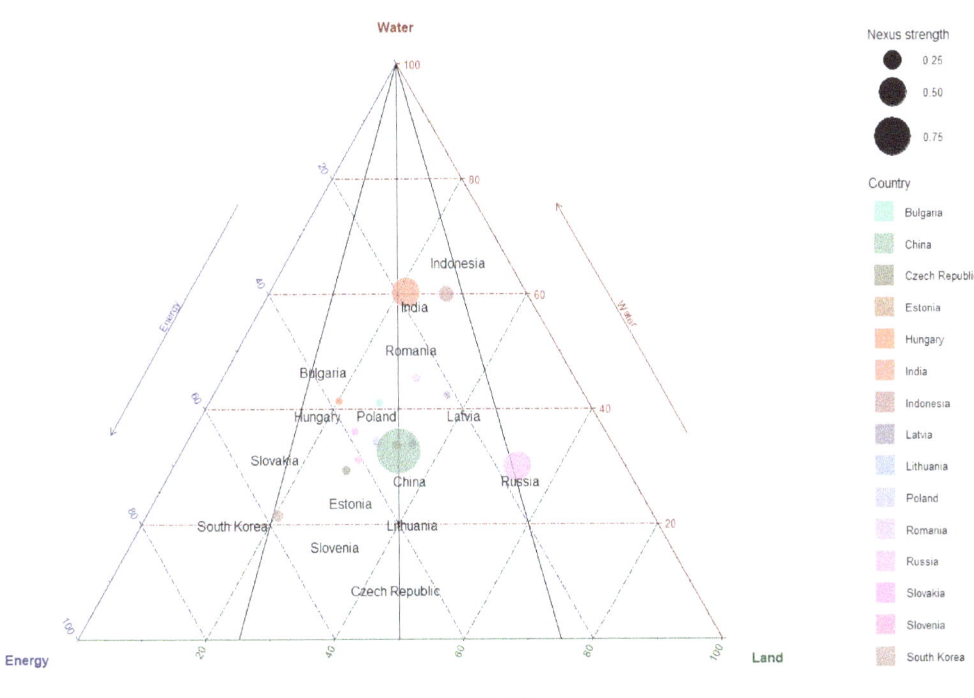

Figure 7. Nexus strength results by country based on the consumption viewpoint. Lines that cut the point position represent aggregated usage of a given resource. Size of the point/or circle inside the triangle shows the combined contribution of the three resources. Location of a point within the triangle provides a series of key information that can be used to compare its resource use combination with other countries.

4. Discussion

Restricted access to vital resources is increasingly seen as a significant impediment to sustainable development. Many technological processes, such as the production of energy and food, encounter resource supply vulnerability issues such as water constraints on energy and food production. Such constraints are usually linked to economic stability, organizational hurdles, political disputes and also the actual availability of supporting natural resources. The nexus concept was suggested in response to such challenges to assist in resource management practices [17].

The nexus focus presented in this work identified key areas of simultaneous resource usage in economic systems. The energy–water–land nexus seems to be strong in China, which is linked particularly to the role of its Others, Electricity and Agriculture industries. Tables A2 and A3 found in the Appendix A present the industrial structure that helps to explain the characteristics of the energy–water–land use flow of China. The second-strongest energy–water–land use flows are in India and Russia. For Russia, Others and Agriculture industries are the main contributors to land and energy use, while for India, Agriculture is the absolute direct contributor to water and land use. The suggested nexus strength indicator that depends on the ternary diagram provided a depiction of the important resource nexuses in economic systems. A point location inside the triangle provides prompt information which can be used to compare and grade resource nexus issues that arise from the transnational economic system. The method used is flexible and can be greatly expanded. The triangle properties, particularly lines and points, not only measure the actual circumstance of a given process, but can predict process behavior as well, based on any change in its driving forces. For example, if there is any sectoral interference, a shift of point position inside the triangle will be noted as well, and one can then test and study the best alternatives. Governments will have a strong tool in the decision-making process on sustainable development for setting policies and selecting alternatives that supersede conventional sectoral interventions.

Currently, the Belt and Road region is at the frontline of undergoing speedy development interventions on a wide scale. Isolated sectoral investment can result in valuing the priorities of one sector in particular over another. Policy actions can be taken in either of the energy, water or land sectors, and we presume that traditionally, decision-making has been solely independent and sector-specific in nature. Thus, the nexus viewpoint should be considered for inter-sectoral negotiations. The current study discussed a balanced nexus structure in the transnational context of the Belt and Road region, and identified important hotspots of simultaneous resource use and associated interlinkages. The use of a sectorally balanced nexus strategy (lowers biases associated with the sectors) serves as a tool to promote discussion to strengthen sectoral collaboration, to potentially accommodate investments that individual sectors would view as sub-optimal and eventually, to boost overall program outcomes.

Amongst all, China is the largest developing country by both population and economic size. Since the BRI was proposed by China, there are concerns that China's trade may lead to natural resource depletion and shifting of detrimental resource effects to neighboring countries. Such concern would certainly plague regional integration and economic cooperation. Thus, in the process of advancing the Belt and Road Initiative, China should develop investment strategies based on the nexus architecture. It is important to support and fund nexus-framed development decisions in the region for better resource management that will certainly help to eliminate misunderstandings. Future research should take on a more dynamic view of scenario development and modeling energy, water and land use flows in the Belt and Road region to provide important information on the resource nexus, so that strategies can be raised by considering the local realities.

Limitations of the Study

This study has some limitations with respect to the method and data. The MRIO model is for the year 2010, so the age of the available data is a significant shortcoming. Additionally, it does not capture trends, a problem that could be solved by using time series data. A number of ways were developed, incorporating multiple spatial scales (for instance, global, national and regional), to capture the regional heterogeneity within the global economy [39,40]. However, increased data inaccuracy is a major disadvantage, due to disaggregation approximations of trade flows from one area in one country to another area in another country. The limitations of IOA are well-documented in the literature [26,41]. For example, data uncertainty due to sectoral aggregation errors. In this research, sectors were aggregated into seven sectors for conformity, which could decrease the accuracy of the results.

As for nexus strength, its development mainly focused on the absolute use of resources, ignoring other aspects related to the nexus debate, i.e., resource availability and price. Additionally, the resource use alone does not necessarily entirely align with the significance of a given nexus issue.

5. Conclusions

Research based on MRIO allows the most detailed and systematic study of resource usage by production and consumption activities at different levels. These activities can trigger the simultaneous use of different resources in a variety of ways, which can be viewed as a kind of resource nexus. This work is placed more effectively to provide new insights into cross-sectoral dynamics and outlines how key resource nexus problems can be identified and given preference in view of alternative and often opposing interests. We established a nexus strength indicator which basically uses ternary diagrams to grade countries based on their combined resources' use and sectoral weighting. Equal sectoral weighting was assigned. In the context of Belt and Road, the findings only provide a snapshot of the transnational resource nexuses' enormous diversity and complexity. However, the overall patterns found can be used to guide future study and resource management activities.

The notion that resources' flow in trade commodities has the ability to challenge environmental policies is supported by various research investigations. The current approach showed that it is possible to evaluate the resource burdens of a region's consumption rather than just production, within its territories. It helped to identify the key regions or industrial sectors that dominate nexus flows, and thus should be prioritized to enhance resource utilization efficiency and lower resource burdens. Further, this study confirmed that drivers of resource consumption can originate from beyond national boundaries. The resource nexus issues are not the same among countries due to disparities in industrial structure, trade policy priorities and resource endowments. Thus, nexus work could disclose different nodes of interest for different countries.

Author Contributions: Conceptualization, G.L.; methodology, A.N., G.L. and F.M.; resources, F.M.; data curation, X.D., Y.H., B.F.G., F.A. and C.M.V.B.A.; writing—original draft preparation, A.N., G.L., F.M. and M.C.; writing—review and editing, A.N., G.L., F.M., A.M.S. and M.C.; supervision, G.L.; project administration, G.L.; funding acquisition, G.L. All authors have read and agreed to the published version of the manuscript.

Funding: This work is funded by the National Natural Science Foundation of China (No. 52070021) and the 111 Project (No. B17005).

Institutional Review Board Statement: Not applicable.

Informed Consent Statement: Not applicable.

Data Availability Statement: Not applicable.

Acknowledgments: The comments of reviewers are acknowledged.

Conflicts of Interest: The authors declare no conflict of interest.

Appendix A

Table A1. 35 sectors aggregated into 7 sectors.

Code	7 Sectors	35 Sectors
1	Agriculture	Agriculture, hunting, forestry and fishing
2	Food and Tobacco	Food, beverages and tobacco
3	Electricity Gas and Water	Electricity gas and water
4	Construction	Construction
5	Transport	Inland transport
		Water transport
		Air transport
		Other supporting and auxiliary transport activities; activities of travel agencies
6	Services	Hotels and restaurants
		Sale, maintenance and repair of motor vehicles and motorcycles; retail sale of fuel
		Wholesale trade and commission trade, except of motor vehicles and motorcycles
		Retail trade, except of motor vehicles and motorcycles; repair of household goods
		Post and telecommunications
		Financial intermediation
		Real estate activities
		Renting and other business activities
		Public admin and defence; compulsory social security
		education
		Health and social work

Table A1. *Cont.*

Code	7 Sectors	35 Sectors
7	Others	Other community, social and personal services Private households with employed persons Mining and quarrying Textile and textile products Leather and footwear Wood and products of wood and cork Pulp paper, printing and publishing Coke-refined petroleum and nuclear fuel Chemical and chemical products Rubber and plastics Other non-metallic minerals Basic metals and fabricated metals Machinery Transport equipment Electrical and optical equipment Manufacturing and recycling

Table A2. Industrial structure of production-based energy–water–land use flow of China in 2010.

	Sector	Total		Local Consumption		International Export		Main International Export Regions	
		SPB	SPB/NPB (%)	LCP	LCP/SPB (%)	IEB	IEB/SPB (%)	Top Three Regions	Ratio (Region/IEB) (%)
Energy (Mtce)	Electricity	1.61×10^6	48.85	1.34×10^6	83.37	2.68×10^5	16.63	India, S.Korea, Russia	10.24
	Others	1.20×10^6	36.44	9.43×10^5	78.48	2.59×10^5	21.52	India, S.Korea, Russia	10.50
	Transport	2.60×10^5	7.88	2.04×10^5	78.55	5.58×10^4	21.45	S.Korea, India, Russia	8.77
	Subtotal	3.07×10^6	93.17	2.49×10^6	81.05	5.82×10^5	18.95	-	10.21
	Total for all sectors (NPBE)	3.30×10^6	100	2.69×10^6	81.57	6.08×10^5	18.43	-	18.43
Water (Mt)	Agriculture	9.04×10^5	91.90	7.91×10^5	87.53	1.13×10^5	12.47	Russia, S.Korea, India	10.32
	Electricity	4.71×10^4	4.79	3.77×10^4	79.93	9.46×10^3	20.07	S.Korea, India, Russia	10.07
	Others	2.11×10^4	2.14	1.52×10^4	71.99	5.91×10^3	28.01	India, S.Korea, Russia	10.29
	Subtotal	9.72×10^5	98.84	8.44×10^5	86.82	1.28×10^5	13.18	-	10.30
	Total for all sectors (NPBW)	9.84×10^5	100	8.55×10^5	86.93	1.29×10^5	13.07	-	13.07
Land (Kha)	Agriculture	5.92×10^5	97.93	5.35×10^5	90.34	5.72×10^4	9.66	Russia, S.Korea, India	10.28
	Others	8.49×10^3	1.40	1.41×10^3	16.56	7.09×10^3	83.44	Russia, India, S.Korea	93.16
	Transport	2.48×10^3	0.41	1.99×10^3	80.40	4.86×10^2	19.60	S.Korea, India, Russia	8.81
	Subtotal	6.03×10^5	99.75	5.39×10^5	89.26	6.48×10^4	10.74	-	19.34
	Total for all sectors (NPBL)	6.05×10^5	100	5.40×10^5	89.26	6.49×10^4	10.74	-	10.74

Note: Taking energy as an example, PB refers to PBE: production-based energy; SPBE: sectoral production-based energy; NPBE: national production-based energy; LCP: production-based energy used for local consumption; IEB: production-based energy embodied in international export; SPB = LCP + IEB; NPB = $\sum S_i PB$, i represents the sector.

Table A3. Industrial structure of consumption-based energy–water–land use flow of China in 2010.

	Sector	Total		Local Production		International Import			
								Main International Import Regions	
		SCB	SCB/NCB (%)	LPC	LPC/SCB (%)	IIB	IIB/SCB (%)	Top Three Regions	Ratio (Region/IEB) (%)
Energy (Ktce)	Others	9.97×10^5	31.01	7.41×10^5	74.27	2.57×10^5	25.73	S. Korea, Russia, India	19.99
	Construction	9.33×10^5	29.01	8.27×10^5	88.63	1.06×10^5	11.37	S. Korea, Russia, India	20.51
	Services	5.82×10^5	18.09	4.92×10^5	84.49	9.02×10^4	15.51	S.Korea, Russia, Indonesia	20.60
	Subtotal	2.51×10^6	78.10	2.06×10^6	81.97	4.53×10^5	18.03	-	20.24
	Total for all sectors (NCBE)	3.22×10^6	100	2.69×10^6	83.66	5.25×10^5	16.34	-	16.34
Water (Mt)	Agriculture	3.76×10^5	35.09	3.55×10^5	94.55	2.05×10^4	5.45	India, Indonesia, Russia	6.30
	Food andTobacco	2.91×10^5	27.17	2.05×10^5	70.69	8.52×10^4	29.31	India, Indonesia, Russia	9.15
	Services	1.48×10^5	13.85	1.13×10^5	76.02	3.55×10^4	23.98	India, Indonesia, Russia	11.67
	Subtotal	8.14×10^5	76.10	6.73×10^5	82.66	1.41×10^5	17.34	-	9.37
	Total for all sectors (NCBW)	1.07×10^6	100	8.55×10^5	79.92	2.15×10^5	20.08	-	20.08
Land (Kha)	Agriculture	3.05×10^5	44.83	2.91×10^5	95.37	1.41×10^4	4.63	Russia, India, Indonesia	3.05
	Food and Tobacco	1.65×10^5	24.19	1.09×10^5	65.99	5.60×10^4	34.01	Russia, India Indonesia	4.21
	Services	8.70×10^4	12.79	6.37×10^4	73.25	2.33×10^4	26.75	Russia, India, Indonesia	7.40
	Subtotal	5.57×10^5	81.81	4.63×10^5	83.22	9.34×10^4	16.78	-	4.83
	Total for all sectors (NCBL)	6.80×10^5	100	5.40×10^5	79.35	1.41×10^5	20.65	-	20.65

Note: Taking energy as an example, CB refers to CBE: consumption-based energy; SCBE: sectoral consumption-based energy; NCBE: national consumption-based energy; LPC: consumption-based energy for local production; IIB: consumption-based energy embodied in international import; SCB = LPC + IIB; NCB = $\sum S_i$CB, i represents the sector.

References

1. White, D.J.; Hubacek, K.; Feng, K.; Sun, L.; Meng, B. The water-energy-food nexus in East Asia: A tele-connected value chain analysis using inter-regional input-output analysis. *Appl. Energy* **2018**, *210*, 550–567. [CrossRef]
2. Le Blanc, D. Towards integration at last? The sustainable development goals as a network of targets. *Sustain. Dev.* **2015**, *23*, 176–187. [CrossRef]
3. Lin, J.Y. "One Belt and One Road" and free trade zones China's new opening-up initiatives. *Front. Econ. China* **2015**, *10*, 585–590.
4. Huang, Y. Understanding China's Belt & Road Initiative: Motivation, framework and assessment. *China Econ. Rev.* **2016**, *40*, 314–321.
5. Brown, L. The Belt and Road Initiative. 2018. Available online: https://www.lehmanbrown.com/wp-content/uploads/2017/08/The-Belt-and-Road-Initiative.pdf (accessed on 4 October 2019).
6. Tian, X.; Geng, Y.; Sarkis, J.; Zhong, S. Trends and features of embodied flows associated with international trade based on bibliometric analysis. *Resour. Conserv. Recycl.* **2018**, *131*, 148–157. [CrossRef]
7. Enderwick, P. The economic growth and development effects of China's One Belt, One Road Initiative. *Strateg. Chang.* **2018**, *27*, 447–454. [CrossRef]
8. Zhang, Y.; Zhang, J.; Tian, Q.; Liu, Z.; Zhang, H. Virtual water trade of agricultural products: A new perspective to explore the Belt and Road. *Sci. Total Environ.* **2018**, *622–623*, 988–996. [CrossRef]

9. Han, L.; Han, B.; Shi, X.; Su, B.; Lv, X.; Lei, X. Energy efficiency convergence across countries in the context of China's Belt and Road initiative. *Appl. Energy* **2018**, *213*, 112–122. [CrossRef]
10. Tian, X.; Hu, Y.; Yin, H.; Geng, Y.; Bleischwitz, R. Trade impacts of China's Belt and Road Initiative: From resource and environmental perspectives. *Resour. Conserv. Recycl.* **2019**, *150*, 104430. [CrossRef]
11. Liu, Y.; Hao, Y. The dynamic links between CO_2 emissions, energy consumption and economic development in the countries along "the Belt and Road". *Sci. Total Environ.* **2018**, *645*, 674–683. [CrossRef]
12. Smajgl, A.; Ward, J.; Pluschke, L. The water-food-energy nexus- realizing a new paradigm. *J. Hydrol.* **2016**, *533*, 533–540. [CrossRef]
13. Endo, A.; Tsurita, I.; Burnett, K.; Orencio, P.M. A review of the current state of research on the water, energy, and food nexus. *J. Hydrol. Reg. Stud.* **2017**, *11*, 20–30. [CrossRef]
14. Wichelns, D. The water-energy-food nexus: Is the increasing attention warranted, from either a research or policy perspective? *Environ. Sci. Pol.* **2017**, *69*, 113–123. [CrossRef]
15. Siciliano, G.; Rulli, M.C.; D'odorico, P. European large-scale farmland investments and the land-water-energy-food nexus. *Adv. Water Resour.* **2017**, *110*, 579–590. [CrossRef]
16. Howells, M.; Rogner, H.-H. Water-energy nexus: Assessing integrated systems. *Nat. Clim. Change* **2014**, *4*, 246–247. [CrossRef]
17. Bazilian, M.; Rogner, H.; Howells, M.; Hermann, S.; Arent, D.; Gielen, D.; Steduto, P.; Mueller, A.; Komor, P.; Tol, R.S.J.; et al. Considering the energy, water and food nexus: Towards an integrated modelling approach. *Energy Policy* **2011**, *39*, 7896–7906. [CrossRef]
18. Ringler, C.; Bhaduri, A.; Lawford, R. The nexus across water, energy, land and food (WELF): Potential for improved resource use efficiency? *Curr. Opin. Environ. Sustain.* **2013**, *5*, 617–624. [CrossRef]
19. Sharmina, M.; Hoolohan, C.; Larkin, A.; Burgess, P.J.; Colwill, J.; Gilbert, P.; Howard, D.; Knox, J.; Anderson, K. A nexus perspective on competing land demands: Wider lessons from a UK policy case study. *Environ. Sci. Policy* **2016**, *59*, 74–84. [CrossRef]
20. Leontief, W. Environmental repercussions and the economic structure: An input-output approach. *Rev. Econ. Stat.* **1970**, *52*, 262–271. [CrossRef]
21. Timmer, M.P.; Dietzenbacher, E.; Los, B.; Stehrer, R.; Vries, G.J.D. An illustrated user guide to the world input-output database: The case of global automotive production. *Rev. Int. Econ.* **2015**, *23*, 575–605. [CrossRef]
22. Miller, R.E.; Blair, P.D. *Input-Output Analysis: Foundations and Extensions*; Cambridge University Press: Cambridge, UK, 2009.
23. Marsh, D.M. The Water-Energy Nexus: A Comprehensive Analysis in the Context of New South Wales. Ph.D. Thesis, Faculty of Engineering and Information Technology Sydney, University of Technology, Sidney, Australia, 2008.
24. Chen, S.; Chen, B. Urban energy–water nexus: A network perspective. *Appl. Energy* **2016**, *184*, 905–914. [CrossRef]
25. Liu, Y.; Wang, S.; Chen, B. Water-land nexus in food trade based on ecological network analysis. *Ecol. Indicat.* **2019**, *97*, 466–475. [CrossRef]
26. Wiedmann, T. A review of recent multi-region input-output models used for consumption-based emission and resource accounting. *Ecol. Econ.* **2009**, *69*, 211–222. [CrossRef]
27. Daher, B.T.; Mohtar, R.H. Water-energy-food (WEF) Nexus Tool 2.0: Guiding integrative resource planning and decision-making. *Water Int.* **2015**, *40*, 748–771. [CrossRef]
28. Bach, H.; Bird, J.; Clausen, T.J.; Jensen, K.M.; Lange, R.B.; Taylor, R.; Viriyasakultorn, V.; Wolf, A. *Transboundary River Basin Management: Addressing Water, Energy and Food Security*; Mekong River Commission: Vientiane, Laos, 2012.
29. Mohtar, R.H.; Daher, B.T. Water, Energy, and Food: The ultimate Nexus. In *Encyclopedia of Agricultural, Food, and Biological Engineering*, 2nd ed.; Heldman, D., Moraru, C.I., Eds.; CRC Press: Boca Raton, FL, USA, 2012.
30. Bizikova, L.; Roy, D.; Swanson, D.; Venema, H.D.; McCandless, M. *The Water-Energy-Food Security Nexus: Towards a Practical Planning and Decision-Support Framework for Landscape Investment and Risk Management*; International Institute for Sustainable Development: Winnipeg, MB, Canada, 2013.
31. Karan, E.; Asadi, S.; Mohtar, R.; Baawain, M. Towards the optimization of sustainable food-energy-water systems: A stochastic approach. *J. Clean. Prod.* **2018**, *171*, 662–674. [CrossRef]
32. Dargin, J.; Daher, B.; Mohtar, R.H. Complexity versus simplicity in water energy food nexus (WEF) assessment tools. *Sci. Total Environ.* **2019**, *650*, 1566–1575. [CrossRef] [PubMed]
33. Kahrl, F.; Roland-Holst, D. China's water-energy nexus. *Water Policy* **2008**, *10*, 51–65. [CrossRef]
34. Voinov, A.; Cardwell, H. The Energy-Water Nexus: Why Should We Care? *J. Cont. Water Res. Edu.* **2009**, *143*, 17–29. [CrossRef]
35. Wang, S.; Chen, B. Energy-water nexus of urban agglomeration based on multiregional input-output tables and ecological network analysis: A case study of the Beijing-Tianjin-Hebei region. *Appl. Energy* **2016**, *178*, 773–783. [CrossRef]
36. Zimmerman, R.; Zhu, Q.; Dimitri, C. Promoting resilience for food, energy, and water interdependencies. *J. Environ. Stud. Sci.* **2016**, *6*, 50–61. [CrossRef]
37. Giannetti, B.F.; Barrella, F.A.; Almeida, C.M.V.B. A combined tool for environmental scientists and decision makers: Ternary diagrams and emergy accounting. *J. Clean. Prod.* **2006**, *14*, 201–210. [CrossRef]
38. Wu, X.D.; Guo, J.L.; Han, M.Y.; Chen, G.Q. An overview of arable land use for the world economy: From source to sink via the global supply chain. *Land Use Policy* **2018**, *76*, 201–214. [CrossRef]
39. Bachmann, C.; Roorda, M.J.; Kennedy, C. Developing a multi-scale multi-region input-output model. *Econ. Syst. Res.* **2014**, *27*, 172–193. [CrossRef]

40. Wenz, L.; Willner, S.N.; Radebach, A.; Bierkandt, R.; Steckel, J.C.; Levermann, A. Regional and sectoral disaggregation of multi-regional input-output tables—A flexible algorithm. *Econ. Syst. Res.* **2014**, *27*, 194–212. [CrossRef]
41. Lenzen, M.; Wood, R.; Wiedmann, T. Uncertainty analysis for multi-region input-output models—A case study of the UK's carbon footprint. *Econ. Syst Res.* **2010**, *22*, 43–63. [CrossRef]

 *energies*

Article

Potential Energy Savings from Circular Economy Scenarios Based on Construction and Agri-Food Waste in Italy

Patrizia Ghisellini [1,*], Amos Ncube [2,*], Gianni D'Ambrosio [3], Renato Passaro [1] and Sergio Ulgiati [4,5]

[1] Department of Engineering, University of Naples "Parthenope", 80143 Naples, Italy; renato.passaro@uniparthenope.it

[2] International PhD Programme/UNESCO Chair "Environment, Resources and Sustainable Development" Department of Science and Technology, University of Naples "Parthenope", 80143 Naples, Italy

[3] Department of Civil Engineering, University of Salerno, 84080 Salerno, Italy; giannid.mod@gmail.com

[4] Department of Science and Technology, University of Naples "Parthenope", 80143 Naples, Italy; sergio.ulgiati@gmail.com

[5] School of Environment, Beijing Normal University, Beijing 100875, China

* Correspondence: patrizia.ghisellini@uniparthenope.it (P.G.); amocube@gmail.com (A.N.)

Abstract: In this study, our aim was to explore the potential energy savings obtainable from the recycling of 1 tonne of Construction and Demolition Waste (C&DW) generated in the Metropolitan City of Naples. The main fraction composing the functional unit are mixed C&DW, soil and stones, concrete, iron, steel and aluminium. The results evidence that the recycling option for the C&DW is better than landfilling as well as that the production of recycled aggregates is environmentally sustainable since the induced energy and environmental impacts are lower than the avoided energy and environmental impacts in the life cycle of recycled aggregates. This LCA study shows that the transition to the Circular Economy offers many opportunities for improving the energy and environmental performances of the construction sector in the life cycle of construction materials by means of internal recycling strategies (recycling C&DW into recycled aggregates, recycled steel, iron and aluminum) as well as external recycling by using input of other sectors (agri-food by-products) for the manufacturing of construction materials. In this way, the C&D sector also contributes to realizing the energy and bioeconomy transition by disentangling itself from fossil fuel dependence.

Keywords: energy savings; circular economy; construction and demolition waste; recycled aggregates; agri-food by-products

Citation: Ghisellini, P.; Ncube, A.; D'Ambrosio, G.; Passaro, R.; Ulgiati, S. Potential Energy Savings from Circular Economy Scenarios Based on Construction and Agri-Food Waste in Italy. *Energies* **2021**, *14*, 8561. https://doi.org/10.3390/en14248561

Academic Editors: Biswajit Sarkar and Gabriele Di Giacomo

Received: 10 November 2021
Accepted: 13 December 2021
Published: 19 December 2021

Publisher's Note: MDPI stays neutral with regard to jurisdictional claims in published maps and institutional affiliations.

1. Introduction

The main research context of the present study is the Construction and Demolition Waste (C&DW) management system of the Metropolitan City of Naples (Italy). This section starts by introducing the relevant environmental and energy impacts of the C&D sector as a whole (Section 1.1), highlighting the need for transitioning to a Circular Economy (CE) (Section 1.2). The goal of this study is described in Section 1.3.

1.1. The Environmental Impacts of C&D Sector

The New Circular Economy Action Plan [1] suggests the urgence of taking actions towards the implementation of CE, particularly in some key product value chains such as C&D (a list of the acronyms used is provided at the end of the manuscript) and agri-foods. In the European Union (EU), around 460 million t/year of C&DW are generated [2], while food waste amounts to 88 million t/year (20% of total food production). The lack of sustainability practices in these sectors largely contributes to the worsening of climate change [3] and other environmental problems [1]. The construction sector in particular is the largest consumer of natural resources [4,5] and this figure is expected to continue in the future [6,7] since urban areas are growing and contributing to the increase of the demand of construction materials and products [8,9]. Sand and gravel are the raw materials most used after water on Earth and their use largely exceed their regeneration rate [10], needed by natural processes to concentrate the raw material [11], and thus is not sustainable [12]. The direct environmental impacts at the extractive sites of such materials are also huge [13,14] (such as to the flora, fauna, habitats, landscape, biodiversity, water bodies) [15] and can be partially mitigated by the adoption of cleaner and more sustainable practices also in compliance with the legislation when available, as in the EU [16].

1.2. Circular Economy Opportunities for C&D Sector

Currently, at the global level it is calculated that about 20–30% of C&DW is recycled or reused. Thus, a change in this pattern is an imperative given the scarcity of natural resources and the associated just above-mentioned environmental impacts due to their extraction [5]. The transition to the CE with a focus on the reduction of the generation of C&DW [16,17] and the increase of their recycling would reduce the dependence on primary resources and improve the efficiency in their use. It will be also beneficial to mitigate the fossil energy demand [18,19] and the related environmental impacts such as global warming [20], simultaneously contributing to the achievement of climate neutrality by 2050 as envisaged in the EU Green Deal [1] and very recently confirmed by EU parliament and the G-20 Rome meeting. In the EU, the production and use of energy accounts for a large share (75%) of GHGs emissions [1]. Moreover, the CE practices for the construction sector also offer the opportunity of improving its environmental performances through the creation of synergistic relationships with other sectors such as agri-food [21] and the use as input of its by-products (e.g., hemp by-products) [22] for the production of construction materials.

1.3. Goal of the Present Study

In this explorative study we mainly evaluate the potential energy savings coming from the recycling of the current annual flows of C&DW available in the Metropolitan City of Naples (Southern Italy). As found by previous literature, the reintroduction of secondary materials from C&DW streams in a new production cycle generates energy savings from avoided landfill disposal as well as limited extraction of raw materials [23–30]. The extent of the life cycle energy savings depends on the recycling potential of the secondary raw materials to substitute the virgin materials of the new products [31]. For example, steel scraps from C&D can be re-converted into valuable materials similar to the virgin materials, whereas in the case of recycled aggregates (RA) their value is currently lower compared to the natural substitutes, resulting in less energy savings [31]. However, in the future should the CE model be more extensively applied to the C&DW sector, the RA could become more suitable substitutes of natural aggregates (NA) [32,33]. This study contributes to the evaluation of CE scenarios in C&DW management that potentially may be beneficial to the achievement of the following United Nations' Sustainable Development Goals: 11 (Sustainable cities and communities), 12 (Responsible consumption and production), and 13 (Climate action) [34].

The present study develops over five sections. In Section 2, we briefly summarize previous studies on the field, whereas in Section 3 (Material and Methods), the main features of the investigated system, the type of data used, and the stages of this Life Cycle Assessment (LCA) study are presented. Section 4 presents the main results, its limitations and proposals for future research avenues, and Section 5 concludes by presenting the main findings, the added value of the present study and their political and managerial implications.

2. Previous Literature on LCA of C&DW Management Systems

So far LCA as a method has been extensively used to analyse the environmental impacts and benefits (including the energy benefits) deriving from the adoption of the CE framework in the C&D sector [5,18,20,23,35–37]. Entire C&DW management systems located in different geographical areas (Italy, Finland) have been investigated by means of LCA [24] or in combination with other tools such as GIS as in [30] or methods such as Life Cycle Costing and Material Flow Accounting [38]. Further analytical frameworks have been also proposed to study C&DW management systems in a more comprehensive sustainability perspective such as by [39], integrating environmental and resource-related impacts, and social and economic impacts.

The energy aspects are key factors that affect the environmental competitiveness of recycled aggregates compared to NA [13]. Studies have found that energy consumption for the extraction and production of NA is higher (1664.11 MJ) compared to the amount used for the recycling of C&DW (246.41 MJ). The largest contribution to the Cumulative Energy Demand (CED) is due to the non-renewable energy category because of the prevalent use of fossil fuels in the processes [28].

Many LCA studies have also found that the transport stage is significant in the life cycle of RA and their collection and re-use should be considered within a limited distance [10,19,20,23,28,40–42]. This highlights that the main market both for the recycling and the delivery of RA should be local. As a result, e.g., the planning of recycling facilities should take into account the relevance of the transport distances for the sustainability of C&DW recycling option and the associated energy, environmental and economic costs [28,43].

With regard to LCA studies analysing entire C&DW management systems, [25] found that the avoided impacts of the life cycle of C&DW in the province of Torino (Italy) are higher than the energy and environmental impacts generated in the life cycle of C&DW. The net energy savings resulted 250 MJ/t whereas the total net contribution to global warming amounted to about 14 kg CO_2 eq. [25]. Reference [38] reported higher avoided environmental impacts (-360 kg CO_2 eq.) for the life cycle of C&DW in Finland including the pre-treatment stage, treatment (landfilling), recovery/utilization, transportation, and avoided production, whereas [19], by modelling three scenarios (current scenario, landfilling scenario and best-case scenario), found that only the latter yielded avoided energy impacts equal to -24 MJ-eq./tonne of managed C&DW whereas the contribution to climate change was -1.78 kg CO_2 eq. Finally, [39] also considered three scenarios: baseline, linear with total disposal of C&DW in landfilling and best practice scenario based on the adoption of selective demolition and an increased amount of high-quality RA produced in stationary recycling plants compared to the baseline scenario. Their indicators in the best practice scenario show that the management of 1 t of C&DW can save 18 kg CO_2 -eq./t and about 6 kg oil-eq./t.

3. Material and Methods

In this section we summarize the main features of the C&DW system under investigation as well as of the Life Cycle Assessment (LCA) model developed in the present study. LCA, as a well-known tool for evaluating the environmental aspects and potential impacts of products, processes and services, was chosen as the main method of analysis and performed according to the standard ISO 14040:2006 [44].

3.1. The Investigated C&DW Generating System

The C&W management system of the Metropolitan City of Naples is considered in this study. The Metropolitan City of Naples is one of the five provinces of Campania Region (Southern Italy) (Figure 1). Its total surface covers a small area (1179 km^2, 8.6%) of the whole regional territory but hosts more than the half of the total regional population. The population density is very high (2630 inhabitants/km^2) both compared to the other provinces of Campania Region and Italy. In administrative terms, the Metropolitan City of Naples was established under the Italian Law No. 56/2014 replacing the Province of Naples from 1 January 2015 while maintaining the same land area.

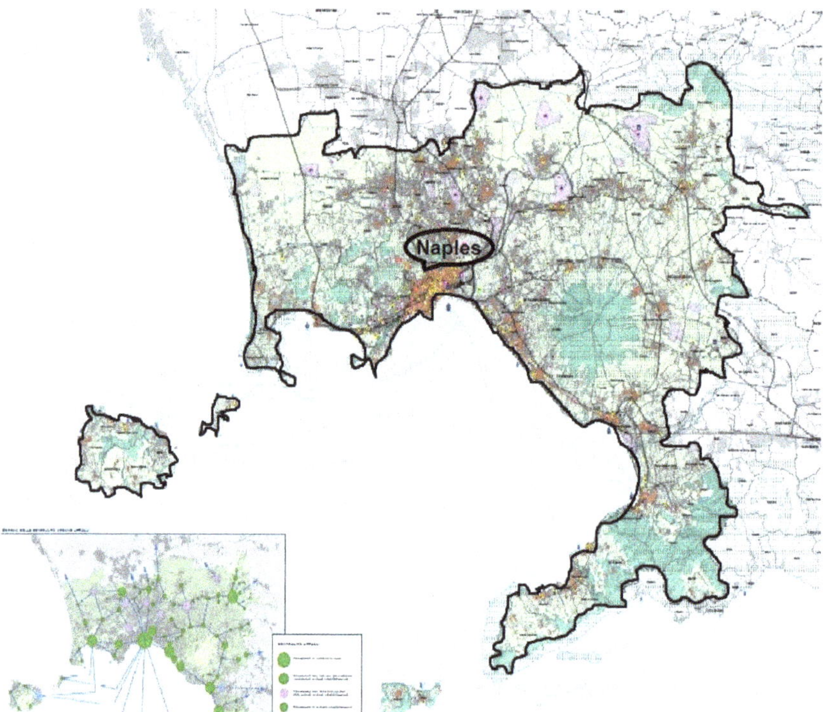

Figure 1. The location of the Metropolitan City of Naples (in Campania Region, Southern Italy). Adapted from [45]. Note: the small box in the Figure 1 depicts the main urban centres of the Metropolitan City of Naples with green circles. Naples is the most important city in the area and has the largest circle compared to the other towns.

With regard to C&DW, the available primary data evidence that its production amounted to 9.13×10^5 tonnes in the year 2017 consisting of non-hazardous C&DW (9.02×10^5 tonnes) and hazardous C&DW (1.12×10^5 tonnes). Figure 2 shows the composition of the generated non-hazardous C&DW in the Metropolitan City of Naples. The main fractions composing the total amount are mixed C&DW (47.37%), soil and stones (24.81%), iron and steel (7.03%), concrete (6.69%), and bituminous mixtures (5.25%).

After the collection on the construction or demolition sites, the C&DW are sent to the available recycling plants in the Metropolitan City of Naples. The data evidence that, in the year 2017, they were almost entirely treated under the management option "R5" (87% of the total amount), that entails the recycling/recovery of other inorganic substances, whereas minor fractions (10%) were treated under the option "R4", that regards the recycling/recovery of metallic compounds. A low fraction (3%) was stocked at the end of the year (31 December 2017). Hazardous C&DW were a minor fraction of the total annual C&DW (1%) and after the generation they were mainly disposed of under the category "D15", involving a preliminary disposal of C&DW before other kinds of disposal options. After that, only a small fraction (973 tonnes) of the total amount of hazardous C&DW produced annually remains in the Metropolitan City, as most of them are sent to other Italian Regions.

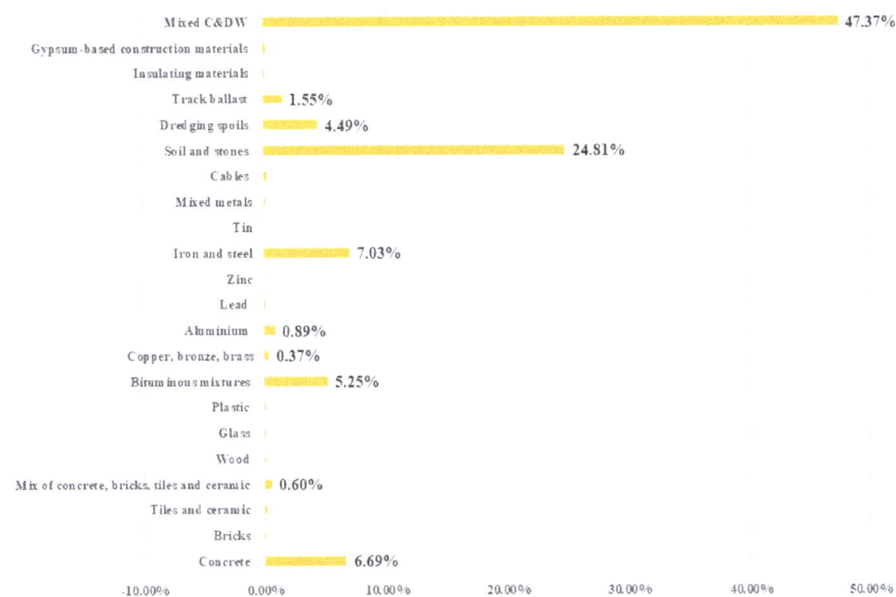

Figure 2. Main fractions composing the amount of C&DW produced in the year 2017 in the Metropolitan City of Naples.

3.2. Life Cycle Assessment Method

The LCA as a technique has been developed since the sixties to better understand and address the environmental impacts of products, services and activities [44,46,47] in a wide range of sectors [48,49], including construction [50–53] and demolition [54,55] activities. The ISO 14040 (2006) [44], that is the main normative framework for the LCA, suggests its use for many purposes:

- Improvement of the environmental performance of products throughout their life cycle;
- Support to decision-makers in industry, government or non-government organizations (e.g., strategic planning, priority setting, product or process design or redesign);
- Selection of relevant indicators of environmental performance, including measurement techniques;
- Marketing (e.g., implementation of an ecolabelling scheme of type I (ISO 14024) such as the Eco-label), or making an environmental claim (e.g., the environmental labelling of type II regulated by the ISO 14021) or adhering to an environmental product declaration (e.g., the environmental labelling of type III within the ISO 14025 standards).

The LCA takes into account the environmental aspects and the potential environmental impacts of a product (e.g., the use of natural resources and the environmental consequences of their use) in a holistic manner given that it considers the whole life cycle of a product from raw material extraction, through production, use, end-of-life treatment, recycling and final disposal (i.e., cradle-to-grave). In so doing, the LCA stimulates industrial activities to look beyond the traditional focus on production sites and manufacturing processes, so to include the environmental impacts of a product in all the other stages, including the end-of-life stages and the return to the original or new production cycle, by means of the reuse of products or components [56–58] or the recycling of materials [59]. This contributes to closing the production and consumption cycle as suggested by the CE framework while maximizing resource reuse (also avoiding their future extraction) and the reduction of waste disposed of in landfills [25,60].

The procedural framework for performing an LCA consists of four phases that comprises: the definition of the goal and scope of the LCA study, the life cycle inventory analysis (LCI), the life cycle impact assessment (LCIA), the life cycle interpretation, reporting and critical review of the LCA, limitations of the LCA, the relationship between the LCA phases, and conditions for use of value choices and optional elements (ISO 14040: 2006) [44].

In waste management, the LCA is useful in the comparison of the environmental impacts of products made of natural and recycled materials since it provides the opportunity to expand the system boundaries beyond the waste management processes [61]. In this perspective, it is also applied to identify the best management options for waste products available in the waste hierarchy (e.g., reuse, recycling, waste to energy and landfilling), being considered a very good scientific alternative to the latter [27].

3.2.1. Goal and Scope

The LCA methodology is applied in this study with the aim of evaluating the energy savings coming from the implementation of recycling scenarios for the different fractions of non-hazardous C&DW generated in the Metropolitan City of Naples in the year 2017.

The present study further integrates previous works of the research group [62,63], having the goal of providing scientific support and useful feedback to the Public Administration of Campania Region that is in charge of the management of C&DW. These latter are classified as special waste in Italy, and are a specific matter of regional authorities, that by means of regional plans, decide the main strategies for such kind of special waste. The functional unit considered in this study is 1 tonne of recycled non-hazardous C&DW.

The system boundaries include the stages and associated processes to the recycling of the main fractions composing the total non-hazardous C&DW (mixed C&DW, soil and stones, iron and steel, aluminium, concrete and bituminous mixtures) (Figure 2). Therefore, the stages considered in this LCA study for the recycling scenario are the following:

- Collection and transportation of the generated C&DW to the recycling plants of the Metropolitan area;
- Recycling of the most relevant materials (mixed waste, iron and steel, Aluminium, soil and stones, concrete) into recycled aggregates of different types (A, B, C) and recycled metals;
- Delivery of the RA and secondary metals and their reintroduction in the production cycle (it was assumed to occur in the local market so as to reduce as much as possible the contribution of this stage);
- Avoided landfilling;
- Avoided extraction and production of virgin materials.

The above first three steps require energy and materials for collection and processing in order to make the recycled materials available to the user. These costs and related impacts are referred to in the following section of this study as "induced", in so meaning that they are needed to implement the recycling process. However, the recycled products allow additional savings in that the landfill and mine operations are avoided. We will refer with the term "avoided" to these much larger costs and impacts that will be no longer needed thanks to the recycling processes, in so pointing out the huge benefit of C&DW recovery.

Figure 3 considers the boundaries of the system and the main unit of process. With regard to the output of the recycling stage, due to the lack of data of the quality of the recycled aggregates, we assumed that all the concrete C&DW could be recycled into recycled aggregates of higher quality (Type A) that can be used in concrete production (UNI EN 12620 Standard). We assumed that the other C&DW fractions could be recycled into aggregates of type B and type C in conformity to the UNI EN 13242 standard. Our assumptions are based on the data of the ARPAC Campania from which result that almost the whole amount of non-hazardous C&DW inert fraction generated annually is recovered under the category R5 (recycling/recovery of other inorganic substances) as described in the annual reports by the Italian Institute for Environmental Protection and Research (ISPRA).

In addition to the evaluation of the recycling scenario for C&DW into recycled aggregates of different types, this study also considers expanding the analyses to the production of concrete to indicate (as an example) the end-use of recycled aggregates in the Metropolitan City of Naples. In that, a comparison of concrete produced from natural, recycled and green aggregates (using agro-industry by-products) is proposed. We assumed the use of hemp-integrated aggregates (aggregates enriched with hemp by-products) for the production of green concrete in agreement with our goal of exploring synergies between the construction and the agri-food sector. In this case we applied the allocation procedure for the partitioning of the energy impacts on the basis of the fact that "when a process has two co-products, the allocation is performed to both of them, generally based on their energy content or their mass or their fraction of economic value" [44,64].

The cumulative energy demand (CED) method [65] was chosen in the present study as LCA impact assessment method to assess the energy consumption and savings related to the recycling of 1 tonne of C&D waste in the Metropolitan City of Naples. Considering a zero-burden approach, CED represents all the direct and indirect energy input flows including the collection and transportation of C&D waste to the recycling plant [66]. According to [67], CED has been criticized as a single-score life cycle impact assessment method and in order to counter this constraint, this paper chose to incorporate the ReCiPe MidPoint and Endpoint method [68] pointing towards decision-making to include environmental impact indicators affecting human health, resources and ecosystems scores. The SimaPro version 9.1.1 [69] software tool is used to both the CED and ReCiPe impact scores.

We complement this study with a further assessment where we evaluated and compared the energy impacts (CED) of conventional concrete with two alternative concretes made of RA and hemp by-products in order to explore the sustainability of this latter material. There is an increasing interest in reintroducing the hemp crop in Italy and in the Campania Region due to the wide range of applications in industry that this crop could have. This latter analysis can be considered preliminary to future research works of the research team of the authors.

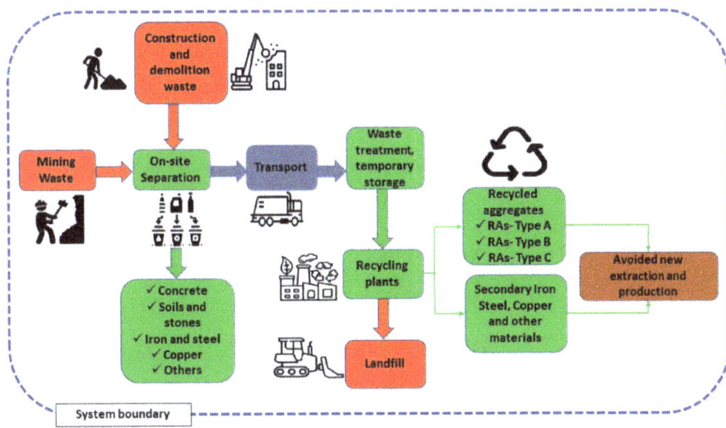

Figure 3. System boundaries of the LCA study.

3.2.2. Life Cycle Inventory (LCI)

This second phase in the LCA consists of an inventory of input/output data of the system under investigation and then involves the collection of the data that are necessary for achieving the goal of the study (ISO 14040: 2006) [44].

The data in this LCA study consist of both primary and secondary data. The primary data regard the annual flows of C&DW generated in the Metropolitan City of Naples in the year 2017 in all projects of construction and demolition of buildings or infrastructures. The data were kindly provided by the Campania Regional Agency for the Environmental Protection (ARPAC).

The secondary data collected regarded the transport stage of the C&DW waste from the construction sites to the modelled recycling plants: Ecoinvent 3.8 database [70] and previous literature [24,31,39]. We assumed to cover a distance of 30 km which aligns with the distance considered by [39]. This latter study was applied to the Campania Region which hosts the Metropolitan City of Naples as one of the five provinces and the area of investigation in this LCA. The data of the treatment of C&D waste at the recycling plant were adopted from [19] based on a number of recycling facilities in the Lombardia Region in Northern Italy.

The landfill option was adopted from the Ecoinvent 3.8 [70] database for a sanitary landfill treatment of inert waste (Europe without Switzerland).

Tables 1 and 2 show the specific inventories (input and output) relating to the recycling of C&D waste, avoided extraction and production of virgin construction materials and finally the production of concrete from natural, recycled and agri-food (hemp–concrete) aggregates. Table 1 includes as input 1 tonne of recycled C&DW composed of mixed C&DW (47.37%), soil and stones (24.81%), iron and steel (7.03%), concrete (6.69%) and bituminous mixtures (5.25%). Table 2 does not include the input flow of C&D waste considering a zero-burden approach but instead includes resources for collection and treatment.

For the comparison of the different types of concrete (made of NA, RA and hemp by-products), we collected the data from the study by [71] related to the production of conventional and recycled concrete as well as from [22] for the production of hemp concrete.

Table 1. Inventory data for 1 tonne of C&DW collected and recycled in the Metropolitan City of Naples.

1	Processes	Amount	Unit	CED (MJ)
	Collection and recycling of C&D waste (functional unit)	1	tonne	
	Avoided landfilling			
	Inert waste (Europe without Switzerland) ⎮ landfill (Ecoinvent 3.8)	1	tonne	
	Materials/fuels (Input)			
	Diesel, low sulphur	0.68	kg	38.58
	Ferromanganese, high-coal, 74.5% Mn (GLO) ⎮ market for ⎮ APOS, S	0.02	kg	0.44
	Transport, freight, lorry >32 metric ton, EURO5 (RER) ⎮ market for transport, freight, lorry >32 metric ton, EURO5 ⎮ APOS, S	30	tkm	45.52
	Water	3.7	kg	0.03
	Lubricating oil (RER) ⎮ market for lubricating oil ⎮ APOS, S	0.001	kg	0.07
	Synthetic rubber (GLO) ⎮ market for ⎮ APOS, S	0.0043	kg	0.38
	Electricity, medium voltage (IT) ⎮ market for ⎮ APOS, S	1.13	kWh	11.58
	Total CED			96.59
	Outputs			
	Recycled aggregates Type A	66.9	kg	
	Recycled aggregates Type B	336.28	kg	
	Recycled aggregates Type C	504.52	kg	
	Recycled Iron and Steel	70.3	kg	
	Recycled aluminium	22.10	kg	
2	Potentially avoided landfilling and mining and production of virgin construction material	1	tonne	
	Avoided landfilling of inert material	1	tonne	
	Avoided steel production	70.3	kg	
	Avoided aluminium production	22.1	kg	
	Avoided production of other virgin construction materials	504.42	kg	
	Avoided extraction of gravel	336.28	kg	
	Concrete production	66.9	kg	

Table 2. Inventory data for the production of conventional concrete and the alternative options made of recycled aggregates and hemp by-products.

Input and Output	Amount	Units
Concrete from natural aggregates *	1	m^3
Materials/fuels (input) *		
Cement, Portland (Europe without Switzerland) ∣ market for ∣ APOS, S	300	kg
Gravel, crushed (RoW) ∣ market for gravel, crushed ∣ APOS, S	1890	kg
Water, deionized (Europe without Switzerland) ∣ market for water, deionized ∣ APOS, S	105	kg
Adhesive mortar (GLO) ∣ market for ∣ APOS, S	3.3	kg
Transport, freight, lorry 7.5–16 metric ton, EURO5 (RER) ∣ market for transport, freight, lorry 7.5-16 metric ton, EURO5 ∣ APOS, S	50	tkm
Concrete from recycled aggregates *	1	m^3
Materials/fuels (input) *		
Cement, Portland (Europe without Switzerland) ∣ market for ∣ APOS, S	320	kg
Water, deionized (Europe without Switzerland) ∣ market for water, deionized ∣ APOS, S	130	kg
Concrete mixing factory (CH) ∣ construction ∣ APOS, S	4.57×10^{-7}	p
Lubricating oil (GLO) ∣ market for ∣ APOS, S	1.19×10^{-2}	kg
Steel, low-alloyed, hot rolled (GLO) ∣ market for ∣ APOS, S	2.38×10^{-2}	kg
Synthetic rubber (GLO) ∣ market for ∣ APOS, S	7.13×10^{-3}	kg
Electricity/heat		
Electricity, medium voltage (IT) ∣ market for ∣ APOS, S	4.36	kWh
Heat, district or industrial, natural gas (RER) ∣ market group for ∣ APOS, S	1.04	MJ
Recycled aggregates	1890	kg
Green concrete from Agri-industry (Hemp by-products) aggregates	1	m^3
Materials/fuels (Input) (**) and (*)		
Water, deionized (Europe without Switzerland) ∣ market for water, deionized ∣ APOS, S	130	kg
Concrete mixing factory (CH) ∣ construction ∣ APOS, S	4.57×10^{-7}	p
Lubricating oil (GLO) ∣ market for ∣ APOS, S	1.19×10^{-2}	kg
Steel, low-alloyed, hot rolled (GLO) ∣ market for ∣ APOS, S	2.38×10^{-2}	kg
Synthetic rubber (GLO) ∣ market for ∣ APOS, S	7.13×10^{-3}	kg
Sun hemp plant, harvested (GLO) ∣ market for sun hemp plant, harvested ∣ APOS, S	1570	kg
Cement, pozzolana and fly ash 36–55% (Europe without Switzerland) ∣ market for cement, pozzolana and fly ash 36–55% ∣ APOS, S (*)	320	kg

(*) [71]; (**) [22].

3.2.3. Life Cycle Impact Assessment

As the third phase in an LCA study, the impact assessment allows to determine the potential contribution on the environment and human health generated by a product or service in its life cycle. The inputs and outputs of the inventory phase are assigned to specific impact categories concerning internationally recognized environmental effects as significant (classification), so as to be able to quantify, through specific characterization methods, the total contribution that the product or service generates to each of the environmental effects considered. In that, the purpose of this phase is elaborating the information resulting from the LCI and better understand their environmental significance (ISO 14040: 2006). The results of this phase are presented in detail in the following Section 4.

4. Results

This section shows the results obtained after processing the inventory data (reported in Tables 1 and 2) of the recycling scenario for the main fractions of C&DW by means of the LCA SimaPro 9.1.1. software tool [69]. In the second part of this section, we show the results of an explorative analysis where we compare the concrete blocks made of NA and RA as well as of hemp by-products.

4.1. Energy and Environmental Impacts of the Recycling Scenario for C&DW

Table 3 shows the results in terms of energy related characterized CE impacts associated with the functional unit (1 tonne of collected and recycled C&DW). The transport stage and the recycling plant stage, both due to the use of diesel, are the most significant energy upstream factors as shown by the higher values compared to the other inputs. The transport and recycling stages mainly contribute to the non-renewable fossil energy category (91.31 MJ) within the total CED. This leads to determine that the life cycle of 1 tonne of C&DW mainly generate impacts related to the non-renewable fossil category with small contributions by the other non-renewable (nuclear and biomass) and renewables (biomass, wind, solar and geothermal) categories.

These results are clearly evidenced in Figure 4 that shows the percentage contribution of each input to the different energy impact categories (fossil, hydro, nuclear, etc). The last column is the total CED, indicating that transport stage and diesel used in the recycling plant contribute to about 90% of the total CED impacts (non-renewable and renewable sources). Electricity and diesel (non-renewable fossil energy) contribute significantly to the energy demand of the recycling facility, due to the mechanical operations for sorting waste and their treatment for the production of RA.

Table 3. Characterized induced CED impacts associated with the collection and recycling of 1 tonne of C&DW.

CED Impact Categories	Unit	Transport	Ferromang.	Water	Lubricat. Oil	Diesel	Synthetic Rubber	Electricity	Total CED
Non-renew. Fossil	MJ	44.13	0.25	0.02	0.06	38.34	0.34	8.17	91.31
Non-renew. Nuclear	MJ	0.82	0.05	0.01	0.00	0.14	0.02	1.26	2.29
Non-renew. Biomass	MJ	0.00	0.00	0.00	0.00	0.00	0.00	0.00	0.00
Renew. Biomass	MJ	0.17	0.01	0.00	0.00	0.03	0.01	0.50	0.72
Renew. (w. Solar, geo.)	MJ	0.08	0.01	0.00	0.00	0.01	0.00	0.56	0.66
Renew. Water	MJ	0.31	0.13	0.00	0.00	0.05	0.01	1.10	1.61
Total induced CED impacts	MJ	45.52	0.44	0.03	0.07	38.58	0.38	11.58	96.59

Note: Renew. (w. solar, geo), renewables (wind, solar, geothermal).

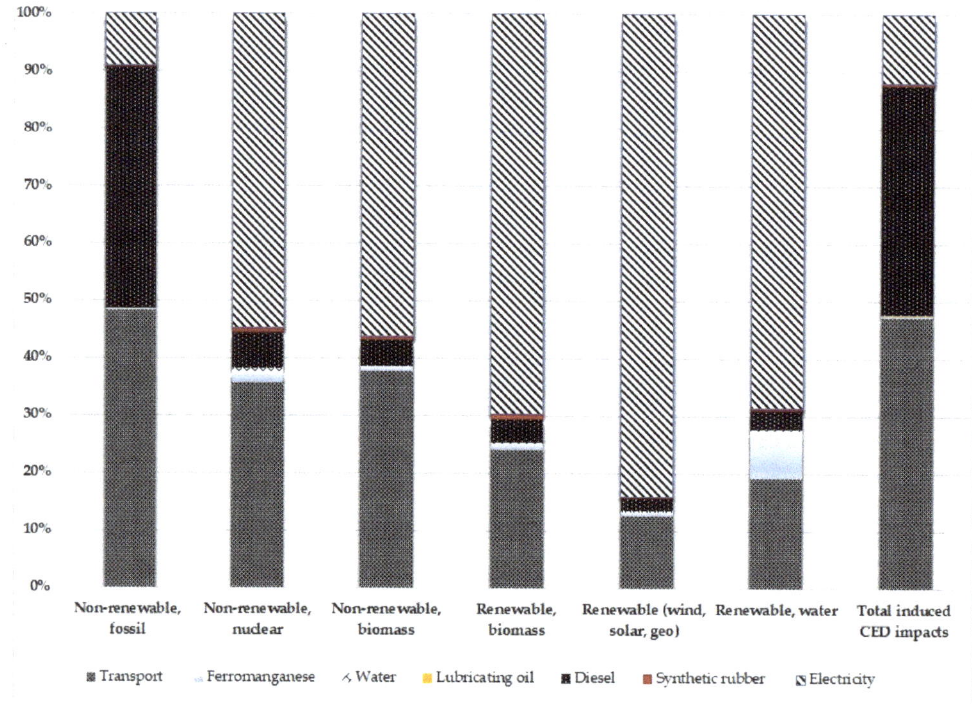

Figure 4. Percentage values of induced CED impacts associated to collection and recycling of 1 tonne of C&DW (from Table 3). Note: Renewable category comprises wind, solar and geothermal (wind, solar, geo).

In the year 2018, the Italian electricity mix was composed of 45% natural gas (a fossil fuel), followed by hydroelectricity for 16.5% and other renewable energy sources accounting for less than 25% combined (biomass, solar and wind). If the energy transition is realized, in the light of the need for reducing the contribution to climate change and greenhouse emissions, there is a possibility to completely replace fossil fuels with renewable fuels (at least for the production of electricity) enabling the reduction of the impacts caused by non-renewable fossils [72,73]. In order to reduce the contribution to global warming it would be important to understand how to replace fossil fuels with renewables in the light of the recent IPCC report on climate change. On the other hand, the avoided extraction and production of virgin construction material replaced by secondary materials will favour the transition to CE thus improving overall energy savings.

Table 4 shows the avoided characterized CED impacts in the life cycle of 1 tonne of C&DW. The high share of prevented impacts (1181.13 MJ) comes from the avoidance of steel production in all the CED categories (non-renewables and renewables). Moreover, avoided aluminium and avoided virgin materials also led to non-negligible avoided CED impacts. The same impacts are shown in Figure 5, as percentage values in each category.

The last column of the Table 4 shows the net energy savings arising from the difference between the induced and avoided CED impacts. In total they amount to −1628.98 MJ. The highest contribution to the total is due to the savings realized in the non-renewable fossil component of the CED (−1498.40 MJ).

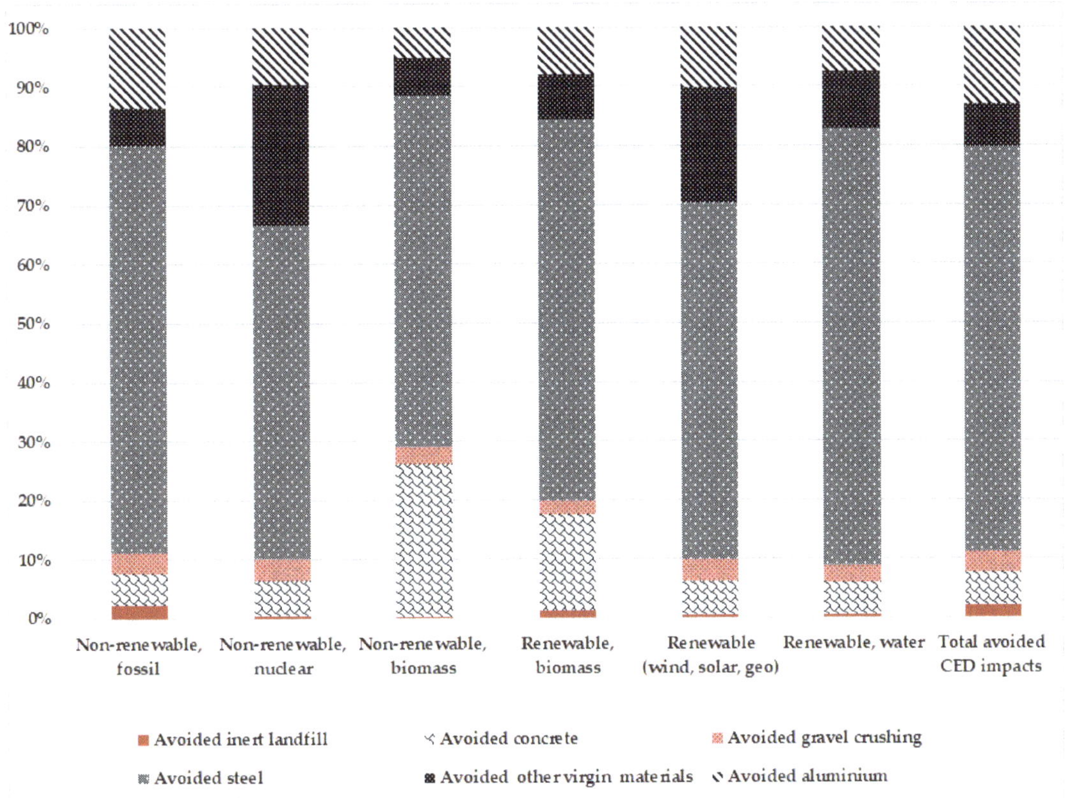

Figure 5. Percentage values of avoided CED impacts associated to the collection and recycling of 1 tonne of C&DW.

Table 4. Avoided versus induced characterized CED impacts associated to the collection and recycling of 1 tonne of C&DW.

CED IMPACT Categories	Unit	Avoided Inert Landfill	Avoided Concrete	Avoided Gravel Crushing	Avoided Steel	Avoided Other Virgin Materials	Avoided Aluminium	Total Avoided CED Impacts	Total Induced CED Impacts	Net Energy Savings
Non-renew. Fossil	MJ	−37.19	−85.88	−54.46	−1095.94	−102.23	−214.01	−1589.71	91.31	−1498.40
Non-renew. Nuclear	MJ	−0.32	−4.06	−2.59	−38.26	−16.10	−6.57	−67.89	2.29	−65.60
Non-renew. Biomass	MJ	0.00	−0.02	0.00	−0.05	−0.01	0.00	−0.09	0.00	−0.09
Renew. Biomass	MJ	−0.33	−3.73	−0.55	−14.90	−1.76	−1.86	−23.12	0.72	−22.40
Renew. (w, Solar, geo)	MJ	−0.04	−0.43	−0.29	−4.56	−1.48	−0.78	−7.59	0.66	−6.92
Renew. Water	MJ	−0.17	−2.06	−1.13	−27.42	−3.58	−2.81	−37.17	1.61	−35.56
Total Av. CED impacts	MJ	−38.05	−96.18	−59.03	−1181.13	−125.15	−226.03	−1725.58	96.59	−1628.98

Notes: Non-renew. (non-renewable); Renew. (renewable); Renew. (w, sol, geo), renewable (wind, solar, geothermal). Total Av. CED Impacts (total avoided CED impacts).

The contribution of steel in total avoided CED impacts is also well highlighted in Figure 5 showing the percentage values of all avoided factors in the life cycle of 1 tonne of C&DW.

As a complement to Table 4, Table 5 summarizes the LCA induced environmental characterized impacts associated with the collection and recycling of 1 tonne of C&DW. The latter contributes to global warming by realizing in total 3.74 kg CO_2 equiv. with the transport stage mainly contributing with 2.73 kg CO_2 equiv. Lower absolute values of GHG emissions are released by the diesel and electricity used in the recycling plants. The use of fossil fuels in the transport and recycling stages translates into environmental impacts in the fossil resource scarcity category. Percentage impacts for this process are also shown in Figure 6, for easier identification of the most contributing steps and flows.

For the sake of clearer identification of the main contributing inflows to the LCA impacts, Figure 6 expresses selected environmental impact categories highlighting transport, electricity and diesel as dominating input flows which are carrying a significant proportion of the environmental burden associated with the collection and recycling of 1 tonne of C&D waste.

Table 6 evidences the avoided environmental impacts resulting in the life cycle of 1 tonne of C&DW. The avoidance of landfilling generates environmental benefits in terms of avoided GHG emissions of 2.56 kg CO_2 equiv. The environmental benefits of steel recycling are relevant as they avoid the production of primary steel and the associated release of GHG emissions (-145.29 kg CO_2 equiv.).

The difference from induced (Table 5) and avoided (Table 6) environmental components result in a negative net contribution to global warming (-181.13 kg CO_2 equiv.) and to fossil resource scarcity (-32.56 kg oil eq.) evidencing the environmental benefits of recycling.

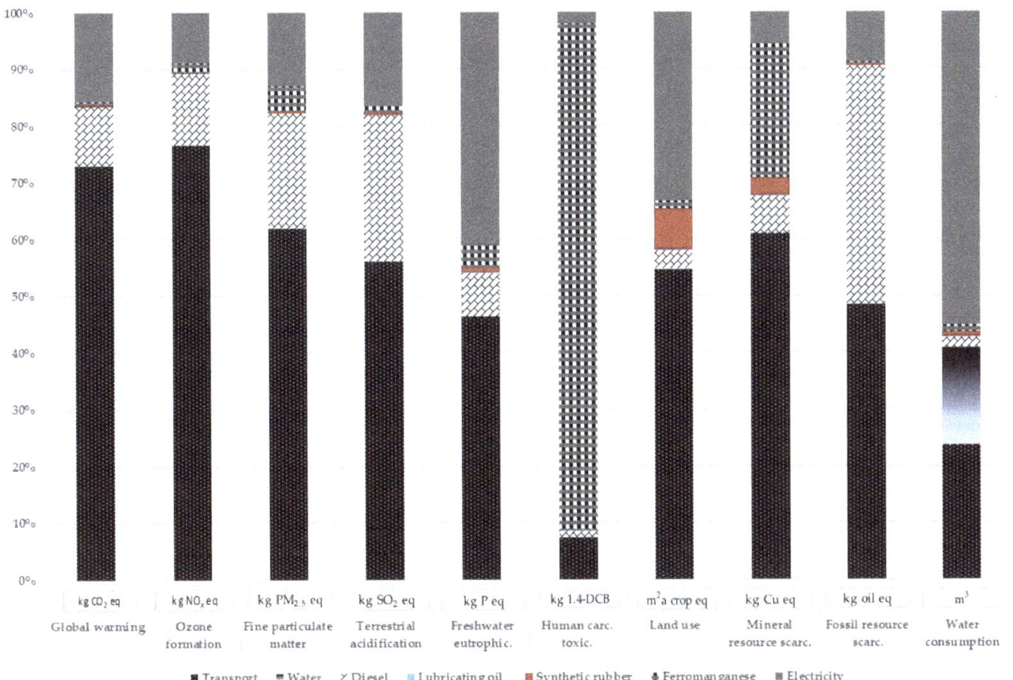

Figure 6. Percentage values of environmental induced impacts coming from the collection and recycling of 1 tonne of C&DW (from Table 5).

Table 5. LCA induced environmental characterized impacts associated to the collection and recycling of 1 tonne of C&DW.

Impact Categories	Unit	Transport	Water	Diesel	Lubric. Oil	Synthetic Rubber	Ferromang.	Electricity	Total ind. env. imp.
Global Warming	kg CO_2 eq.	2.73	0.00	0.39	0.00	0.01	0.02	0.59	3.74
Ozone Formation	kg NOx eq.	0.01	0.00	0.00	0.00	0.00	0.00	0.00	0.01
Fine Partic. Matter	kg $PM_{2.5}$ eq.	0.00	0.00	0.00	0.00	0.00	0.00	0.00	0.01
Terrestrial Acidific.	kg SO_2 eq.	0.01	0.00	0.00	0.00	0.00	0.00	0.00	0.01
Freshwater Eutroph.	kg P eq.	0.00	0.00	0.00	0.00	0.00	0.00	0.00	0.00
Human carc. Toxicity	kg 1,4-DCB	0.05	0.00	0.01	0.00	0.00	0.61	0.01	0.68
Land Use	m^2a crop eq.	0.36	0.00	0.02	0.00	0.05	0.01	0.22	0.65
Miner. Resour. Scarc.	kg Cu eq.	0.01	0.00	0.00	0.00	0.00	0.00	0.00	0.01
Fossil Resour. Scarc.	kg oil eq.	0.96	0.00	0.84	0.00	0.01	0.01	0.18	1.99
Water Consumption	m^3	0.01	0.00	0.00	0.00	0.00	0.00	0.01	0.02

Table 6. Environmental avoided characterized impacts in the life cycle of 1 tonne of C&DW.

Impact Categories	Unit	Avoided Inert Landfill	Avoided Concrete	Avoided Gravel Crushing	Avoided Steel	Avoided Other v. Materials	Avoided Aluminium	Total Av. env. Impacts	Net Environ. Impacts
Global Warming	kg CO_2 eq.	−2.56	−10.60	−3.91	−145.29	−7.75	−14.75	−184.87	−181.13
Ozone Formation	kg NOx eq.	−0.03	−0.03	−0.02	−0.31	−0.04	−0.09	−0.53	−0.52
Fine Partic. Matter	kg $PM_{2.5}$ eq.	−0.01	−0.01	−0.01	−0.23	−0.02	−0.03	−0.30	−0.30
Terrestrial Acidific.	kg SO_2 eq.	−0.01	−0.03	−0.02	−0.37	−0.04	−0.06	−0.52	−0.51
Freshwater Eutroph.	kg P eq.	0.00	0.00	0.00	−0.05	0.00	0.00	−0.06	−0.06
Human carc. Toxicity	kg 1,4-DCB	−0.05	−0.47	−0.18	−21.01	−0.54	−0.35	−22.59	−21.90
Land Use	m^2a crop eq.	−0.49	−2.55	−0.62	−23.47	−1.54	−2.00	−30.67	−30.01
Miner. Resour. Scarc.	kg Cu eq.	0.00	−0.11	−0.02	−5.51	−0.07	−0.99	−6.71	−6.70
Fossil Resour. Scarc.	kg oil eq.	−0.81	−1.87	−1.19	−23.77	−2.23	−4.68	−34.55	−32.56
Water Consumption	m^3	0.00	−0.09	−0.48	−0.92	−0.20	−0.04	−1.74	−1.72

Figure 7 (with percentage values derived from Table 6) highlights very clearly the highest shares of avoided steel, aluminium and virgin materials production in all the environmental impact categories. Moreover, a non-negligible share results from avoided gravel crushing in the environmental category "water consumption".

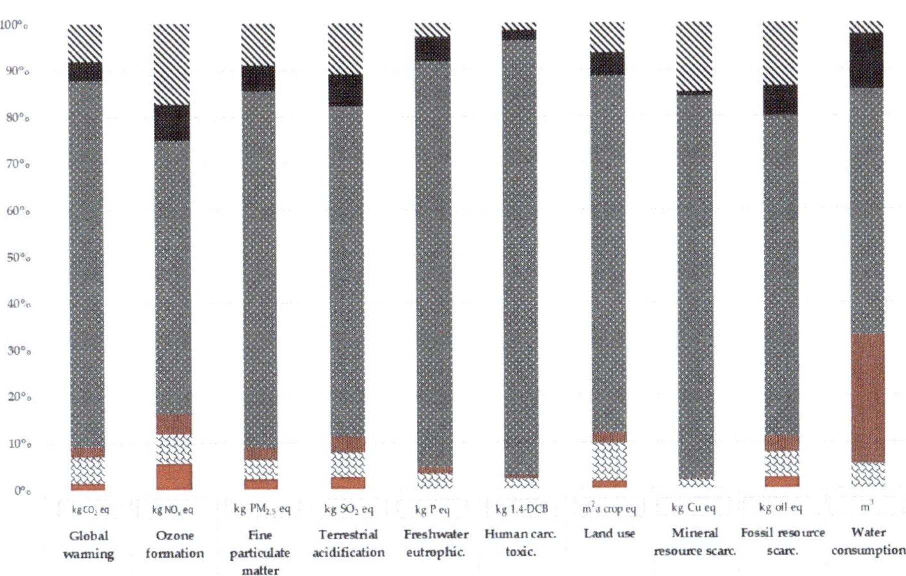

Figure 7. Percentage values of environmental avoided impacts in the life cycle of 1 tonne of C&DW.

4.2. Comparison of Conventional, Recycled and Green Concrete Aggregates

After understanding the performances of the recycling plant in processing and treating 1 tonne of C&D waste, and the avoided extraction and mining of virgin construction materials, the next step considers expanding the analyses to the production of concrete to indicate (as an example) the end-use of recycled aggregates in Naples. A comparison of concrete produced from natural, recycled and green aggregates (using agro-industry by-products) is proposed and presented in Table 7 and Figure 8. All concrete production systems include raw materials production such as cement, additive, hemp production (in the case of green concrete) and water supply to produce 1 m³ of concrete as an output. Table 7 shows the energy costs to produce 1 m³ of concrete of different characteristics and production process. The first one, conventional concrete (made with natural aggregates), requires 1963.67 MJ of energy, out of which 1635.53 MJ is fossil sources, 217.51 MJ is nuclear source, 42.08 MJ is biomass source, 21.43 MJ from wind, solar and geothermal sources, and finally 46.62 MJ from hydro sources. The total is carried out vertically and provides the CED calculated by the LCA software. The second kind of concrete, from recycled aggregates, of course requires less energy (total: 1401.02 MJ) because the raw material is not primary mineral but recycled one and therefore there are no mining energy costs. The non-renewable demand is less, while the other typologies are more or less the same. Finally, the third typology (green concrete) is produced by means of agro-industrial hemp by-products. Its total demand is lower, depending on the allocation of the energy costs, and has a larger fraction of renewable energy demand from biomass compared to natural and recycled aggregate concretes. Concerning green concrete, a sensitivity test was performed by allocating by 30%, 20%, and 10%, independently on the choice of mass, energy or economic based allocation.

Table 7. Energy characterized CED impacts for production of 1 m^3 of conventional, recycled and green concretes aggregates.

Impact Categories	Unit	Natural agg. Concrete	Recycled agg. Concrete	Green Concrete ***	Green Concrete **	Green Concrete *
Non-renewable, Fossil	MJ	1635.53	1138.80	766.92	757.02	747.12
Non-renewable, Nuclear	MJ	217.51	165.94	110.06	109.62	109.18
Non-renewable, Biomass	MJ	0.50	0.52	0.37	0.35	0.33
Renewable, Biomass	MJ	42.08	40.75	385.46	266.35	147.25
Renewable, (Wind, Solar, geo)	MJ	21.43	18.01	12.38	12.33	12.27
Renewable, Water	MJ	46.62	37.00	26.31	26.12	25.93
Total CED impacts	MJ	1963.67	1401.02	1301.50	1171.78	1042.07

*** Green concrete made of hemp by-products (allocation to hemp by-products 30%); ** green concrete made of hemp by-products (allocation to hemp by-products 20%); * green concrete made of hemp by-products (allocation to hemp by-products 10%).

Table 7 and Figure 8 show that the total CED characterized impacts decrease from values for natural aggregates concrete down to lower values for green concretes, due to the replacement of the fossil energy component by means of different percentages of biomass source.

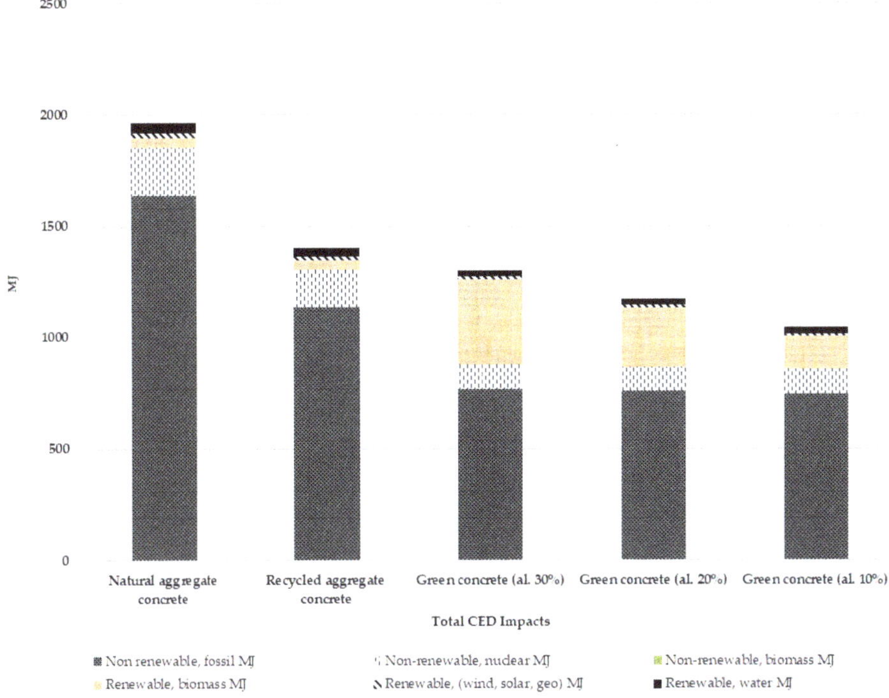

Figure 8. Comparison of CED characterized impacts for the different concrete types: conventional (with natural aggregates) and alternatives (with recycled aggregates and hemp by-products with different allocation percentages).

4.3. Discussion

The results of this LCA study show that the avoidance of landfilling (that in the waste hierarchy is the less preferable option for waste management) by means of the recycling of non-hazardous C&DW fractions into aggregates of different types and secondary materials (iron, steel and aluminium) has the potential of providing many energetic and environmental benefits contributing to reduce the dependence of the sector on fossil energy and associated environmental impacts. The performances of recycling scenarios can be further improved by reducing the share of fossil energy use in the recycling plants by means of electricity from renewable sources (e.g., the installation of PV panels) as found by previous studies [40].

The results agree with previous LCA studies that have analysed the environmental and energy impacts of entire C&DWM systems (national, regional or provincial) such as [24,30]. However, in [24], the avoided energy and environmental impacts of the recycling of C&DW are higher than the energy and environmental impacts of landfilling (for almost all impact categories), only in the best-case scenario. In the best-case scenario the authors [24] assumed that all the C&DW are sent to recycling; all the recycling plants are powered by electricity; transport distances have been reduced at the minimum value of their range with the exception of NAs selling distance that was unchanged; 90% of the produced RAs are considered of high quality and the related replacement coefficient has been maximized so it was set equal to 1 (10% of the produced RAs are still considered of low quality because of the presence of fine non-removable material in the C&DW [24]. The only category that performs worst in the best-case scenario compared to landfilling scenario is freshwater ecotoxicity. Other studies evaluating the recycling of C&DW compared to other options such as waste to energy and landfilling found that recycling is a better option compared to landfilling [27,74–79] even if it is dependent on the transport distances [25,36,80].

In the present study, the above benefits are definitely already achieved for iron, steel and aluminium that have well developed markets, whereas for RA, as evidenced in our previous research, the market is still underdeveloped, and the demand is low [48]. The primary data collected about the annual generation and recovery of non-hazardous C&DW evidence that they are almost recovered for the whole amount in the Metropolitan City of Naples, but their value is still underestimated both from an environmental and purely economic point of view due to the very low demand [63,81]. This is in contrast with previous studies where, e.g., the market price of NA is lower than the price of RA [82,83].

The next steps in our research will be to further improve the knowledge on the recycling stage in the Metropolitan City of Naples in order to rely on primary data about the recycling plants and related processes and products and their market. This would overcome one of the limits in this LCA study due to the reliance on secondary data from Ecoinvent database and previous LCA literature. Another limit is due to our assumption about the replacement ratio of RA with NA that we assumed to be 1:1 which is not currently the real case in the Metropolitan City of Naples due to the lack of confidence on RA.

Finally, the explorative analysis in this LCA study involving the comparison of alternative concrete blocks made of virgin materials, recycled aggregates and agri-food by-products from hemp crop show the potential of further improving the environmental sustainability of the construction sector by using alternative concretes. From our results, 1 m^3 of green concrete made with hemp by-products requires an energy cost in terms of CED ranging from 1301.50 to 1042.07 MJ/m^3 that is much lower than the energy cost of conventional concrete made of virgin materials (1963.67 MJ/m^3). There is an increasing interest in Italy on construction products and materials made of agri-food by-products [64,84,85]. In this view it is worth highlighting that the available certified construction products in the Italian market made of hemp by-products are designed to be recyclable and biodegradable at the end-of-life [22,64,86,87], contributing to the opportunity of a better alignment of the construction sector to the principles of CE [88].

4.4. Policy Implications

The results of this study confirm the importance, in this initial phase of transition to CE, of the political support to favour the substitution of NA with RA whenever possible in non-structural applications so as to reduce the huge environmental impacts of NA. The political support in the creation of circular supply chains and networks is needed, to reduce the uncertainties and risks embedded in the use of circular products and in general of the adoption of the CE model. Currently, in the Metropolitan City of Naples, the main barrier to the use of RA is the lack of confidence by the designers or contractors [63].

It is important to underline that if the RA would be considered as perfect substitutes, the annual amount of generated C&DW, assuming their complete recycling, might even not be enough to cover the demand for aggregates for non-structural applications This is according to our calculation and previous research including interviews to stakeholders in the Metropolitan City of Naples [63]).

Hopefully, in the Metropolitan City and Campania Region, the current transition to the CE, also supported by the adoption of the Environmental Minimum Criteria decree [89,90], would be a driver for boosting the use and production of certified recyclable construction materials and products such as those bearing the "Remade in Italy" [91]. This latter certification scheme, in turn, will encourage the traceability and transparency of the life cycle of RA, further integrating the information provided by the CE marking and declaration of performance with those related to the environmental quality of the RA in terms of recycled content and Italian origin [91].

If the Environmental Minimum Criteria is extended beyond the public buildings, to cover private buildings, the effects could be much higher. Given the lack of confidence by the stakeholders of the sector on the use of RA, only within a strict legislative framework, their use could increase and progress.

5. Conclusions

This explorative LCA study aimed to evaluate the energy savings coming from the implementation of recycling scenarios for the different fractions of non-hazardous C&DW generated in the year 2017 in the Metropolitan City of Naples (Southern Italy). We also included the results of other environmental impact categories such as global warming, fossil resources scarcity and land use for a more complete environmental assessment. The main results are highlighted in the following:

○ The construction sector as the biggest consumer of natural resources, by means of the adoption of CE recycling scenarios (as showed in this LCA study), has the potential of contributing to tackling the current environmental challenges also caused by the fossil energy use for mining and manufacturing of construction materials;

○ The results show that prolonging the value of construction and demolition materials by means of their recycling has the potential of realizing environmental and energetic savings compared to the disposal in landfill in line with the waste hierarchy.

○ Recycling of C&DW into RA should be encouraged at the political level to favour their use. The political support should occur in an integrated framework along with the other CE strategies (e.g., reduce, reuse) throughout the waste hierarchy.

○ In a circular product design perspective, the recycling of C&DW into RA is an intended strategy and not an end-of-pipe solution, as it is still now, and then its adoption in the C&DW sector would be important for further progressing their recyclability including the quality of RA and increase the trust in their use.

○ The circular designer may also decide to replace the use of technical conventional materials with bio-based construction materials and this study can be also useful for that purpose as it shows how the energy and environmental performances of concrete change according to the feed stock materials (natural aggregates, recycled aggregates, hemp by-products).

○ Finally, the funding of research projects is essential for educating professionals that have the technical and knowledge skills on the CE model in order to be applied in the C&D sector and favour its technological renewal in line with the CE principles [92].

Author Contributions: Conceptualization, P.G., A.N. and S.U.; methodology, P.G., A.N. and S.U.; software, A.N.; data curation, P.G. and G.D.; writing—original draft preparation, P.G.; writing—review and editing, P.G., A.N., R.P. and S.U.; visualization, P.G.; supervision, R.P. and S.U.; project administration, R.P. and S.U.; funding acquisition, R.P. and S.U. All authors have read and agreed to the published version of the manuscript.

Funding: The research described in this paper received funding from the European Commission's research programmes Horizon 2020-SC5-2020-2 scheme, Grant Agreement 101003491 (JUST Transition to the Circular Economy project) and Horizon 2020-Marie Sklodowska-Curie Actions-Innovative Training Networks-2018 programme (Grant Number: 814247) (Realizing the Transition to Circular Economy project); Patrizia Ghisellini and Sergio Ulgiati also gratefully acknowledge the China-Italy High Relevance Bilateral Project funded by the Ministry of Foreign Affairs and International Cooperation (MAECI), General Directorate for the Promotion of the Country System (Grant No. PGR05278).

Institutional Review Board Statement: Not applicable.

Informed Consent Statement: Not applicable.

Data Availability Statement: The data about the annual generation and management of C&DW in the Metropolitan City of Naples in the year 2017 were kindly provided by the Environmental Protection Agency of Campania Region. The authors greatly acknowledge the Agency for providing the data for this study.

Conflicts of Interest: The authors declare no conflict of interest. The funders had no role in the design of the study; in the collection, analyses, or interpretation of data; in the writing of the manuscript, or in the decision to publish the results.

Nomenclature

APOS	At Point of Substitution
C&DW	Construction and Demolition Waste
CE	Circular Economy
CED	Cumulative energy demand
EU	European Union
GHG	Greenhouse gas
IPCC	Intergovernmental Panel on Climate Change
ISO	International Organization for Standardization
LCA	Life Cycle Assessment
MJ	MegaJoules
NA	Natural aggregates
RA	Recyled aggregates
Non-renew.	Non-renewable
Renew.	Renewable
Renew. (w, solar, geo)	Renewable (wind, solar, geothermal)
Total Av. CED impacts	Total avoided CED impacts

References

1. European Commission, 2020. Communication from the Commission to the European Parliament, the Council, the European Economic and Social Committee and the Committee of the Regions. A new Circular Economy Action Plan for a Cleaner and More Competitive Europe, 2020; COM/2020/98 Final. Available online: https://eur-lex.europa.eu/legal-content/EN/TXT/?qid=1583933814386&uri=COM:2020:98:FIN (accessed on 8 November 2021).
2. European Cement Association (Cembureau). Construction and Demolition Waste. Available online: https://www.cembureau.eu/policy-focus/sustainable-construction/construction-demolition-waste/ (accessed on 22 November 2021).
3. Ritchie, H. Our World in Data. Sector by Sector: Where Do Global Greenhouse Gas Emissions Come From? Available online: https://ourworldindata.org/ghg-emissions-by-sector (accessed on 22 November 2021).
4. Liu, Y.; Li, H.; Huang, S.; An, H.; Santagata, R.; Ulgiati, S. Environmental and economic-related impact assessment of iron and steel production. A call for shared responsibility in global trade. *J. Clean. Prod.* **2020**, *269*, 122239. [CrossRef]
5. World Economic Forum. Can the Circular Economy Transform the World's Number One Consumer of Raw Materials? Available online: https://www.weforum.org/agenda/2016/05/can-the-circular-economy-transform-the-world-s-number-one-consumer-of-raw-materials (accessed on 8 November 2021).
6. Meglin, R.; Kytzia, S.; Habert, G. Regional circular economy of building materials: Environmental and economic assessment combining Material Flow Analysis, Input-Output Analysis and Life Cycle Assessment. *J. Ind. Ecol.* **2021**, 1–15. [CrossRef]
7. Ginga, C.P.; Ongpeng, J.M.C.; Daly, M.K.M.; Klarissa, M. Circular economy on construction and demolition waste: A literature review on material recovery and production. *Materials* **2020**, *13*, 2970. [CrossRef] [PubMed]
8. Jain, S.; Singhal, S.; Pandey, S. Environmental life cycle assessment of construction and demolition waste recycling: A case of urban India. *Resour. Conserv. Recycl.* **2020**, *155*, 104642. [CrossRef]
9. Ma, M.; Tam, V.W.Y.; Le, K.N.; Li, W. Challenges in current construction and demolition waste recycling: A China study. *Waste Manag.* **2020**, *118*, 610–625. [CrossRef] [PubMed]
10. UNEP (United Nations Environment Programme). March 2014. Sand, Rarer Than One Thinks. Available online: https://wedocs.unep.org/bitstream/handle/20.500.11822/8665/GEAS_Mar2014_Sand_Mining.pdf?sequence=3&isAllowed=y (accessed on 27 October 2021).
11. Brown, M.T.; Buranakarn, V. Emergy indices and ratios for sustainable material cycles and recycle options. *Resour. Conserv. Recycl.* **2003**, *38*, 1–22. [CrossRef]
12. Pearce, D.W.; Turner, R.K. *Economics of Natural Resources and the Environment*, Italian ed.; Harvester Wheatsheaf: London, UK, 1991; Il Mulino: Bologna, Italy, 1991; pp. 1–362.
13. Villagrán-Zaccardi, Y.A.; Marsh, A.T.M.; Sosa, M.E.; Zega, C.J.; De Belie, N.; Bernal, S.A. Complete re-utilization of waste concretes–Valorisation pathways and research needs. *Resour. Conserv. Recycl.* **2021**, *177*, 105955. [CrossRef]

14. Gavriletea, M.D. Environmental Impacts of Sand Exploitation. Analysis of Sand Market. *Sustainability* **2017**, *9*, 1118. [CrossRef]

15. Cochran, J.K.; Bokuniewicz, H.J.; Yager, P.L. *Encyclopedia of Ocean Sciences*, 3rd ed.; Academic Press: Cambridge, MA, USA, 2019; p. 4306. ISBN 978-0-12-813082-7.

16. European Commission. Environmental Impacts along the Supply Chain. Available online: https://rmis.jrc.ec.europa.eu/?page=environmental-impacts-along-the-supply-chain-3dfccf (accessed on 22 January 2021).

17. Ambrosini, C. Prevenire è Meglio Che Smaltire, Il Caso Dell'edilizia Italiana, Atlante dell'Economia Circolare. Available online: https://economiacircolare.com/prevenire-e-meglio-che-smaltire-il-caso-delledilizia-italiana/ (accessed on 28 October 2021).

18. Quattrone, M.; Angulo, S.C.; John, V.M. Energy and CO_2 from high performance recycled aggregate production. *Resour. Conserv. Recycl.* **2014**, *90*, 21–33. [CrossRef]

19. Halkos, G.; Petrou, K.N. Analysing energy efficiency of EU Member States: The potential of energy recovery from waste in the circular economy. *Energies* **2019**, *12*, 3718. [CrossRef]

20. Yu, Y.; Yazan, D.M.; Bhochhibhoya, S.; Volker, L. Towards Circular Economy through Industrial Symbiosis in the Dutch construction industry: A case of recycled concrete aggregates. *J. Clean. Prod.* **2021**, *293*, 126083. [CrossRef]

21. Ambrosini, C. Ripensare L'edilizia Attraverso L'economia Circolare, Lo Studio Della Luiss, Atlante dell'Economia Circolare. Available online: https://economiacircolare.com/ripensare-ledilizia-attraverso-leconomia-circolare-lo-studio-della-luiss/ (accessed on 28 October 2021).

22. Arrigoni, A.; Pelosato, R.; Melià, P.; Ruggieri, G.; Sabbadini, S.; Dotelli, G. Life cycle assessment of natural building materials: The role of carbonation, mixture components and transport in the environmental impacts of hempcrete blocks. *J. Clean. Prod.* **2017**, *149*, 1051–1061. [CrossRef]

23. Farina, I.; Colangelo, F.; Petrillo, A.; Ferraro, A.; Moccia, I.; Cioffi, R. LCA of concrete with construction and demolition waste. In *Advances in Construction and Demolition Waste Recycling: Management, Processing and Environmental Assessment*; Pacheco-Torgal, F., Ding, Y., Colangelo, F., Tuladhar, R., Koutamanis, A., Eds.; Woodhead Publishing, Elsevier: Sawston, UK, 2020; pp. 520–594.

24. Borghi, G.; Pantini, S.; Rigamonti, L. Life cycle assessment of non-hazardous Construction and Demolition Waste (CDW) management in Lombardy Region (Italy). *J. Clean. Prod.* **2018**, *184*, 815–825. [CrossRef]

25. Bowea, M.D.; Powell, J.C. Developments in life cycle assessment applied to evaluate the environmental performance of construction and demolition wastes. *Waste Manag.* **2016**, *50*, 151–172.

26. Penteado, C.S.G.; Rosado, L.P. Comparison of scenarios for the integrated management of construction and demolition waste by life cycle assessment: A case study in Brazil. *Waste Manag. Res.* **2016**, *34*, 1026–1035. [CrossRef] [PubMed]

27. Vossberg, C.; Mason-Jones, K.; Cohen, B. An energetic life cycle assessment of C&D waste and container glass recycling in Cape Town, South Africa. *Resour. Conserv. Recycl.* **2014**, *88*, 39–49.

28. Simion, I.M.; Fortuna, M.E.; Bonoli, A.; Gavrilescu, M. Comparing environmental impacts of natural inert and recycled construction and demolition waste processing using LCA. *J. Environ. Eng. Landsc. Manag.* **2013**, *21*, 273–287. [CrossRef]

29. Yuan, H.P.; Shen, L.-Y.; Li, Q.-M. Emergy analysis of the recycling options for construction and demolition waste. *Waste Manag.* **2011**, *31*, 2503–2511. [CrossRef] [PubMed]

30. Blengini, G.A.; Garbarino, E. Resources and waste management in Turin (Italy): The role of recycled aggregates in the sustainable supply mix. *J. Clean. Prod.* **2010**, *18*, 1021–1030. [CrossRef]

31. Blengini, G.A. Life cycle of buildings, demolition and recycling potential: A case study in Turin, Italy. *Build. Environ.* **2009**, *44*, 319–330. [CrossRef]

32. Garcia-Gonzalez, J.; Rodriguez-Robles, D.; De Belie, N.; Morán-del Pozo, J.M.; Guerra-Romero, M.I.; Juan-Valdés, A. Self-healing concrete with recycled aggregates. In *Advances in Construction and Demolition Waste Recycling, Management, Processing and Environmental Assessment*; Pacheco-Torgal, F., Ding, Y., Colangelo, F., Tuladhar, R., Koutamanis, A., Eds.; Woodhead Publishing, Elsevier: Sawston, UK, 2020; pp. 355–377. ISBN 978-0-12-819055-5.

33. Tam, V.W.Y.; Soomro, M.; Evangelista, A.C.J. A review of recycled aggregate in concrete applications (2000–2017). *Constr. Build. Mater.* **2018**, *172*, 272–292. [CrossRef]

34. United Nations. Department of Economic and Social Affairs, Sustainable Development. The 17 Goals. Available online: https://sdgs.un.org/goals (accessed on 23 November 2021).

35. Hossain, M.U.; Ng, S.T.; Antwi-Afari, P.; Amor, B. Circular economy and the construction industry: Existing trends, challenges and prospective framework for sustainable construction. *Renew. Sustain. Energy Rev.* **2020**, *130*, 109948. [CrossRef]

36. Colangelo, F.; Forcina, A.; Farina, I.; Petrillo, A. Life cycle assessment (LCA) of different kinds of concrete containing waste for sustainable construction. *Buildings* **2018**, *8*, 70. [CrossRef]

37. Ghisellini, P.; Ripa, M.; Ulgiati, S. Exploring environmental and economic costs and benefits of a circular economy approach to the construction and demolition sector. A literature review. *J. Clean. Prod.* **2018**, *178*, 618–643. [CrossRef]

38. Dahlbo, H.; Bachér, J.; Lähtinen, K.; Jouttijärvi, T.; Suoheimo, P.; Mattila, T.; Sironen, S.; Myllymaa, T.; Saramäki, K. Construction and demolition waste management e a holistic evaluation of environmental performance. *J. Clean. Prod.* **2015**, *107*, 333–341. [CrossRef]

39. Iodice, S.; Garbarino, E.; Cerreta, M.; Tonini, D. Sustainability assessment of Construction and Demolition Waste management applied to an Italian case. *Waste Manag.* **2021**, *128*, 83–98. [CrossRef] [PubMed]

40. Faleschini, F.; Zanini, M.A.; Pellegrino, C.; Pasinato, S. Sustainable managment and supply of natural and recycled aggregates in a medium-size integrated plant. *Waste Manag.* **2016**, *49*, 146–155. [CrossRef]

41. Rosado, L.P.; Vitale, P.; Penteado, C.S.G. Life cycle assessment of natural and mixed recycled aggregate production in Brazil. *J. Clean. Prod.* **2017**, *151*, 634–642. [CrossRef]
42. Pantini, S.; Rigamonti, L. Effectiveness and efficiency of construction and demolition waste in Lombardy: A life cycle based evaluation, 2019. In Proceedings of the 5th Conference on Final Sinks, Vienna, Austria, 8–11 December 2019; Available online: http://www.icfs2019.org/wp-content/uploads/2019/11/Se08-03_Rigamonti_Effectiveness-And-Efficiency-Of-Construction-And-Demolition-Waste-Recycling-In-Lombardy-A-Life-Cycle-Based-Evaluation.pdf (accessed on 5 November 2021).
43. Yuan, H. A SWOT analysis of successful construction waste management. *J. Clean. Prod.* **2013**, *39*, 1–8. [CrossRef]
44. ISO 14040:2006 Environmental Management—Life Cycle Assessment—Principles and Framework. Available online: https://www.iso.org/standard/37456.html (accessed on 5 August 2021).
45. Metropolitan City of Naples, Tavole di Progetto (In Italian). Available online: https://www.cittametropolitana.na.it/tavole-di-progetto (accessed on 23 November 2021).
46. Baldo, G.L.; Marino, M.; Rossi, S. *Analisi del Ciclo di Vita LCA. Materiali, Prodotti, Processi*; Edizioni Ambiente: Milano, Italy, 2005; pp. 1–289.
47. Ciacci, L.; Passarini, F. Life cycle assessment (LCA) of environmental and energy systems. *Energies* **2020**, *13*, 5892. [CrossRef]
48. Al-Khori, K.; Al-Ghami, S.G.; Boulfraud, S.; Koç, M. Life cycle assessment for integration of solid oxide fuel cells into gas processing operations. *Energies* **2021**, *14*, 1–19.
49. Nastro, R.A.; Leccisi, E.; Tuscanesi, M.; Liu, G.; Trifuoggi, M.; Ulgiati, S. Exploring avoided environmental impacts as well as energy and recource recovery from micribial desalination cell treatment of brine. *Energies* **2021**, *14*, 1–16.
50. Buyle, M.; Braet, J.; Audenaert, A. Life cycle assessment in the construction sector: A review. *Renew. Sustain. Energy Rev.* **2013**, *26*, 379–388. [CrossRef]
51. Buyle, M.; Braet, J.; Audenaert, A. Life Cycle Assessment of an Apartment Building: Comparison of an Attributional and Consequential Approach. *Energy Procedia* **2014**, *62*, 132–140. [CrossRef]
52. Cabeza, L.F.; Rincón, L.; Vilariño, V.; Pérez, G.; Castell, A. Life cycle assessment (LCA) and life cycle energy analysis (LCEA) of buildings and the building sector: A review. *Renew. Sustain. Energy Rev.* **2014**, *29*, 394–416. [CrossRef]
53. Asdrubali, F.; Roncone, M.; Grazieschi, G. Embodied energy and embodied GWP of windows: A critical review. *Energies* **2021**, *14*, 1–17.
54. Vilches, A.; Garcia-Martinez, A.; Sanchez-Montanes, B. Life cycle assessment (LCA) of building refurbishment: A literature review. *Energy Build.* **2017**, *135*, 286–301. [CrossRef]
55. Vitale, P.; Arena, N.; Di Gregorio, F.; Arena, U. Life cycle assessment of the end of-life phase of a residential building. *Waste Manag.* **2017**, *60*, 311–321. [CrossRef]
56. Thormark, C. Environmental analysis of a building with reused building materials. *Int. J. Low Energy Sust. Build.* **2000**, *1*, 1–18.
57. Da Rocha, C.G.; Sattler, M.A. A discussion on the reuse of building components in Brazil: An analysis of major social, economical and legal factors. *Resour. Conserv. Recycl.* **2009**, *54*, 104–112. [CrossRef]
58. Diyamandoglu, V.; Fortuna, L.M. Deconstruction of wood-framed houses: Material recovery and environmental impact. *Resour. Conserv. Recycl.* **2015**, *100*, 21–30. [CrossRef]
59. Silva, R.V.; De Brito, J.; Dhir, R.K. Availability and processing of recycled aggregates within the construction and demolition supply chain: A review. *J. Clean. Prod.* **2017**, *143*, 598–614. [CrossRef]
60. Huang, B.; Wang, X.; Kua, H.; Geng, Y.; Bleischwitz, R.; Ren, J. Construction and demolition waste management in China through the 3R principle. *Resour. Conserv. Recycl.* **2018**, *129*, 36–44. [CrossRef]
61. Ekvall, T.; Assefa, G.; Björklund, A.; Eriksson, O.; Finnveden, G. What life cycle assessment does and does not do in assessments of waste management. *Waste Manag.* **2007**, *27*, 989–996. [CrossRef]
62. Ghisellini, P.; Cristiano, S.; Santagata, R.; Gonella, F.; Dumontet, S.; Ulgiati, S. *Report on the Construction and Demolition Waste Management and the Potential for Circular Options in the Metropolitan City of Naples. China-Italy Bilateral Project, Analysis on the Metabolic Process of Urban Agglomeration and the Cooperative Strategy of Circular Economy*; Dipartimento di Scienze e Tecnologie, Università degli Studi Napoli "Parthenope": Naples, Italy, 2019; Available online: http://www.urbancirculareconomy.it/wp-content/uploads/2019/09/Annex-7.pdf (accessed on 10 November 2021).
63. Cristiano, S.; Ghisellini, P.; D'Ambrosio, G.; Xue, J.; Nesticò, A.; Gonella, F.; Ulgiati, S. Construction and demolition waste in the Metropolitan City of Naples, Italy: State of the art, circular design, and sustainable planning opportunities. *J. Clean. Prod.* **2021**, *293*, 125856. [CrossRef]
64. Zampori, L.; Dotelli, G.; Vernelli, V. Life Cycle Assessment of Hemp Cultivation and Use of Hemp-Based Thermal Insulator Materials in Buildings. *Environ. Sci. Technol.* **2013**, *47*, 7413–7420. [CrossRef]
65. Frischknecht, R.; Wyss, F.; Büsser Knöpfel, S.; Lützkendorf, T.; Balouktsi, M. Cumulative energy demand in LCA: The energy harvested approach. *Int. J. Life Cycle Assess.* **2015**, *20*, 957–969. [CrossRef]
66. Puig, R.; Fullana-i-Palmer, P.; Baquero, G.; Riba, J.R.; Bala, A. A Cumulative Energy Demand indicator (CED), life cycle based, for industrial waste management decision making. *Waste Manag.* **2013**, *33*, 2789–2797. [CrossRef] [PubMed]
67. Huijbregts, M.A.J.; Hellweg, S.; Frischknecht, R.; Hendriks, H.W.M.; Hungehbühler, K.; Hendriks, A.J. Cumulative energy demand as predictor for the environmental burden of commodity production. *Environ. Sci. Technol.* **2010**, *44*, 2189–2196. [CrossRef]

68. National Institute for Public Health and the Environment. *ReCiPe 2016 v1.1 A Harmonized Life Cycle Impact Assessment Method at Midpoint and Endpoint Level, Report I: Characterization*; RIVM Report 2016-0104a; National Institute for Public Health and the Environment: Bilthoven, The Netherlands, 2016; Available online: https://pre-sustainability.com/legacy/download/Report_ReCiPe_2017.pdf (accessed on 6 December 2021).

69. SimaPro. SimaPro, Version 9.1.1., Full Update Instruction. Available online: https://simapro.com/wp-content/uploads/2020/10/FullUpdateInstructionsToSimaPro911.pdf (accessed on 24 November 2021).

70. Ecoinvent 3.8 Database. Available online: https://ecoinvent.org/the-ecoinvent-database/data-releases/ecoinvent-3-8/ (accessed on 24 November 2021).

71. Knoeri, C.; Sany_e-Mengual, E.; Althaus, H.-J. Comparative LCA of recycled and conventional concrete for structural applications. *Int. J. Life Cycle Assess.* **2013**, *18*, 909–918. [CrossRef]

72. Gargiulo, A.; Carvalho, M.L.; Girardi, P. Life cycle assessment of Italian Electricity Scenarios to 2030. *Energies* **2020**, *13*, 3852. [CrossRef]

73. Cellura, M.; Cosenza, M.A.; Guarino, F.; Longo, S.; Mistretta, M. Life cycle assessment of electricty generation scenarios in Italy. In *Life Cycle Assessment of Energy Systems and Sustainable Energies Technologies*; Springer: Cham, Switzerland, 2018; pp. 3–15.

74. Roussat, N.; Méhu, J.; Dujet, C. Indicators to assess the recovery of natural resources contained in demolition waste. *Waste Manag. Res.* **2009**, *27*, 159–166. [CrossRef]

75. Dewulf, J.; Van der Vorst, G.; Versele, N.; Janssens, A.; Van Langenhove, H. Quantification of the impact of the end-of-life scenario on the overall resource consumption for a dwelling house. *Resour. Conserv. Recycl.* **2009**, *53*, 231–236. [CrossRef]

76. Carpenter, A.; Jambeck, J.R.; Gardner, K.; Weitz, K. Life cycle assessment of end-of-life management options for construction and demolition debris. *J. Ind. Ecol.* **2012**, *17*, 396–406. [CrossRef]

77. Kucukvar, M.; Egilmez, G.; Tatari, O. Evaluating environmental impacts of alternative construction waste management approaches using supply chainlinked life-cycle analysis. *Waste Manag. Res.* **2014**, *32*, 500–508. [CrossRef]

78. Ortiz, O.; Pasqualino, J.C.; Castells, F. Environmental performance of construction waste: Comparing three scenarios from a case study in Catalonia, Spain. *Waste Manag.* **2010**, *30*, 646–654. [CrossRef]

79. Ding, T.; Xiao, J.; Tam, V.W.Y. A closed-loop life cycle assessment of recycled aggregate concrete utilization in China. *Waste Manag.* **2016**, *56*, 367–375. [CrossRef]

80. Ferronato, N.; Lizarazu, G.E.G.; Portillo, M.A.G.; Moresco, F.; Conti, F.; Torretta, V. Environmental assessment of construction and demolition waste recycling in Bolivia: Focus on transportation distances and selective collection rates. *Waste Manag. Res.* **2021**, 1–13. [CrossRef] [PubMed]

81. Giorgi, S.; Lavagna, M.; Campioli, A. Circular economy and regeneration of building stock in the Italian context: Policies, partnership and tools. *IOP Conf. Ser. Earth Environ. Sci.* **2019**, *225*, 012065. [CrossRef]

82. Ghisellini, P.; Ulgiati, S. Economic assessment of circular patterns and business models for reuse and recycling of construction and demolition waste. In *Advances in Construction and Demolition Waste Recycling: Management, Processing and Environmental Assessment*; Pacheco-Torgal, F., Ding, Y., Colangelo, F., Tuladhar, R., Koutamanis, A., Eds.; Woodhead Publishing, Elsevier: Sawston, UK, 2020; pp. 31–50. ISBN 978-0-12-819055-5.

83. Wijayasundara, M.; Mendis, P.; Crawford, R.H. Integrated assessment of the use of recycled concrete aggregate replacing natural aggregate in structural concrete. *J. Clean. Prod.* **2018**, *174*, 591–604. [CrossRef]

84. Capucci, M.G.; Ruffini, V.; Barbieri, V.; Siligardi, C.; Ferrari, A.M. Life cycle assessment of a wall made with agro-concrete blocks with wheat husk. In Proceedings of the EM4SS'21–Engineered Materials for Sustainable Structures, Modena, Italy, 26–28 April 2021; p. 101.

85. Annibaldi, V.; Cucchiella, F.; D'Adamo, I.; Gastaldi, M.; Rotilio, M. Recycled materials for circular economy in construction sector. A review. In Proceedings of the EM4SS'21–Engineered Materials for Sustainable Structures, Modena, Italy, 26–28 April 2021; p. 95.

86. Ghisellini, P.; Passaro, R.; Ulgiati, S. The role of product certification in the transition towards the circular economy for construction sector. In Proceedings of the EM4SS'21–Engineered Materials for Sustainable Structures, Modena, Italy, 26–28 April 2021; p. 103.

87. Legambiente, 2016. Rapporto dell'Osservatorio Recycle Legambiente, Cento Materiali per una Nuova Edilizia. Available online: https://www.legambiente.it/sites/default/files/docs/cento_materiali_rapporto_osservatorio_recycle.pdf (accessed on 5 November 2021).

88. Equilibrium, Natural Beton (Italian Company and Product). Available online: https://www.equilibrium-bioedilizia.it/it/prodotto/natural-beton (accessed on 24 November 2021).

89. Decree 56/2017 establishing "Minimal Criteria for the Design Services and Works for the Construction, Restoring, and Refurbishing of Public Buildings. (In Italian). Available online: https://www.mite.gov.it/sites/default/files/archivio/allegati/GPP/dlgs_19_04_2017_56.pdf (accessed on 24 November 2021).

90. The Italian Decree October 2017 "Minimum Environmental Criteria for the Assignment of Design Services and Works for the New Construction, Renovation and Maintenance of Buildings and for the Management of Public Administration Sites". (In Italian). Available online: https://www.mite.gov.it/sites/default/files/archivio/allegati/GPP/allegato_tec_CAMedilizia.pdf (accessed on 24 November 2021). (In Italian)

91. ReMade in Italy. Frantoio Fondovalle (Italian Company). Available online: https://www.remadeinitaly.it/portfolio-tag/frantoio-fondovalle/ (accessed on 24 November 2021).

92. Stahel, W.R. Circular Economy. *Nature* **2016**, *531*, 435–438. [CrossRef] [PubMed]

Article

Social and Environmental Assessment of a Solidarity Oriented Energy Community: A Case-Study in San Giovanni a Teduccio, Napoli (IT)

Serena Kaiser [1], Mariana Oliveira [1,*], Chiara Vassillo [2], Giuseppe Orlandini [3] and Amalia Zucaro [4]

1   International PhD Programme, UNESCO Chair "Environment, Resources and Sustainable Development", Department of Science and Technology, Parthenope University of Naples, Centro Direzionale, Isola C4, 80143 Naples, Italy; serena.kaiser001@studenti.uniparthenope.it
2   Department of Agriculture, University of Naples Federico II, 80138 Naples, Italy; cvassillo@gmail.com
3   Independent Researcher, 80137 Naples, Italy; orlandinigiuseppe@yahoo.it
4   ENEA, Department for Sustainability, Resource Efficiency Division, Research Centre of Portici, 80055 Portici, Italy; amalia.zucaro@enea.it
*   Correspondence: mariana.oliveira@studenti.uniparthenope.it

Abstract: Renewable energy communities (RECs) are alternatives toward sustainable production and consumption pathways. In 2020, Italy implemented the EU Directive 2018/2001, defining a common framework for promoting energy from renewable sources. The "Famiglia di Maria", a foundation dealing with social issues in San Giovanni a Teduccio, Napoli (Italy), in collaboration with "Legambiente" and "Con il Sud" Foundations, released the first Solidarity Oriented Renewable Energy Community project in Italy. Therefore, by applying social life cycle assessment (s-LCA) and life cycle assessment (LCA) methodologies, this study aims to: (i) promote the dissemination of RECs in the Italian and European contexts, (ii) suggest REC scenarios for the best social and environmental solutions, and (iii) support the policymakers for sustainable local development. Some key results show that the solidarity-oriented project has already produced mature outcomes about community cohesion. In contrast, technical skills and awareness about environmental issues still need to be further developed and shared among the stakeholders. Finally, social and environmental indicators converge on the self-consumption model as a feasible alternative for energy justice, community empowerment, and economic and market competition independence.

Keywords: s-LCA; LCA; energy communities; empowerment; energy justice

Citation: Kaiser, S.; Oliveira, M.; Vassillo, C.; Orlandini, G.; Zucaro, A. Social and Environmental Assessment of a Solidarity Oriented Energy Community: A Case-Study in San Giovanni a Teduccio, Napoli (IT). *Energies* **2022**, *15*, 1557. https://doi.org/10.3390/en15041557

Academic Editor: Martin Junginger

Received: 24 December 2021
Accepted: 15 February 2022
Published: 20 February 2022

Publisher's Note: MDPI stays neutral with regard to jurisdictional claims in published maps and institutional affiliations.

1. Introduction

Centralised services have shown their weakness, especially during the COVID-19 pandemic. For instance, the hospital-based health system has stalled because of too much concentration of services demand, most often causing inadequacy of territorial assistance [1,2]. Therefore, the reorganisation of centralised services, including energy production, needs to be addressed. The interest for "reterritorialisation" based on sustainable energy production, small scale self-production, and renewable sources is growing [3]. Several countries are designing future energy plans, including balancing centralised facilities and distributed energy systems [4]. In this context, the new concept of "prosumers" arises. According to Lang et al. [5], prosumers are "individuals who consume and produce value, either for self-consumption or consumption by others and can receive implicit or explicit incentives from organisations involved in the exchange".

Energy Communities (ECs) are becoming a compelling opportunity among the solutions currently proposed to overcome energy production (electricity, heat, and gas) from fossil fuels. ECs promote renewable sources in local territories and decentralised energy production. ECs represent a socio-economic alternative in which the collective dimension becomes prominent, thus creating options for social change beyond, and in addition to, environmental protection. Moreover, ECs can be adopted as a solid antidote to energy poverty, aiming to enhance citizens' participation and control over centralised decision-making, creating opportunities for empowerment and energy justice [6,7]. Finally, ECs are naturally one of the most virtuous solutions for the energy transition in Europe [8].

Different authors have thoroughly investigated the energy subject from a socio-political perspective. A review discloses the concepts of energy democracy, relations between energy and political power, and possible scenarios for the democratisation of renewable energy development [9]. This study offers a comprehensive critical outlook on building community-based renewable energy, assuming that renewables represent a possibility, but not a certainty, to a democratic energy future [9]. In a smart community in the UK, Burchell et al. [10] provide a specific investigation about energy-saving. Through interviews with the participants, the authors discuss the concept of community and the importance of non-commercial projects. Another case study [11] focuses on innovation factors and hybridisation phenomena in a French initiative, highlighting the importance of a new shared identity based on energy projects as a source of new job opportunities and economic wealth and the possibility of reshaping territories from below (from the citizens perspective). Creamer et al. [12] provide a review from a spatial and geographical perspective to express the importance of intermediary organisations, which can play a fundamental role among the State, private organisations, and communities. Bomberg and McEwen [13] analyse mobilising factors preceding the creation of community energy groups by a qualitative study on 100 Scottish groups, explaining community actions motivated by many immaterial and symbolic resources. An Italian review gives energy socio-political and community-oriented perspective [14]. From a markedly political standpoint, a materialistic historical point of view, provided by a Marxist critical thinking, inspires alternatives to the capitalistic model and the fossil-based energy production [15]. Finally, some other studies about blockchain technology formulate suggestions for communities to manage economic transactions without intermediaries, thus providing options for broader autonomy [16].

It would be impossible to sufficiently understand ECs without considering the legal framework for their implementation. Moreover, different political backgrounds provide additional opportunities and affect the operation of ECs in specific territories. In this context, both European and Italian legal frameworks should be considered. The first one would be unworkable without national implementation.

The EU framework is based on the EU Directive 2001/2018, "On the promotion of the use of energy from renewable sources" [17] and Directive 944/2019 "On common rules for the internal market for electricity and amending Directive 2012/27/EU" [18]. Both directives are part of the "Clean Energy Package" (CEP) [19] and define Energy Communities as a juridical subject based on open and voluntary participation, whose priority is not financial profit but environmental, economic, and social benefits for members and territories. In detail, the EU Directive 2001/2018 [17] deals with adequately incentivised administrative frameworks to stimulate the transition from fossil fuels to renewables and defines the Renewable Energy Communities (RECs), whereas the Directive 944/2019 [18] specifies the Energy Communities of Citizens (ECCs). The main differences between the RECs and ECCs are the energy management and the location of power generation facilities. RECs manage electricity, gas, and heat, and the members need to be close to the power production plants, whereas ECCs work exclusively for electricity production without any specific requirements regarding the proximity between the consumers and the power generating facilities [17,18].

The legal context implements the EU Directives in Italy and draws the Italian operative framework. The first regulation about ECs is the Decree-Law 162/2019, so-called "Milleproroghe Decree" [20], which was converted into the Law 8/2020 [21], establishing many opportunities for the promotion of renewable sources of energy. These regulations allow the installation of: (i) power generation plants on private houses to produce energy for self-consumption [20,21]; (ii) collective power generation plants, also managed by a third party, to produce energy for groups of people living in the same building [20,21]; and (iii) power generation plants for people not living in the same building (including RECs and ECCs), in which direct self-consumption is not allowed. Thus, in the last case, the produced energy must be sold to the local grid and managed by an external service company [19–21].

In 2019, the first ECC was implemented in Bologna (Italy) [19], but the first Solidarity Oriented REC was launched in October 2021, in Southern Italy (San Giovanni a Teduccio, Napoli). The Solidarity Oriented REC has as the primary beneficiary the group of families living in the neighbourhood that will be monetarily rewarded from the electricity produced by the photovoltaic panels installed on the rooftop of a local building.

In this work, to evaluate the social and environmental sustainability of the Solidarity Oriented REC of San Giovanni a Teduccio, the life cycle assessment methods (s-LCA and LCA for social and environmental evaluation, respectively) were applied to: (i) promote the dissemination of RECs in the Italian and European contexts, (ii) suggest REC scenarios for the best social and environmental options, and (iii) support the policymakers for sustainable local development.

2. Materials and Methods

2.1. The System—Territorial Context

The Solidarity Oriented REC is located in San Giovanni a Teduccio, VI district of the municipality of Napoli (Southern Italy). This territory has a historical environmental and social exploitation background since industrialisation occurred during the 1950s. Companies settled in the area (primarily food industries and refineries) instead of creating local economic wealth (except for some job opportunities), destroyed the natural capital of the territory, polluting both the land and the sea [22]. As it often happens, areas with environmental exploitation are also socially and economically depressed [23].

An overview of the social profile of the San Giovanni a Teduccio district is shown in Table 1 and Figure 1.

Table 1. The unemployment rate in VI District and Napoli *.

Indicator	Napoli	VI District
Unemployment rate, age 20–24	69%	69%
Unemployment rate, age 25–29	47%	52%
Unemployment rate, age 30–34	34%	41%
Unemployment rate, age 35–39	26%	34%
Unemployment rate, age 40–44	21%	28%
Unemployment rate, age 45–49	15%	20%
Unemployment rate, age 50–54	11%	16%
Unemployment rate, age 55–59	12%	16%
Unemployment rate, age 60–64	10%	17%
Unemployment rate, age from 65 on	10%	23%
Inactivity rate	57%	62%
University graduates looking for a job for every 100 inhabitants looking for a job, age 15–34	7%	3%
Middle school graduates looking for a job, for every 100 inhabitants looking for a job, age 15–34	46%	52%

* Data come from a 2001 census [24].

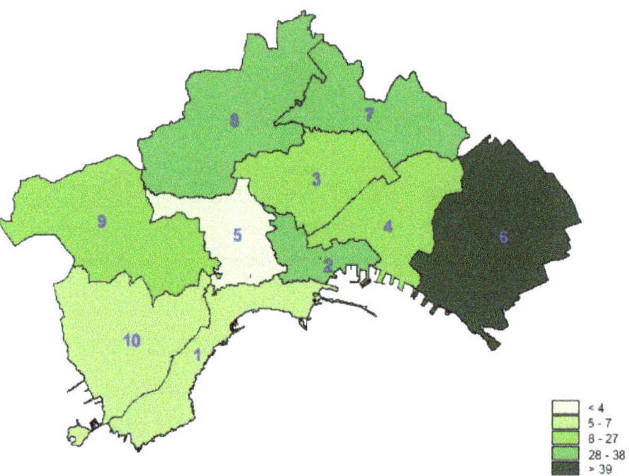

Figure 1. Number of babies born from mothers younger than 20 years old in each district of Napoli in 2008. Numbers on the figure are related to the numerical denominations of districts, whereas different colours refer to the different numbers of children born [24].

The unemployment rates and the education indicators (Table 1) show the social unrest that has led to increasing delinquency in this territory over the years. Figure 1 shows the number of children born from mothers younger than 20 years old, highlighting for the VI district the highest value in the city of Napoli (more than 39 children born from mothers more youthful than 20 years old in 2008). In addition, some other indicators complete the social overview. In 2012, 25.7% of minors (age 0–18) were in foster care, 23.2% of minors (age 3–18) were in day-care centres, 17.7% of minors (age 8–16) were in special territorial educational programmes, and 16.3% of adults were in external penal execution offices [24].

2.2. The Solidarity Oriented REC Project and Stakeholders

The Solidarity Oriented REC of San Giovanni a Teduccio is hosted and implemented by the "Famiglia di Maria" Foundation, based on a project developed by the environmentalist association "Legambiente Campania" and funded by the "Con il Sud" Foundation. On the rooftop of the "Famiglia di Maria" Foundation building, photovoltaic (PV) panels were installed to produce electricity to be sold to the Italian electricity grid, providing an income to the beneficiaries (families). The original project involves 40 families connected to the same street power pack and energy box junction (these aspects explain the technical limitation). However, the project faced bureaucratic obstacles from the city administration and started with only 15 families. Because the other families (25 families) are expected to be included soon, LCA analysis was applied in this study considering 40 families of the original project. However, a preliminary s-LCA has identified all relevant stakeholders, interviewing seven representatives of these 15 families (face-to-face interviews), the technical partner, and the foundations involved.

The identification of stakeholders is based on several variables, such as liability, influence, proximity, and representation [25–27], to create an easy interaction and encourage communication and comprehension among them [25]. Therefore, all identified stakeholders are:

1. Families;
2. Local community;
3. "Famiglia di Maria" Foundation;
4. "Con il Sud" Foundation;
5. Environmentalist association "Legambiente";
6. Private technical partner: "3eee" Company;
7. Public institutions (national);
8. Public institutions (local).

The families represent the main stakeholder. Due to current regulation restrictions, they produce and sell renewable electricity. However, when the Italian ECs legislation becomes less obstructive, they might produce and self-consume the generated electricity, selling the surplus to the grid, thus becoming prosumers in the literal sense. In addition, families in this project are also part of the local community, including other local inhabitants out of the Solidarity Oriented REC. The "Famiglia di Maria" Foundation is a local organisation that addresses many social problems in the neighbourhood. They work with children and women, developing projects against school abandonment and gender violence, among other issues. The Foundation physically hosts the PV panels on the rooftop of its building and plays as an intermediary among all stakeholders. The "Con il Sud" Foundation is the leading financial partner of this REC and entirely funded the project. It is a private non-profit organisation founded in 2006 from the alliance between bank-owned foundations and other non-profit organisations. The primary purpose of this Foundation is the development of social and environmental projects to promote social infrastructure in Southern Italy. The environmentalist association Legambiente was founded in 1980, aiming to develop projects in defence of the environment on a solid scientific basis, thus indicating realistic and feasible solutions. The "3eee" Company is the technical partner of the project that installed the PV panels on the "Famiglia di Maria" Foundation building, which manages the bureaucratic and the accounting aspects between the families and the electricity company. Public institutions were identified as representatives of the national and local legal authorities.

2.3. Assessment Methods

According to the life cycle thinking tools, social and environmental impact assessments were applied to Solidarity Oriented REC of San Giovanni a Teduccio using social life cycle assessment (s-LCA) and life cycle assessment (LCA). The s-LCA and LCA stages are: goal and scope definition, life cycle inventory (LCI), life cycle impact assessment (LCIA), and interpretation of results [28,29]. For both analyses, a cradle-to-gate approach was used. Thus, the selected system boundary (Figure 2) accounts for the physical limits of the investigated REC, including the installation and maintenance of the PV panels and the electricity production and supply to the national grid. The product of the investigated system is the solar electricity produced and sold to the national grid (following the Italian regulations in which direct self-consumption is not allowed [19–21]).

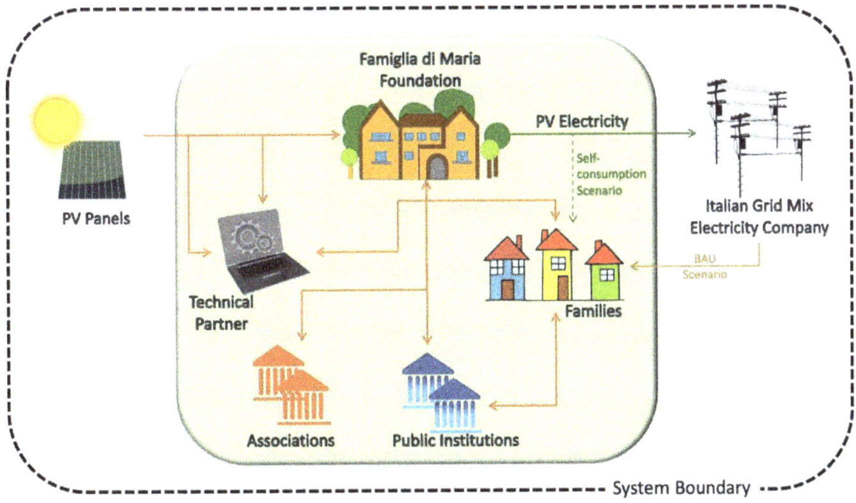

Figure 2. The system boundary of the investigated Solidarity Oriented REC of San Giovanni a Teduccio.

In this study, the s-LCA was performed to provide a preliminary overview of the social impact generated by the REC. Moreover, the LCA was conducted to outline possible environmental benefits thanks to implementing a social-oriented project in a degraded territory.

2.3.1. Social Life Cycle Assessment (s-LCA)

Social life cycle assessment (s-LCA) accounts for social impacts of products and services, highlighting positive and negative impacts, named "opportunities" and "risks", respectively [30]. In this study, the goal and scope of the s-LCA are based on identifying the social impacts of the Solidarity Oriented REC, suggesting good practices for policymakers within the energy transition framework (both from the point of view of energy production and socio-cultural activities). However, considering the social impact from a comprehensive perspective, the definition of a functional unit (FU) can be controversial. Indeed, including only those social impacts from the production of a single product or service can create a "distraction" from the general behaviour of companies: they might perform a specific output in a virtuous way while having a harmful impact within other productive activities [31]. Therefore, in this study, the FU for s-LCA was not considered to provide a broader overview of the social behaviour of the investigated system.

The s-LCA inventory is based on implementing appropriate questionnaires [32] for each stakeholder, completed during face-to-face interviews or remotely. These questionnaires were built based on the selected social indicators identified according to each stakeholder's group characteristics. For the representatives of families, several field visits and live meetings took place to construct and finalise the questionnaire, which was submitted to the families in Italian to break down the language barriers (English translation is provided in Appendix A). For the other stakeholders, face-to-face and remote meetings were both held (Appendix B). Translating the information gathered during this phase into scientific data was challenging and time-demanding due to the enormous amount of collected information (written notes, online forms, and audio recordings). Undeniably, site-specific data (primary data) are more accurate than secondary ones. Still, the interactions between researchers and interviewees, as in any human relationship, might influence the answers, thus negatively affecting the data accuracy [33].

The s-LCI is directly connected to the impact assessment stage. Stakeholders are grouped into different stakeholders categories assessed according to specific impact categories and subcategories [30,34]. Impact categories are related to human rights, working conditions, health and safety, cultural heritage, governance, and socio-economic repercussions. Impact subcategories represent the analytic topic ramifications of the six impact categories for each stakeholder category. In this study, the stakeholders' categories, impact subcategories, and indicator definitions were based on the energy justice-oriented modified version [32]. In Table 2, the stakeholders' categories and related impact subcategories connected to the Solidarity Oriented case study are reported. Impact subcategories indicators are in the results table (Appendix C—based on [32]).

The assessment stage of s-LCA requires the identification of different stakeholders' categories: families are categorised as "Prosumers", merging the concept of producer and consumer, and are also part of the "Local Community". In addition, the "Workers" category is relevant for the foundations and the technical partner, whereas the "Society" category is relevant for all the actors.

In the last stage of s-LCA, the answers to all questionnaires were used to create a preliminary table of results (Appendix C), including numerical values and descriptive sentences. In this study, only some descriptive results were converted into a scale value, whereas others were considered not convertible into numbers [35,36]. The qualitative data converted to numerical scale values express the proportion and the level of occurrence of some impacts, based on the answers to the questionnaires.

Table 2. Stakeholders categories and impact subcategories of Solidarity Oriented REC (adapted from [32]).

Stakeholders Categories	Impact Subcategories
Prosumers	Freedom of choice about sources Feedback mechanisms Costs Quality of the service
Local Community	Delocalisation and migration Community engagement and participation Sense of place and cultural heritage Respect for local culture Access to material resources Access to immaterial resources and information protests
Workers	Child labour Unpaid labour Wage Discrimination Health services Safety Right to unionise Hours of work and time off Freedom of mobility Technology, R&D Ethical principles
Society	Public commitment to sustainability issues Prevention and mitigation of conflicts Contribution to economic development corruption Technology development

2.3.2. Environmental Life Cycle Assessment-LCA

Environmental life cycle assessment (LCA) addresses the environmental impacts of a product or a process from a life cycle perspective, evaluating released emissions and resource extractions into different environmental impact categories. This evaluation technique can improve the environmental performance of manufactured and consumed products by identifying bottlenecks (hotspots) and suggesting possible recommendations to improve the environmental performance. LCA accounts for resources from raw material acquisition through production, use, end-of-life treatment, recycling, and final disposal, throughout a product's life cycle and the environmental consequences of releases, using a "cradle-to-grave" approach. However, with proper justification, the LCA technique can also be used in studies with a "cradle-to-gate" or "gate-to-gate" perspective.

The goal to be reached by LCA in this study is to evaluate the potential environmental benefits of the investigated Solidarity Oriented REC of San Giovanni a Teduccio (cradle-to-gate approach). Therefore, a comparison was carried out between 1 kWh of electricity produced (selected functional unit) by PV panels and 1 kWh of electricity produced by the Italian Electricity Mix. Moreover, considering a timeframe of 1 year, scenarios were evaluated to provide feasible alternatives aiming to improve the investigated Solidarity Oriented REC. The first investigated scenario considers that the solar electricity produced by the REC is entirely sold to the grid (business as usual scenario—BAU). The second scenario is the self-consumption scenario, in which families consume the solar electricity produced by the investigated REC, and only the surplus is sold to the grid (families as prosumers).

The professional software SimaPro v.9.0.0.48 (Pre-Consultants) coupled with the ReCiPe2016 method [37] and the EcoInvent v.3.5 [38] database were used to set up the LCA model of the investigated system and implement the impact assessment calculations.

The PV plant of the energy community is composed of 166 PV panels, flat installed on the rooftop of the building belonging to the "Famiglia di Maria" Foundation (primary data collected in the inventory stage). Each PV panel has a power of 330 w, totalling 54.78 kW of installed power [39], produced with 60 cells of 158.75 mm^2 made of monocrystalline silicon PV panels; the front has a 3.2 mm solar glass, and the back has a polymer sheet supported by a frame of aluminium [40]. PV panels are modules with a limited lifetime (the expected lifetime of a PV panel is 25 years). Currently, PV waste is exponentially growing due to the PV expansion market in the last 20 years [41]. However, end-of-life panels treatment options have advantages and disadvantages from the economic and environmental points of view [42].

The LCA was applied to evaluate the avoided environmental impacts of the electricity produced by PV panels installed at "Famiglia di Maria" Foundation, showing and discussing (interpretation of results phase) the environmental and the additional benefits achieved in the families (social) perspective.

3. Results and Discussion

3.1. Social Life Cycle Assessment (s-LCA)

The starting point of the s-LCA was the identification of stakeholders: families/prosumers, local community, "Famiglia di Maria" Foundation, "Con il Sud" Foundation, environmentalist association "Legambiente", private technical partner "3eee" Company, and public national and local institutions. First, identified stakeholders were grouped and categorised to provide a preliminary overview of the social impacts generated by the evaluated Solidarity Oriented REC of San Giovanni a Teduccio. Next, based on the field visits and live meetings, specific questionnaires for each stakeholder were finalised (see Appendices A and B for additional details).

The answers gathered from the questionnaires' compilation during the interviews provided the results of the implemented s-LCA indicators (Appendix C), summarised in Table 3. These indicators followed the goal and scope definition of the analysis and were selected according to the identified stakeholders' groups. The desired direction of indicators expresses the expected answer (positive or negative; Yes or No) to detect the social impact in terms of risks and opportunities [35].

3.1.1. Families (Prosumers)

The answers collected among the representatives of the families provided information about the territory identity and inhabitants relationship with the neighbourhood. In situations in which researchers are not aware of personal and internal dynamics among individuals, the neutral definition of "household" is recommended. However, after the meetings, interviews, and shared social moments, the gained closeness between researchers and the interviewees allows the authors to use the word "families".

The results from s-LCA show, among the most representative indicators for families (prosumers—stakeholder category), the sense of place and the cultural heritage. Therefore, the question "What is a community for you?" highlights the influence of the project on the families' perception. To this question, five respondents answered "A group of people who take part to the collective wellbeing of their own territory"; two respondents answered "A group of people who join to improve their own and theirs living conditions"; no one answered choosing the third possible option, "A group of people that join together to get a goal" (Appendix A). These answers underlined that the Solidarity Oriented REC produces a vast sense of community related to the improvement of wellbeing and is not limited to specific purposes (e.g., energy production or economic gain). The answers to the question "To what extent does receiving a sum of money at the end of the year influence your decision to join the Energy Community project?" shows that the economic benefits provided by the project represent only a tiny part of the positive impacts of the REC. The neighbourhood's problematic social conditions (high level of energy poverty) encourage ambitious projects that offer economic benefits and cultural and social empowerment.

Furthermore, the empowering processes also need immaterial resources to grow, e.g., social cohesion, awareness, and technical competencies for governance on processes.

Table 3. Selected s-LCA indicators group for Solidarity Oriented REC of San Giovanni a Teduccio.

Stakeholders Categories	Indicators Group	Indicators Type	Desired Direction
Prosumers	Access to information about energy use and sources of electricity	Semi-quantitative (Yes/No)	Yes
	Choices in electricity generation options	Semi-quantitative (Yes/No)	Yes
	Feedback mechanisms to electricity suppliers	Semi-quantitative (Yes/No)	Yes
	Responses and actions after feedback and complaints	Semi-quantitative (Yes/No)	Yes
	Economic rewarding system	Semi-quantitative (Scale 1 to 5)	Positive
	Inequality of electricity costs	Semi-quantitative (Yes/No)	No
	Quality of supplier services (burnouts)	Semi-quantitative (Yes/No)	No
	Penalties and charges related to the project membership	Semi-quantitative (Yes/No)	No
Local Community	Involvement and recognition	Semi-quantitative (Scale 1 to 5)	Positive
	Participation	Quantitative	Positive
	Displacement by population group	Semi-quantitative (Yes/No)	No
	Involuntary relocation	Semi-quantitative (Scale 1 to 5)	Negative
	Land and resources ownership	Quantitative	Positive
	Resources and electricity access	Semi-quantitative (Yes/No)	Yes
	Project activities influence the sense of place and cultural heritage	Semi-quantitative (Scale 1 to 5)	Positive
	Project activities influence health and safety	Semi-qualitative (Poor/High)	High
	Availability of project information	Semi-quantitative (Yes/No)	Yes
	Access to project information	Semi-quantitative (Yes/No)	Yes
	Project policies for local culture preservation and promotion	Semi-quantitative (Yes/No)	Yes
	Social mobilisation and organisation (protests)	Quantitative	Negative
Workers	Child labour	Quantitative	Negative
	Unpaid labour	Quantitative	Negative
	Paid labour—wages periodicity	Semi-quantitative (Yes/No)	Yes
	Paid labour—wages deduction	Semi-quantitative (Yes/No)	No
	Wage gaps by sex, gender, nationality, cultural group, and race	Quantitative	Negative
	Paid labour—wages based on living location	Quantitative	Positive
	Paid labour—minimum wage	Quantitative	Negative
	Paid labour—health insurance	Quantitative	Positive
	Safety—accidents and death	Quantitative	Negative
	Safety—education and training	Semi-quantitative (Yes/No)	Yes
	Safety—appropriate equipment and availability	Semi-quantitative (Yes/No)	Yes
	Labour union—rights	Semi-quantitative (Yes/No)	Yes
	Labour union—affiliation	Semi-quantitative (Yes/No)	Yes
	Working hours	Quantitative	Negative
	Paid leave—holidays and vacations	Quantitative	Positive
	Employment freedom and justice	Semi-quantitative (Yes/No)	Yes
	Access to technology	Semi-quantitative (Yes/No)	Yes
	Access to research and development options	Semi-quantitative (Yes/No)	Yes
	Relationship with violent conflicts, including war	Semi-quantitative (Yes/No)	No
	Corruption and unethical practices	Semi-quantitative (Yes/No)	No
Society	Sustainability and social responsibility—orientation	Semi-quantitative (Yes/No)	Yes
	Sustainability and social responsibility—behaviour	Semi-quantitative (Yes/No)	Yes
	Sustainability and social responsibility—economic contribution to regions and nations	Semi-quantitative (Scale 1 to 5)	Positive
	Sustainability and environmental responsibility—promotion	Semi-quantitative (Yes/No)	Yes

3.1.2. "Famiglia di Maria" Foundation

The strong bonds between the families and the "Famiglia di Maria" Foundation and the Foundation that hosts the project and the territory identified positive elements. Due to the activities organised in the foundation building, the contact between the hosting Foundation (Famiglia di Maria) and the families enables the Foundation to collect feedback coming from the families. Moreover, the performed activities always consider the participants' interests, hobbies, and skills. Thus, the participants can also share competencies during meetings and laboratories, manual activities, and information moments, showing the Foundation's respect for the local cultural heritage (one of the main indicators of this study is: "Project activities influence the sense of place and cultural heritage"). Indeed, all the interviewed family members stated that they would also attend the foundation activities not connected to the REC project, demonstrating that the sense of community born around the project is strong and goes beyond the economic gains and environmental goals.

Results of the assessment of "Project activities influence on health and safety indicator" (Appendix C) highlighted another positive impact of the trustful relationship between the "Famiglia di Maria" Foundation and families. A hub for vaccinations against COVID-19 was organised at the foundation building for the entire local community. The vaccination hub would have probably been equipped even if the Solidarity Oriented REC had not existed. Nevertheless, the project brought many new people to the Foundation. Thus, many individuals have information and access to health services, which are fragile in peripheral neighbourhoods.

3.1.3. Project Perspective from Stakeholders' Interactions

From the interviews and the meetings with the representatives of the "Famiglia di Maria" Foundation, the association "Legambiente", and the technical partner "3eee" Company, many obstacles were faced with starting the operation of the PV panels plant due to bureaucratic burdens. In particular, the local administrative authorities (public local institution stakeholders) raised landscape constraints based on historical buildings regulations limiting the installation of PV panels on rooftops. Therefore, the project schedule was delayed, even if the PV plant was ready to produce electricity. Another identified limit for REC's operability was the contrast between the local institutions and the key promoters of the project ("Famiglia di Maria" Foundation, the association "Legambiente", "3eee" Company. and Con il Sud Foundation) during the start-up phase of the project. Furthermore, the current legal impossibility to self-consume in RECs created questions about alternative legal frameworks in which self-consumption and different project governance of processes become possible.

Results from "3eee" Company data collection (Appendix B) showed some challenging elements. The first one is connected to the absence of trade union membership among workers (identified indicators: "Labour union—rights" and "Labour union—affiliation"), which is understandable considering the tiny dimension of the company. However, the risk of workers' rights was identified (Appendix C). The other challenge is related to the ethical indicator ("Sustainability and social responsibility—promotion" indicators, Appendix C). The company seems to respect standards and select partners and suppliers virtuously, even though no initiative has been undertaken to promote these good practices among partners and society in general (Appendices B and C). Therefore, the participation in the energy community project and the involvement of the "3eee" Company in the educational activities are likely to open possibilities for future proactivity. Additionally, results demonstrate that components of the plant come from sustainable production.

In addition to the s-LCA reported results, another important aspect was observed from the context and interaction among all stakeholders: the project's governance, which limits families' involvement in the decision process, is also a risk. Families' empowerment is still at a starting point, far from the everyday reality of well-established solidarity oriented association. There is still no place for essential decisions from below. The upcoming behaviour of the "Famiglia di Maria" Foundation might determine more involvement of families in the major decisions by reducing the level of management control over decisions in the near future. Meanwhile, related to the mission and main activities of the "Famiglia di Maria" Foundation, some initiatives were implemented about gender discrimination and female inclusion within the Solidarity Oriented REC project framework.

The initiatives against gender violence involve a music laboratory in which a song against violence was produced and recorded. Considering that all the participants in the laboratories are women, this result shows the positive and empowering impact of the project. Regarding female inclusion, women are the main actors in the investigated REC project. As often happens, projects with environmental and energy purposes have high participation by women due to the unfair distribution of job opportunities and, in general, the differences in public life between genders. Moreover, the diffused gender inequality pushes women to commit to family care and, for this reason, women are more available to engage in this kind of project. Gender inequality is even more relevant for disadvantaged territories, where depressed economies enlarge the gap to extremes. Therefore, energy-related projects can represent an empowerment tool in these territories than elsewhere, especially for women [43].

The final remark from this preliminary s-LCA applied to the Solidarity Oriented REC highlights that cohesion is already present as a mature outcome. At the same time, awareness, governance opportunities, and technical competencies about the environmental value of the project are still outcomes to be realised through the course of future events and interactions. Certainly, RECs bring many opportunities to territories and local communities. However, the increase of RECs should not correspond to an intensification of electricity consumption and production because it comes from a clean source. Instead, the rise of RECs is desirable based mainly on achievable social awareness and opportunities for territories.

3.2. Life Cycle Assessment (LCA)

The environmental burdens for each investigated impact category of the electricity produced by the Italian electricity mix and the PV panels installed at the Solidarity Oriented REC of San Giovanni a Teduccio are reported in Table 4.

Table 4. Characterised impacts calculated for the evaluated PV plant compared to the Italian electricity mix—functional unit 1 kWh of electricity produced.

Impact Categories	Abbreviation	Units	Italian Electricity Mix	PV Plant (REC)
Global warming	GWP	kg CO_2 eq	4.31×10^{-1}	1.59×10^{-1}
Stratospheric ozone depletion	ODP	kg CFC11 eq	3.40×10^{-7}	1.02×10^{-7}
Ionising radiation	IRP	kBq Co-60 eq	4.88×10^{-2}	1.00×10^{-3}
Ozone formation, human health	OFHP	kg NOx eq	8.11×10^{-4}	2.85×10^{-4}
Fine particulate matter formation	PMFP	kg $PM_{2.5}$ eq	5.43×10^{-4}	1.67×10^{-4}
Ozone formation, terrestrial ecosystems	OFTP	kg NOx eq	8.25×10^{-4}	2.89×10^{-4}
Terrestrial acidification	TAP	kg SO_2 eq	1.58×10^{-3}	4.95×10^{-4}
Freshwater eutrophication	FEP	kg P eq	1.36×10^{-4}	3.63×10^{-5}
Marine eutrophication	MEP	kg N eq	1.28×10^{-5}	3.70×10^{-6}
Terrestrial ecotoxicity	TETP	kg 1,4-DCB	1.23	1.16×10^{-1}
Freshwater ecotoxicity	FETP	kg 1,4-DCB	4.24×10^{-2}	2.38×10^{-3}
Marine ecotoxicity	METP	kg 1,4-DCB	5.28×10^{-2}	3.22×10^{-3}
Human carcinogenic toxicity	HCTP	kg 1,4-DCB	1.42×10^{-2}	3.06×10^{-3}
Human non-carcinogenic toxicity	HNTP	kg 1,4-DCB	3.95×10^{-1}	7.33×10^{-2}
Land use	LUP	m^2a crop eq	1.71×10^{-1}	6.03×10^{-2}
Mineral resource scarcity	MSP	kg Cu eq	9.24×10^{-4}	1.13×10^{-4}
Fossil resource scarcity	FSP	kg oil eq	1.30×10^{-1}	4.92×10^{-2}
Water consumption	WCP	m^3	8.99×10^{-3}	3.03×10^{-3}

These results show an overall average reduction of impacts of 76% for the electricity generated by the PV plant of the REC compared to the Italian electricity mix. The significant decreases (Figure 3 and Table 4) are shown in ionising radiation (IRP) 98%, followed by 94% in freshwater ecotoxicity (FETP) and marine ecotoxicity (METP), 91% in terrestrial ecotoxicity (TETP), and 88% in mineral resource scarcity (MSP).

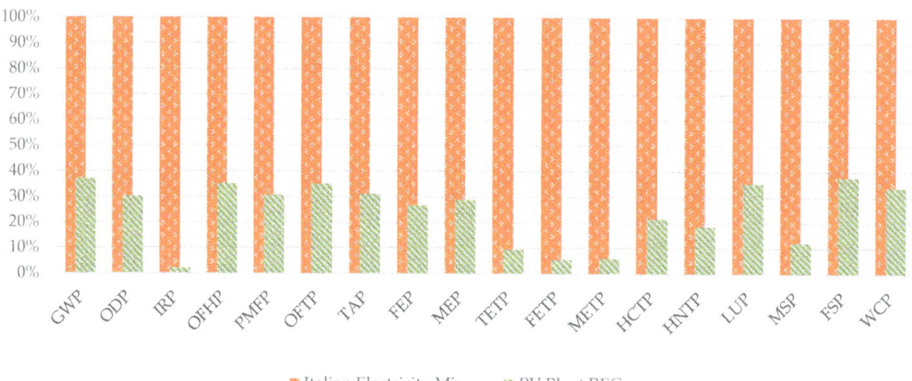

Figure 3. LCA characterisation graph showing the comparison between electricity produced by the Italian electricity mix and the evaluated PV plant (functional unit 1 kWh of electricity produced).

The normalised impacts (Table 5) for the electricity generated by the Italian mix and the PV plant of the Solidarity Oriented REC show that the most impacted categories are the marine ecotoxicity (METP, 0.051 and 0.003 for Italian mix and the PV plant, respectively), freshwater ecotoxicity (FETP, 0.034 and 0.002 for Italian mix and the PV plant, respectively), human carcinogenic toxicity (HCTP, 0.005 and 0.001 for Italian mix and the PV plant, respectively), and human non-carcinogenic toxicity (HNTP, 0.002 and 0.0004 for Italian mix and the PV plant, respectively). These results are in line with pertinent scientific literature reporting reduced impacts for PV installations on toxicity impact categories (human, marine, and freshwater) [44–47].

Table 5. Normalised impacts calculated for the evaluated PV plant compared to the Italian electricity mix (functional unit 1 kWh of electricity produced).

Impact Categories	Abbreviation	Italian Electricity Mix	PV Plant (REC)
Global warming	GWP	5.401×10^{-5}	1.997×10^{-5}
Stratospheric ozone depletion	ODP	5.679×10^{-6}	1.711×10^{-6}
Ionising radiation	IRP	1.015×10^{-4}	2.086×10^{-6}
Ozone formation, human health	OFHP	3.941×10^{-5}	1.383×10^{-5}
Fine particulate matter formation	PMFP	2.123×10^{-5}	6.520×10^{-6}
Ozone formation, terrestrial ecosystems	OFTP	4.644×10^{-5}	1.629×10^{-5}
Terrestrial acidification	TAP	3.857×10^{-5}	1.208×10^{-5}
Freshwater eutrophication	FEP	2.096×10^{-4}	5.597×10^{-5}
Marine eutrophication	MEP	2.779×10^{-6}	8.020×10^{-7}
Terrestrial ecotoxicity	TETP	1.184×10^{-3}	1.122×10^{-4}
Freshwater ecotoxicity	FETP	3.454×10^{-2}	1.941×10^{-3}
Marine ecotoxicity	METP	5.112×10^{-2}	3.125×10^{-3}
Human carcinogenic toxicity	HCTP	5.112×10^{-3}	1.104×10^{-3}
Human non-carcinogenic toxicity	HNTP	2.649×10^{-3}	4.917×10^{-4}
Land use	LUP	2.766×10^{-5}	9.766×10^{-6}
Mineral resource scarcity	MSP	7.694×10^{-9}	9.432×10^{-10}
Fossil resource scarcity	FSP	1.329×10^{-4}	5.019×10^{-5}
Water consumption	WCP	3.371×10^{-5}	1.136×10^{-5}

Characterised and normalised results underline that the electricity generated by PV panels installed in the Solidarity Oriented REC reduces environmental burdens and is potentially considered an environmentally friendly energy source.

The electricity produced by the Solidarity Oriented REC PV panels reduces the environmental impact (Table 6). The absolute benefit value was calculated by subtracting the environmental burdens (values reported in Table 4) of the electricity produced by the PV plant from the electricity supplied by the Italian grid.

Table 6. Annual environmental benefits are sorted in descending order of absolute benefit value.

Impact Categories	Units	Absolute Benefit Value of 1 kWh *	Project Perspective		Families Perspective	
					Self-Consumption Scenario	
			Solidarity Oriented REC	BAU Scenario	Electricity Consumed	Electricity Surplus Sold to the Grid
TETP	kg 1,4-DCB	1.11	1.28×10^5	3.21×10^3	3.11×10^2	2.90×10^3
HNTP	kg 1,4-DCB	3.22×10^{-1}	3.71×10^4	9.28×10^2	9.00×10	8.38×10^2
GWP	kg CO_2 eq	2.72×10^{-1}	3.14×10^4	7.85×10^2	7.61×10	7.09×10^2
LUP	m^2a crop eq	1.10×10^{-1}	1.27×10^4	3.19×10^2	3.09×10	2.88×10^2
FSP	kg oil eq	8.11×10^{-2}	9.36×10^3	2.34×10^2	2.27×10	2.11×10^2
METP	kg 1,4-DCB	4.95×10^{-2}	5.72×10^3	1.43×10^2	1.39×10	1.29×10^2
IRP	kBq Co-60 eq	4.78×10^{-2}	5.52×10^3	1.38×10^2	1.34×10	1.25×10^2
FETP	kg 1,4-DCB	4.00×10^{-2}	4.62×10^3	1.15×10^2	1.12×10	1.04×10^2
HCTP	kg 1,4-DCB	1.11×10^{-2}	1.28×10^3	3.20×10	3.11	2.89×10
WCP	m^3	5.96×10^{-3}	6.88×10^2	1.72×10	1.67	1.55×10
TAP	kg SO_2 eq	1.09×10^{-3}	1.25×10^2	3.13	3.04×10^{-1}	2.83
MSP	kg Cu eq	8.10×10^{-4}	9.35×10	2.34	2.27×10^{-1}	2.11
OFTP	kg NOx eq	5.35×10^{-4}	6.18×10	1.55	1.50×10^{-1}	1.40
OFHP	kg NOx eq	5.26×10^{-4}	6.08×10	1.52	1.47×10^{-1}	1.37
PMFP	kg $PM_{2.5}$ eq	3.76×10^{-4}	4.34×10	1.09	1.05×10^{-1}	9.81×10^{-1}
FEP	kg P eq	9.98×10^{-5}	1.15×10	2.88×10^{-1}	2.79×10^{-2}	2.60×10^{-1}
MEP	kg N eq	9.11×10^{-6}	1.05	2.63×10^{-2}	2.55×10^{-3}	2.37×10^{-2}
ODP	kg CFC11 eq	2.38×10^{-7}	2.74×10^{-2}	6.86×10^{-4}	6.65×10^{-5}	6.19×10^{-4}

* Reduction of families' environmental impacts by consuming electricity from photovoltaic panels instead of electricity from the Italian energy grid mix (absolute value).

The Project and Families perspectives consider all 40 families planned to be involved in the Solidarity Oriented REC (Table 6). The project perspective accounts for the total PV electricity produced in one year by the installed PV plant: 115,434 kWh/year (primary data collected during the interview with a manager of the "3eee" Company [39]). The families perspective accounts for the average annual electricity consumption of a family part of the Solidarity Oriented REC (an average family in the neighbourhood consumes 280 kWh/year [39]). Therefore, the project benefit value for each investigated impact category was calculated to underline the potential environmental advantages after one year of project operation (project perspective: PV electricity entirely sold to the national grid). Shifting to the Families perspective, scenarios were built based on the share of electricity self-consumed and sold to the Italian grid. In the business as usual (BAU) scenario, 100% of the PV electricity produced is sold to the grid. Each family receives the same income from the electricity company for this transaction. In contrast, in the self-consumption scenario, families (40 families) self-consume the electricity produced by the PV plant. The self-consumed electricity amounts to approximately 10% of the total electricity produced by the Solidarity Oriented REC. In this case, only the electricity surplus is sold to the grid, and the families receive two advantages: no expenses for electricity and a small profit are still recorded. As highlighted in recent literature [47], shifting the perspective shows that, even if the environmental benefits between the two REC operational models (entirely sold to the grid and partially self-consumed electricity) are the same, self-consuming energy is also economically convenient. However, self-consumption means acquiring batteries to accumulate electricity. Thus, results might be worse than the current assessed one, on some specific impact categories, due to the need to account for new materials and services to enable self-consuming electricity.

4. Conclusions

This study is characterised by a specificity: whereas the environmental effects are almost exclusively related to electricity production (single product), the social consequences are pervasive and disseminated all over the project of the Solidarity Oriented REC of San Giovanni a Teduccio, denoting a complex object of analysis with many stakeholders and outcomes. This complexity also enables the formulation of political considerations about ECs addressing the fundamental issue of energy justice as a crucial component of environmental justice. Therefore, ECs should be encouraged, and their potential should be studied and disseminated. However, the thirst for the quantitative expansion of renewable energy generation should be turned into a qualitative shift, driving to a resilient transition towards an ethical, shared, empowering, accessible, and clean energy.

The s-LCA investigation demonstrated a great sense of community among the prosumers. However, it is still impossible to state whether the project produces real awareness and empowerment. Nevertheless, the LCA results support the s-LCA outcomes, highlighting the self-consumption scenario as a feasible alternative for energy justice.

Additional considerations were made about the Italian legal framework regarding the self-consumption model: peer to peer sharing systems allow the transactions among producers and consumers without an intermediary in a democratic and consensus-based way. Therefore, the self-consumption model represents an alternative system for communities to: (i) empower themselves, (ii) get complete economic independence from the energy supply companies, and (iii) develop sharing practices outside of the market competition.

Author Contributions: S.K.—conceptualisation, methodology, formal analysis, s-LCA investigation, writing—original draft preparation; M.O.—conceptualisation, methodology, formal analysis, LCA investigation, writing—original draft preparation, writing—review and editing; C.V.—investigation, writing—original draft preparation; G.O.—conceptualisation, methodology; A.Z.—writing—review and editing, reviewing, validation, project administration. All authors have read and agreed to the published version of the manuscript.

Funding: This project has received funding from the European Union's Horizon 2020 research and innovation programme under the Marie Skłodowska-Curie Innovative Training Networks (H2020-MSCA-ITN-2018) scheme, grant agreement number 814247 (ReTraCE).

Institutional Review Board Statement: This research has obtained ethical approval according to the procedures and standards of University Parthenope of Napoli.

Informed Consent Statement: Informed consent was obtained from all subjects involved in the study.

Acknowledgments: A special thank goes to the "Famiglia di Maria" Foundation, the "Con il Sud" Foundation, the Association "Legambiente Campania", the "3eee" Company, and to the women of the Energy Community of San Giovanni a Teduccio.

Conflicts of Interest: The authors declare no conflict of interest.

Appendix A

The s-LCA questionnaire for the families in English (the original questionnaire in Italian is available online at: https://www.survio.com/survey/d/N8X4Q8F0E3P5O1Y2L; accessed on: 16 February 2022).

Families Energy Community San Giovanni

Dear Sir or Madam,

Please take a few minutes of your time to complete the following survey.

1. Gender
   - Woman
   - Man
   - Other

2. How old are you? *Type one or more words* Do you live here in the neighbourhood? Choose an answer

- Yes
- No

3. How did you get to know "Famiglia di Maria"? Choose one or more answers
    - Some people who come to the Foundation have told me about it.
    - For the activities they do with children: I bring my child/children here.
    - From the internet/TV/Newspapers.
    - I didn't know them: they contacted me for the Energy Community.
    - Other.

4. Do you currently attend the Foundation only for the Energy Community project or also for other activities? Choose an answer
    - Only for activities related to the Energy Community.
    - For something else too.

5. What is a community for you? Choose an answer
    - A group of people who come together to achieve a goal.
    - A group of people who come together to improve their own life conditions and those of others.
    - A group of people who participate in the collective wellbeing of their territory.

6. What value does ecology/respect for the environment have for you? Choose an answer
    - It is important, but it is not a priority.
    - It is fundamental and a priority.
    - It is essential, that is, without it, there is no true wellbeing.

7. Is it necessary to produce and consume clean energy today? Choose an answer
    - No, it is just a topic of the moment, but it is indifferent.
    - Yes, because energy is needed for life.
    - Yes, because its production and consumption are among the main factors of atmospheric pollution.

8. Has the way you see respect for the environment and what community means have changed since you were part of the project? Choose an answer
    - Yes.
    - No.

9. Why did you decide to enter the Energy Community project? Choose one or more answers
    - To have the economic advantage of receiving a sum of money at the end of the year.
    - To be part of a project with other people from the neighbourhood: I enjoy being in a group.
    - To do something good for the environment and produce clean energy.
    - To do something good for my neighbourhood is always described as dangerous and degraded.

10. What is an Energy Community, in your opinion? Choose one or more answers
    - One way to demonstrate that the problem of pollution can be solved with clean energy and can serve as an example for giving birth to other communities.
    - One way to show that people can organise themselves even if institutions leave us alone and can serve as an example for other places with problems.
    - Just a nice project to make people talk about a disadvantaged neighbourhood, but things will not change on a general level, neither here nor in other places.
    - It's a nice project to make people talk about a disadvantaged neighbourhood, but it can only change things here, others in other places will not notice it, or at least they will not do anything similar.

11. Do you know other examples of Energy Community in other places? Choose an answer

   - Yes.
   - No.
   - I've heard of it, but I don't know where they are.

Let's rate the different reasons for joining an Energy Community: how many stars would you give to this?

12. Economic advantage: the money we will earn at the end of the year.
13. Being part of a group and being able to meet new people.
14. Doing something important for the environment.
15. Become an example for other difficult places like San Giovanni.
16. Learn more about environmental problems and solutions.
17. Changing my neighbourhood and improving it together with others, because together we can also do many other good things for San Giovanni.
18. Becoming famous and going on TV because now everyone is talking about our project.
19. Now, let's put the same reasons in order as before, from the most important to the least important. Change the order of preference (1—most important, last—least important)

   - Economic advantage: the money we will earn at the end of the year.
   - Becoming famous and going on TV because now everyone is talking about our project.
   - Understanding more about environmental problems and solutions.
   - Being part of a group and meeting new people.
   - Doing something important for the environment.
   - Becoming an example for other difficult places like San Giovanni.
   - Changing my neighbourhood and improving it together with others, because together we can also do many other good things for San Giovanni.

20. What is changing in the neighbourhood thanks to the project? Choose one or more answers in each row

	They know a lot more about the environment thanks to the project.	They are happy with the project.	They are happy with the project, but they know nothing more about the environment than before.	They don't care about the project or the environment in general.	They are unhappy with the project.
The people of the Energy Community					
The people of San Giovanni who are not in the project					
The people outside San Giovanni					

21. Do you think there will be someone in the neighbourhood who will not be happy with the project? Choose an answer

   - Yes, many.
   - Yes, but few.
   - No, they will all be happy with the project.

22. If someone in the neighbourhood is not happy with the project, what could be the reason? Choose one or more answers

   - Because they don't know what it is and they talk without knowing.
   - Because they are not interested in the environment and in changing the neighbourhood.
   - Because they have not been involved and are envious.

- Because they don't want things to change for the better in the neighbourhood.

23. How would you solve the problem of people who are possibly against the project? Choose one or more answers
    - I would like to meet them and explain the importance of the project.
    - I would invite them here to involve them in some activities and show them they are interesting.
    - I would not consider them because I am not interested in explaining to these people.
    - I would explain to someone; some others cannot be convinced.

24. Do you like having all this attention from newspapers and TV? Choose an answer
    - Yes, because what's going on is funny and I like being popular.
    - Yes, because they will talk about our community and other places can do the same in this way.

25. In your opinion, what will change in the near future thanks to the project? Choose one or more answers
    - Anything.
    - Few things, but it is already something.
    - Few things that won't solve anything.
    - Many things can open the doors to better development for San Giovanni.

26. The money you will earn in the project will be: Choose an answer
    - Little stuff, but better than nothing.
    - An important help we need.
    - Little stuff, but an important symbol for change

27. What will you do with the money you will earn in the project? Choose one or more answers
    - I will use them for myself.
    - I will keep them.
    - I will use them for household expenses.

28. How are decisions made in the project? Choose an answer
    - We meet and talk to decide together.
    - The Foundation understands more and decides, then explains the decisions made.
    - The Foundation decides on its own without talking to us; we are so confident that everything will be fine following their decisions.
    - For now, the Foundation decides, but when we better understand the issues of the project, we will always decide together.

29. How do you feel about participating in the project? Choose an answer
    - I feel that I participate more in the life of the neighbourhood and that I can change it.
    - I do not decide anything for the neighbourhood, but I participate in a good project and I am happy to be part of it anyway.
    - Nothing has changed compared to how I felt before joining the project.

30. Would you have any ideas to put into the project to do something else? Choose an answer
    - No, things are fine this way.
    - Yes, I would like to propose doing something else, but I know that it is impossible.
    - Yes, I would also like to propose something else and I think that I will be able to make this proposal in the future.

31. Choose one of the following sentences that express your thoughts: Choose an answer
    - This project is only positive for me, the neighbourhood and everyone and nothing worry me neither for the present nor for the future.

- This project is only positive for me, the neighbourhood and everyone, but it worries me that we will not be able to get it started due to bureaucratic problems with the municipality.
- This project is only positive for me, the neighbourhood and everyone, but it worries me that we will not be able to make it work after the departure due to some neighbourhood residents who will create problems.
- This project is good for me, the neighbourhood and everyone, but I'm afraid we won't be able to make it work once it starts.

Thank you for participating!
Serena Kaiser,
Department of Science and Technology "Parthenope" University

Appendix B

Synthesis of the Meetings, Interviews and Questionnaires—Other Stakeholders

1. "Famiglia di Maria" Foundation

The meetings with the Foundation "Famiglia di Maria" have been both live meetings and remote interviews. Therefore, it has been possible to talk to the President and other operators. The main aspects touched were the project's origins, the story of the Foundation and the relation with the territory and the people, and the activities within and outside the Energy Community project.

About the families, the Foundation has explained how they have been chosen and why and the motivation of the families to join the project.

Elements about feedback mechanisms and privacy have been explained.

The Foundation has always been an intermediary between the authors and the families.

A questionnaire has been used, but it has only been a base for wider conversations. However, it can be useful to introduce it, to have elements about the key points of the meetings:

- Who came up with the idea to form an Energy Community and why?
- According to which criteria were the families chosen in the neighbourhood to be involved?
- Are families expected to adhere to the values, principles, and aims of the project in addition to the requirements through which they were chosen? If so, how has this been observed?
- What socio-economic background do the families come from?
- Were the families involved in any part of the design?
- In involved families, have there been people more active and responsible for the dialogue with you and the decision to join? (More adults/older/young people/males/females . . . ?)
- Were there families who refused to join? If so, why?
- What were, in your opinion, the main reasons that drove families to join the project?
- Do you perceive that other reasons have been added to the initial reasons as a result of the accession?
- What kind of consequences are expected once the project starts? (economic, social, cultural . . .) (this question belongs to the first period of the research when the project had not started yet).
- Have you thought of introducing a feedback mechanism for families as a tool to express their impressions about the project?
- Is the privacy of the participants protected?
- Can families have access, at any time, to the documents and any tool that is useful for a transparent understanding of the objectives and functioning of the project?
- Has the project created/will it create any employment opportunities on the territory?
- When will the cultural activities start? (question from before the project started).
- What kind of cultural activities are you planning to develop? (question from before the project started).
- What will the topics of the cultural activities be?

- Will participants in cultural activities be involved only as learners/audience, or can they propose cultural sharing and exchange activities, topics they are interested in, requests from below, etc . . . ?
- Is there any plan, at a certain stage in the project's development, e.g., during/after cultural activities, to initiate or—if already active—to encourage the participation of families in the decision-making process concerning the development of the project?
- In your opinion, what are the main benefits that will arise in the territory, even beyond the families directly involved in the project?
- Are you planning to link the project only to environmental issues or extend it to the possibility of dealing with other socially sensitive issues?
- Is there a perception that this project could prevent and/or mitigate not only economic and energy poverty, but also violence, crime, and local tensions?
- Is anyone outside your territory interested in your practices and has contacted you? If so, who and why?
- Would you be interested in networking with other similar realities?

2. "3eee" Company

In addition to meetings with the Foundation "Famiglia di Maria", several meetings and conversations with the "3eee" Company were held, both live and remote. As in the previous case, a set of questions had been prepared and can be useful to be presented here, even though they only represent a base for wider conversations.

- How many workers are there in your company?
- How many are men, how many are women? (In the absence of precise data, give an approximate answer)
- In which jobs are men more represented, in which are women?
- How many workers were employed to install the photovoltaic panels used for the project of Energy Community of San Giovanni a Teduccio?
- How many panels were assembled for the project?
- Would it be possible to know how many working hours were needed to install each panel (if not, do you know the total amount of working hours, so that you can calculate the hours for a single panel)?
- Where were the panels (or the different components) produced?
- Do you know the company that makes it?
- Which of these statements does your company represent? (you can tick multiple boxes):
  - The majority of workers are employed on an indefinite basis;
  - The majority of workers are employed on a fixed-term basis;
  - Some workers are hired on a project basis (what number out of the total?);
  - Some employees are interns (how many compared to the number of employees hired? ___ on____).
- If there are interns, how many have been hired in the last period (about five years)? (___ recruited out of a total of _____ trainees);
- Given the same skills and qualifications, which of these factors are, in your opinion, the most important for recruitment to the company? (*Give a mark from 0 to 5 in parentheses for each element*)
  - Age [ ];
  - General [ ];
  - Previous experience [ ];
  - Nationality [ ];
  - Disability [ ];
  - Knowledge of foreign languages [ ].
- Could you express YES/NO to the following statements?
  - Safety at the workplace is respected (installations, emergency procedures . . .)
  - I have received and am receiving information and training on risk and safety YES NO;

- The spaces are suitable and comfortable YES NO;
- The demands of the company are clear YES NO;
- I have been pressured or experienced unpleasant situations because of the person who runs the company YES NO;
- I have sometimes been pressured or experienced unpleasant situations by other colleagues YES NO;
- I can take enough breaks YES NO;
- The work rates are sustainable YES NO;

My relationship with executives is:

- good YES NO;
- dialectical YES NO;
- mutually collaborative YES NO;
- respectful YES NO;

My relationship with other colleagues is:

- pleasant YES NO;
- competitive YES NO;
- of team YES NO;
- fair YES NO.

- How would you define the relationship between your work and your pay?
  - Adequate; Slightly inadequate; Totally inadequate.
- How would you define the distribution of workload in the company?
  - Adequate; Slightly inadequate; Totally inadequate.
- How would you define the distribution of the burden of responsibility within the company?
  - Adequate; Slightly inadequate; Totally inadequate.
- How would you define the company's respect for different trade union memberships?
  - There is respect for every union membership, without any difference;
  - There is respect for all memberships, but some trade unions are better considered;
  - Employees are discouraged from joining trade unions in general;
  - Workers are discouraged from joining certain trade unions.
- Regarding the S. Giovanni a Teduccio Energy Community project, could you give an order of importance to the reasons that led, in your opinion, the company to participate, among the following options? (Insert a number in the brackets next to each reason, expressing an order of importance):
  - Environmental purposes (__);
  - Involvement of the inhabitants for the empowerment and awareness of local communities (__);
  - Consequences on society at a general level (__);
  - Positive economic achievements for project participants (__);
  - Positive economic achievements for the company;
  - Positive marketing purpose for the company (__).
- Has anyone from the company met the families involved in the project?
- Will you also participate in the environmental training phase?
- Will you participate in other phases of the project?
- If so, which ones?
- Are you currently involved in other projects with relevant social objectives?
- If so, could you tell us briefly?

Appendix C

Social Life Cycle Assessment Results

Table A1. s-LCA results

Stakeholders Categories	Indicators Group	Indicators Subgroup—Questions	Indicators Type	Desired Direction/Answer	Results
Prosumers	Access to information about energy use and sources of electricity	Do electricity prosumers have free access to objective information about energy use and sources of electricity?	Semi-quantitative (Yes/No)	Yes	Yes
	Choices in electricity generation options	Do electricity prosumers have a choice in generation methods used by their utility?	Semi-quantitative (Yes/No)	Yes	Yes
	Feedback mechanisms to electricity suppliers	Do prosumers have a mechanism to provide feedback to their utility?	Semi-quantitative (Yes/No)	Yes	Yes
	Responses and actions after feedback and complaints	Does the electric utility (project management) act to address prosumers' feedback or complaints?	Semi-quantitative (Yes/No)	Yes	Yes
	Economic rewarding system	To which extent does receiving a sum of money at the end of the year influence your decision to join the Energy Community project?	Semi-quantitative (Scale 1 to 5)	Positive	4
	Inequality of electricity costs	Does the cost of electricity relative to household income significantly differ across populations served by the utility?	Semi-quantitative (Yes/No)	No	No
	Quality of supplier services (burnouts)	Does the number of brownouts over time differ across populations the utility serves?	Semi-quantitative (Yes/No)	No	No. Brownouts, so far, can't be considered as caused by the connection to the system.
	Penalties and charges related to the project membership	Are there charges and possible penalties connected to the membership in the project?	Semi-quantitative (Yes/No)	No	No
Local Community	Involvement and recognition	The extent to which the local community was involved and recognised in the decision to begin operations in an area	Semi-quantitative (Scale 1 to 5)	Positive	3: very much involved but not all the local community groups that were able to be involved and groups already attending the Foundation's activities.
	Participation	Quantification of the number of meetings with individual community groups or leaders prior to the project's decision-making that could affect the local community	Quantitative	Positive	Once a week between the Foundation and the families; occasionally between "3EEE" Company (technical partner) and the families. Very frequently among the "Legambiente" Association, the foundation "Famiglia di Maria" and the families. Occasionally among the foundation "Con il Sud", the foundation "Famiglia di Maria" and the families.
	Displacement by population group	Is the percentage of the local community that is displaced different by population group in the area?	Semi-quantitative (Yes/No)	No	Not relevant because no displacement is connected to the project.
	Involuntary relocation	The extent to which relocation of local community members is involuntary	Semi-quantitative (Scale 1 to 5)	Negative	Not relevant because no relocation is connected to the project

Table A1. *Cont.*

Stakeholders Categories	Indicators Group	Indicators Subgroup—Questions	Indicators Type	Desired Direction/Answer	Results
Local Community (cont.)	Land and resources ownership	Quantification of the percentage of the resources in an area, including land, used by the company that is owned by members of the local community	Quantitative	Positive	0%
	Resources and electricity access	Does the local community still retain access to raw materials extracted at a site or have access to the final product (electricity) generated at a site?	Semi-quantitative (Yes/No)	Yes	Yes (only about electricity generated: there are no raw materials to be extracted).
	Project activities influence the sense of place and cultural heritage	Extent to which the activities of the project either positively or negatively affect the local community's sense of place and cultural heritage	Semi-quantitative (Scale 1 to 5)	Positive	5—all the activities aimed at enhancing cultural heritage and bonds among local community members and the territory.
	Project activities influence health and safety	Quantification of the health and safety impacts on local community members by the activities of the project	Semi-qualitative (Poor/High)	High	High—highly positive due to initiatives performed (and currently performing) within the project, in particular: access to COVID-19 vaccinations; initiatives against gender violence.
	Availability of project information	Is project information available in all local languages?	Semi-quantitative (Yes/No)	Yes	Yes
	Access to project information	Is project information easily accessible for local community members?	Semi-quantitative (Yes/No)	Yes	Yes, but mainly for community members who are also members of the project or for individuals attending activities at the "Famiglia di Maria" Foundation for other reasons.
	Project policies for local culture preservation and promotion	Does the project have and enact policies that show respect for local culture, including observance of cultural events?	Semi-quantitative (Yes/No)	Yes	Yes. Furthermore, many cultural events are directly organised by the "Famiglia di Maria" Foundation, that is the most active on the territory among the partners of the project
	Social mobilisation and organisation (protests)	Quantification of the number and duration of protests the project and the number of protesters that are from the local community	Quantitative	Negative	0% for all of the partners
Workers	Child Labour	Percentage of labour that is child labour	Quantitative	Negative	0% for all partners
	Unpaid Labour	Percentage of labour that is unpaid	Quantitative	Negative	There are activities within the project that are paid only with reimbursement, but this is due to the solidarity oriented nature of the project, and the material work of installation of the plant ("3eee" Company) has been regularly paid
	Paid labour—wages periodicity	Are employees paid at known and regular intervals?	Semi-quantitative (Yes/No)	Yes	Yes, for all the organisations involved in the project.
	Paid labour—wages deduction	Are there deductions on employees' wages that were enacted for reasons beyond an employee's control?	Semi-quantitative (Yes/No)	No	No, for all the organisations involved in the project.

Table A1. *Cont.*

Stakeholders Categories	Indicators Group	Indicators Subgroup—Questions	Indicators Type	Desired Direction/Answer	Results
Workers (Cont.)	Wage gaps by sex, gender, nationality, cultural group, and race	Quantification of wage gaps by sex, gender, nationality, cultural group, and race	Quantitative	Negative	Impossible to estimate due to the different nature of the subjects participating in the project: employees do not belong to a single company
	Paid labour—wages based on living location	Percentage of workers earning a living wage based on their location	Quantitative	Positive	Not relevant: no different location connected to the project
	Paid labour—minimum wage	Percentage of workers earning the legal minimum wage	Quantitative	Negative	0%
	Paid labour—health insurance	Percentage of workers with benefits such as health insurance	Quantitative	Positive	Not assessed
	Safety—accidents and death	Quantification of the number of workplace accidents resulting in injuries or death over a period	Quantitative	Negative	None
	Safety—education and training	Are appropriate safety education and training provided to employees?	Semi-quantitative (Yes/No)	Yes	Yes
	Safety—appropriate equipment and availability	Is the appropriate safety equipment for workers' activities consistently available and accessible to employees?	Semi-quantitative (Yes/No)	Yes	Yes
	Labour union—rights	Do workers have the right to unionise?	Semi-quantitative (Yes/No)	Yes	Yes, but they do not, for the following reasons: In the "3eee" Company, because of the very small dimensions of the company; In the "Famiglia di Maria" Foundation, in the "Con il Sud" Foundation and in the Association "Legambiente", because they are solidarity oriented foundations and an environmentalist association.
	Labour union—affiliation	Are employees unionised?	Semi-quantitative (Yes/No)	Yes	No (see above)
	Working hours	Quantification of the average and maximum number of hours worked per week by workers at different levels	Quantitative	Negative	Impossible to assess because of the difference among partners.
	Paid leave—holidays and vacations	Quantification of the number of holidays and other paid time off available to workers annually	Quantitative	Positive	Impossible to assess because of the difference among partners

Table A1. *Cont.*

Stakeholders Categories	Indicators Group	Indicators Subgroup—Questions	Indicators Type	Desired Direction/Answer	Results
Workers (Cont.)	Employment freedom and justice	Are workers free to end their employment and not tied by debt to a company, lack of mobility, monopoly of employment in the region by the company, or the company holding onto their legal documentation?	Semi-quantitative (Yes/No)	Yes	Yes
	Access to technology	Is the technology used accessible and affordable to developing countries?	Semi-quantitative (Yes/No)	Yes	Yes
	Access to research and development options	Are research and development results disseminated without barriers or monetary charges?	Semi-quantitative (Yes/No)	Yes	Yes
	Relationship with violent conflicts, including war	Are the companies and actors involved connected to violent conflicts, including war?	Semi-quantitative (Yes/No)	No	No
	Corruption and unethical practices	Have the companies and actors been sued or fined for or known to be involved in corruption and unethical practices?	Semi-quantitative (Yes/No)	No	No
Society	Sustainability and social responsibility—orientation	Have the companies shown behaviour that can be considered sustainability and social responsibility oriented about the choice of partners, suppliers, and relationships along the supply chain?	Semi-quantitative (Yes/No)	Yes	Yes
	Sustainability and social responsibility—behaviour	Have the companies shown proactive behaviour in terms of initiatives to promote sustainability and social responsibility?	Semi-quantitative (Yes/No)	Yes	No
	Sustainability and social responsibility—economic contribution to regions and nations	To what extent has the activities along the life cycle of the electrical energy system contributed to economic progress for different geographic regions or nations?	Semi-quantitative (Scale 1 to 5)	Positive	3 for the implementation territory; Not assessed for other territories.
	Sustainability and environmental responsibility—promotion	Are the companies promoting low-carbon energy systems over conventional fossil energy systems at their respective stages in the life cycle?	Semi-quantitative (Yes/No)	Yes	Yes

References

1. Plagg, B.; Piccoliori, G.; Oschmann, J.; Engl, A.; Eisendle, K. Primary Health Care and Hospital Management During COVID-19: Lessons from Lombardy. *Risk Manag. Healthc. Policy* **2021**, *14*, 3987–3992. [CrossRef] [PubMed]
2. Belleri, G.; Il Territorio Abbandonato. Covid, Politiche Regionali e Cure Primarie | Lombardia Sociale 2020, Program e Governance, Punti di Vista. Available online: http://www.lombardiasociale.it/2020/04/24/il-territorio-abbandonato/?doing_wp_cron=1643316844.5760290622711181640625 (accessed on 27 January 2022).
3. Bolognesi, M.; Magnaghi, A. Verso le comunità energetiche (in Italian: Towards energy communities). *Sci. Del. Territ.* **2020**, 142–150. [CrossRef]
4. Rikkonen, P.; Tapio, P.; Rintamäki, H. Visions for small-scale renewable energy production on finnish farms—A Delphi study on the opportunities for new business. *Energy Policy* **2019**, *129*, 939–948. [CrossRef]
5. Lang, B.; Dolan, R.; Kemper, J.; Northey, G. Prosumers in times of crisis: Definition, archetypes and implications. *J. Serv. Manag.* **2020**, *32*, 176–189. [CrossRef]
6. Caramizaru, A.; Uihlein, A. *Energy Communities: An Overview of Energy and Social Innovation*; EUR 30083 EN; Publications Office of the European Union: Luxembourg, 2020. [CrossRef]
7. Sovacool, B.K.; Dworkin, M.H. Energy justice: Conceptual insights and practical applications. *Appl. Energy* **2015**, *142*, 435–444. [CrossRef]
8. European Union. *Models of Local Energy Ownership and the Role of Local Energy Communities in Energy Transition in Europe.* QG-01-18-933-EN-N; European Union: Brussels, Belgium, 2018. [CrossRef]
9. Burke, M.J.; Stephens, J.C. Political power and renewable energy futures: A critical review. *Energy Res. Soc. Sci.* **2018**, *35*, 78–93. [CrossRef]
10. Burchell, K.; Rettie, R.; Roberts, T. Community, the very idea!: Perspectives of participants in a demand-side community energy project. *People Place Policy Online* **2014**, *8*, 168–179. [CrossRef]
11. Yalçın-Riollet, M.; Garabuau-Moussaoui, I.; Szuba, M. Energy autonomy in Le Mené: A French case of grassroots innovation. *Energy Policy* **2014**, *69*, 347–355. [CrossRef]
12. Creamer, E.; Eadson, W.; van Veelen, B.; Pinker, A.; Tingey, M.; Braunholtz-Speight, T.; Markantoni, M.; Foden, M.; Lacey-Barnacle, M. Community energy: Entanglements of community, state, and private sector. *Geogr. Compass* **2018**, *12*, e12378. [CrossRef]
13. Bomberg, E.; McEwen, N. Mobilizing community energy. *Energy Policy* **2012**, *51*, 435–444. [CrossRef]
14. Osti, G.; Pellizzoni, L. *Energia e Innovazione tra Flussi Globali e Circuiti Locali (in Italian—Energy and Innovation Between Global Flows and Local Circuits)*; EUT Edizioni Università di Trieste: Trieste, Italy, 2018.
15. Bellamy, B.R.; Diamanti, J. *Materialism and the Critique of Energy*; MCM' Publishing: Chicago, IL, USA, 2018.
16. Andoni, M.; Robu, V.; Flynn, D.; Abram, S.; Geach, D.; Jenkins, D.; McCallen, P.; Peacock, A. Blockchain technology in the energy sector: A systematic review of challenges and opportunities. *Renew Sustain. Energy Rev.* **2019**, *100*, 143–174. [CrossRef]
17. European Commission. Directive (EU) 2018/2001 of the European parliament and of the council of 11 December 2018 on the promotion of the use of energy from renewable sources. *Off. J. Eur. Union* **2018**, *328*, 82–209.
18. European Commission. Directive (EU) 2019/944 of the European parliament and of the council of 5 June 2019 on common rules for the internal market for electricity and amending Directive. *Off. J. Eur. Union* **2019**, *158*, 125–199.
19. ENEA. *La Comunità Energetica—Vademecum 2021 (in Italian: The Energy Community—Vademecum 2021)*; ENEA: Rome, Italy, 2021; Volume 24. [CrossRef]
20. Italian Government. *Disposizioni Urgenti in Materia di Proroga di Termini Legislativi, di Organizzazione Delle Pubbliche Amministrazioni, Nonché di Innovazione Tecnologica D.L. 162/2019/A.C. 2325*; Gazzetta Ufficiale della Repubblica Italiana: Rome, Italy, 2020. (In Italian)
21. Italian Government. *LEGGE 28 febbraio 2020, n. 8. Conversione in Legge, con Modificazioni, del Decreto-Legge 30 Dicembre 2019, n. 162, Recante Disposizioni Urgenti in Materia di Proroga di Termini Legislativi, di Organizzazione delle Pubbliche Amministrazioni, Nonché di in*; Gazzetta Ufficiale della Repubblica Italiana: Rome, Italy, 2020.
22. Caruso, V. Territorio e deindustrializzazione. *Gli anni settanta e le origini del declino economico di Napoli est. Meridiana* **2019**, *96*, 209–230.
23. Hoff, M.D.; Polack, R.J. Social dimensions of the environmental crisis: Challenges for social work. *Soc. Work* **1993**, *38*, 204–211.
24. Ministero della Giustizia. *Profilo di Comunità Municipalità 6—Distretto 32 Barra—San Giovanni a Teduccio—Ponticelli 2010–2012*; Ministero della Giustizia: Napoli, Italiy, 2012. (In Italian)
25. Freeman, R.E. A stakeholder theory of the modern corporation. In *The Corporation And its Stakeholders*; Clarkson, M., Ed.; University of Toronto Press: Toronto, ON, Canada, 2016; pp. 125–138. [CrossRef]
26. Freeman, R.E.; Harrison, J.S.; Wicks, A.C.; Parmar, B.L.; De Colle, S. *Stakeholder Theory: The State of the Art*; Cambridge University Press: New York, NY, USA, 2010.
27. Vassillo, C.; Restaino, D.; Santagata, R.; Viglia, S.; Vehmas, J.; Ulgiati, S. Barriers and solutions to the implementation of energy Efficiency. A survey about stakeholders' diversity, motivations and engagement in Naples (Italy). *J. Environ. Account Manag.* **2019**, *7*, 229–251. [CrossRef]
28. ISO, 14040. *Environmental Management—Life Cycle Assessment—Principles and Framework*; ISO: Geneva, Switzerland, 2006.
29. ISO, 14044. *Life cycle Assessment—Requirements and Guidelines*; ISO: Geneva, Switzerland, 2006.

30. UNEP. *Guidelines for Social Life Cycle Assessment of Products and Organizations 2020*; Benoît Norris, C., Traverso, M., Neugebauer, S., Ekener, E., Schaubroeck, T., Russo Garrido, S., Berger, M., Valdivia, S., Lehmann, A., Finkbeiner, M., et al., Eds.; United Nations Environment Programme (UNEP): Nairobi, Kenya, 2020.
31. Zamagni, A.; Feschet, P.; De Luca, A.I.; Iofrida, N.; Buttol, P. Social life cycle assessment. In *Sustainability Assessment of Renewables-Based Products: Methods and Case Studies*; John Wiley & Sons, Ltd.: Chichester, UK, 2015; pp. 229–240. [CrossRef]
32. Fortier, M.O.P.; Teron, L.; Reames, T.G.; Munardy, D.T.; Sullivan, B.M. Introduction to evaluating energy justice across the life cycle: A social life cycle assessment approach. *Appl. Energy* **2019**, *236*, 211–219. [CrossRef]
33. Jørgensen, A.; Le Bocq, A.; Nazarkina, L.; Hauschild, M. Methodologies for social life cycle assessment. *Int. J. Life Cycle Assess* **2007**, *13*, 96. [CrossRef]
34. UNEP. *The Methodological Sheets for Sub-Categories in Social Life Cycle Assessment (S-LCA)*; United Nations Environment Programme (UNEP): Nairobi, Kenya, 2013.
35. Alicea, R.J.B.; Fu, K. Social life—Cycle assessment (S-LCA) of residential rooftop solar panels using challenge-derived framework. *Energy Sustain. Soc.* **2022**, *12*, 7. [CrossRef]
36. Tsalidis, G.A. Integrating individual behavior dimension in social life cycle assessment in an energy transition context. *Energies* **2020**, *13*, 5984. [CrossRef]
37. Huijbregts, M.A.J.; Steinmann, Z.J.N.; Elshout, P.M.F.; Stam, G.; Verones, F.; Vieira, M.D.M.; Hollander, A.; Zijp, M.; val Zelm, R. *ReCiPe 2016 v1.1. A Harmonized Life Cycle Impact Assessment Method at Midpoint and Endpoint Level Report I: Characterization*; Rijksinstituut voor Volksgezondheid en Milieu RIVM: Bilthoven, The Netherlands, 2016.
38. EcoInvent v3.5. The EcoInvent Database n.d. Available online: https://www.ecoinvent.org/database/database.html (accessed on 29 September 2020).
39. 3eee. Portfolio Page—PRIMA Comunità Energetica E Solidale D'italia (in Italian: First Energy and Solidarity Community in Italy 2021). 2021. Available online: https://www.3eee.it/portfolio_page/prima-comunita-energetica-e-solidale-ditalia/ (accessed on 15 June 2021).
40. ALEO SOLAR GMBH. Data Sheet—P23 320–330W 2020. Available online: https://www.aleo-solar.com/app/uploads/2016/01/P23_320-330W_EN_web.pdf (accessed on 10 October 2021).
41. IEA. *Snapshot of Global PV Markets 2014—Report IEA-PVPS T1-37: 2020*; IEA: Paris, France, 2020.
42. Ansanelli, G.; Fiorentino, G.; Tammaro, M.; Zucaro, A. A life cycle assessment of a recovery process from end-of-life photovoltaic panels. *Appl. Energy* **2021**, *290*, 116727. [CrossRef]
43. Tagne, R.F.T.; Dong, X.; Anagho, S.G.; Ulgiati, S. Technologies, challenges and perspectives of biogas production in China and Africa. *Energy Sustain. Dev.* **2021**, *23*, 14799–14826. [CrossRef]
44. Muteri, V.; Cellura, M.; Curto, D.; Franzitta, V.; Longo, S.; Mistretta, M.; Parisi, M.L. Review on life cycle assessment of solar photovoltaic panels. *Energies* **2020**, *13*, 252. [CrossRef]
45. Bracquene, E.; Peeters, J.R.; Dewulf, W.; Duflou, J.R. Taking evolution into account in a parametric LCA model for PV panels. *Procedia CIRP* **2018**, *69*, 389–394. [CrossRef]
46. Oliveira, M.; Cocozza, A.; Zucaro, A.; Santagata, R.; Ulgiati, S. Circular economy in the agro-industry: Integrated environmental assessment of dairy products. *Renew Sustain. Energy Rev.* **2021**, *148*, 111314. [CrossRef]
47. Rossi, F.; Heleno, M.; Basosi, R.; Sinicropi, A. LCA driven solar compensation mechanism for renewable energy communities: The Italian case. *Energy* **2021**, *235*, 121374. [CrossRef]

Article

Modeling Dynamic Multifractal Efficiency of US Electricity Market

Haider Ali [1], Faheem Aslam [1] and Paulo Ferreira [2,3,4,*]

1 Department of Management Sciences, Comsats University, Islamabad 45550, Pakistan; haideralinaqi55@gmail.com (H.A.); faheem.aslam@comsats.edu.pk (F.A.)
2 VALORIZA—Research Center for Endogenous Resource Valorization, 7300-555 Portalegre, Portugal
3 Department of Economic Sciences and Organizations, Polytechnic Institute of Portalegre, 7300-555 Portalegre, Portugal
4 CEFAGE-UE, IIFA, University of Évora, 7000-809 Évora, Portugal
* Correspondence: pferreira@ipportalegre.pt

Abstract: The dramatic deregulatory reforms in US electricity markets increased competition, resulting in more complex prices compared to other commodities. This paper aims to investigate and compare the overall and time-varying multifractality and efficiency of four major US electricity regions: Mass Hub, Mid C, Palo Verde, and PJM West. Multifractal detrended fluctuation analysis (MFDFA) is employed to better quantify the intensity of self-similarity. Large daily data from 2001 to 2021 are taken in order to make a more conclusive analysis. The four electricity market returns showed strong multifractal features with PJM West having the highest multifractality (corresponding to lowest efficiency) and Mass Hub having the lowest multifractality (i.e., highest efficiency). Moreover, all series exhibited mean reverting (anti-persistent) behavior in the overall time period. The findings of MFDFA rolling window suggest Palo Verde as the most volatile index, while a significant upward trend in the efficiency of Mass Hub and PJM West is observed after the first quarter of 2014. The novel findings have important implications for policymakers, regulatory authorities, and decision makers to forecast electricity prices better and control efficiency.

Keywords: electricity; efficiency; multifractal detrended fluctuation analysis; multifractality; MLM; rolling window

Citation: Ali, H.; Aslam, F.; Ferreira, P. Modeling Dynamic Multifractal Efficiency of US Electricity Market. *Energies* **2021**, *14*, 6145. https://doi.org/10.3390/en14196145

Academic Editors: Sergio Ulgiati, Hans Schnitzer and Remo Santagata

Received: 14 August 2021
Accepted: 24 September 2021
Published: 27 September 2021

Publisher's Note: MDPI stays neutral with regard to jurisdictional claims in published maps and institutional affiliations.

1. Introduction

For decades, the electricity industry has undergone drastic reforms in the United States (US) [1]. After World War II, this industry was regarded as a natural monopoly and was regulated as a state-owned utility [2]. In the 1990s, the industry was revolutionized from a natural monopoly to a free market through a wave of major regulatory reforms. Competition was injected into power generation and distribution segments of the system only [3,4]. The idea behind reforming the electricity market was to provide consumers with more choice so that the market would be driven by their preferences. For example, a consumer may choose an electricity retailer that uses 'green' sources such as solar, hydro, wind, etc., while another may choose the lowest cost provider irrespective of their sources [5]. Currently, most parts of the US feature open markets for new electricity generators, free trade between (giant) consumers and producers, and competitive pricing formation. Transmission services, on the other hand, are separate from generation and distribution segments and remained a monopoly under regulation due to their characteristics. Therefore, the transmission sector has been unable to attract the necessary investment due to its lack of development incentives [6].

Electricity reforms have not progressed uniformly in all US states. Some states continue to have a natural monopoly in electricity production, while others have been deregulated [5]. Under both regimes i.e., regulated and deregulated, the regulator producer is still in charge of production, but there is competition in the electricity supply. In deregulated states, electricity is distributed to suppliers who then sell it in a competitive market, while in regulated states, power is supplied directly to the consumers. Producers, distributers, and major industrial consumers who are more able to forecast electricity prices may change their consumption/supply strategies to outperform the competition in terms of cost and benefit. Furthermore, electricity prices absorb all the shocks of supply and demand. This results in spikes that exacerbate volatility because there are no electricity stockpiles to buffer shocks [7]. Hence, both producers and suppliers require better forecasts on both the demand and supply sides of the business model under both regimes. Now, the challenge is to develop strong forecasting models that accurately forecast demand and allow producers to determine the optimal output levels.

One major challenge in forecasting demand and supply is that electricity prices frequently present autocorrelation, heteroscedasticity, and nonlinearity [8]. Hence, assessment should be done using an accurate forecasting model [9–11]. Several studies combine mathematical approaches with artificial intelligence to predict electricity prices better (see [12] for a detailed review), as these methods have the potential to examine the complex inter-relationships between inputs and outputs. For example, Lin, Gow [13] proposed an Enhanced Radial Basis Function Network (ERBFN) by combining the Radial Basis Function Network (RBFN) and Orthogonal Experimental Design (OED) to forecast electricity market prices in order to minimize the price volatility risk. Keles et al. [14] present the forecasting methodology based on artificial neuronal networks (ANN) and find it well fitting for electricity prices with lowest possible errors. Agrawal et al. [15], on the other hand, employ New England electricity market data to introduce a novel forecasting model primarily centered on relevance vector machine (RVM). Luo and Weng [16] propose a more precise forecasting method by diversifying data sources such as highly correlated power data.

Electricity consumption is known to be influenced by the weather (wind speed, temperature, precipitation, and so on) as well as business activities (peaks, weekdays, weekends, holidays, etc.) [12]. Global climate change and the production burdens on old infrastructure also impact the supply and demand sides [5]. As a result, these characteristics lead to complex, highly volatile price dynamics which are not observed so much in other commodity markets [17], such as showing seasonality at different frequencies i.e., daily, weekly, annually, and sudden short-lived and generally unanticipated price spikes [18]. These complex characteristics make multifractality a particularly interesting way to look at the market efficiency of an electricity time series.

The concept of market efficiency has its roots in the efficient market hypothesis (EMH) of Fama [19], which plays an important role in modern financial economics. According to EMH, all available information is incorporated in the prices of efficient financial markets, which makes them impossible to predict as they behave randomly. Therefore, investors are unable to earn abnormal profits through arbitrage opportunities. However, due to external events of market friction and noisy traders, financial market prices fluctuate from their fair market values [20]. Empirically, it is shown that EMH fails to explain many irregularities/complexities of financial market time series. These irregularities include nonlinearity, long-range dependence [21–23], fat tails [24,25], volatility clustering [26], chaos [27], asymmetry [28], and self-similarity [29]. Mandelbrot [30] names these irregular structures as 'fractals' and introduced the concept of multifractality. Then, Peters [31] used the theory of fractals to propose the fractal market hypothesis (FMH), which strongly rejected the EMH. According to Peters [31], fractional Brownian motion produces more accurate financial market projections since it accounts for some well-proven irregularities.

Previous studies employed various multifractal econometric approaches such as wavelet transform modulus maxima, entropy methods, and others, but these studies found spurious indications of multifractality [32]. The Hurst [33] Rescaled Range method (R/S), on the other hand, has gained significance in the past two decades. However, it produces significant errors if time series are not stationary and have short-term memory. Then, Lo [34] developed a revised version of R/S to address its flaws and to account for short-term dependencies. These approaches appear to be best suited to examining long dependence correlation for stationary time series only [35]. Considering this fact, Kantelhardt et al. [36] proposed the multifractal detrended fluctuation analysis (MFDFA), which is an extension of the mono-fractal DFA [37]. As financial time series possess multifractal characteristics, using the single scaling exponent of the DFA to examine financial time series could produce spurious results [38–41]. Therefore, the MFDFA is stronger and performs better than the above-mentioned methods. MFDFA can explore the dynamics of market efficiency, long memory properties, degree of persistency, and the forecasting of financial markets. Hence, this approach can reliably characterize the multifractality of non-linear financial time series.

Researchers have applied MFDFA in a variety of financial time series such as stock markets [42–45], foreign exchanges [46,47], cryptocurrency market [48,49], gold [50], futures market [51,52], green bonds [53], and the carbon emission trading market [54]. For example, Aslam et al. [43] used MFDFA to examine the multifractal characteristics of MSCI in emerging Asian markets. The findings revealed the highest multifractality levels for India and Malaysia, while the Chinese and South Korean markets showed the lowest multifractality. Diniz-Maganini et al. [47] find that currencies which follow a Free Float regime have low multifractality i.e., higher efficiency than those currencies that follow a Managed Float regime. Telli and Chen [50] apply MFDFA and find the different multifractal characteristics of bitcoin return series from gold. More recently, there has been a significant increase in research examining the multifractality of financial markets during COVID-19. For instance, Choi [55] employed MFDFA to examine the multifractality of various sectors in US markets during COVID-19. The findings reveal the lowest levels of efficiency in the utilities sector, while the consumer discretionary sector has the highest efficiency. Aslam et al. [46] notice a significant decrease in forex markets' efficiency during COVID-19. Aslam et al. [44] study the impact of COVID-19 on the markets of Central Eastern Europe and find a significant fluctuation in persistence behavior during the pandemic.

Various studies quantify/rank efficiency after examining MFDFA through the Magnitude of Long-Memory index (MLM) also known as Market Deficiency Measure (MDM) [56]. MLM ensures the robustness of the results and has become a significant tool for regulators and policymakers. For example, Li et al. [57] used this MLM measure to quantify the efficiency of six foreign exchange markets and found Chinese Renminbi (CNY) currency to be the most inefficient. Shahzad et al. [58] also employed this metric to examine clean energy indices. Recently, it has also been used to quantify the efficiency of financial markets during COVID-19 [28,48,49].

Considering the fact that electricity time series are non-linear and complex, compared to other commodities, this paper contributes to the literature on the efficiency and forecasting of the electricity market in three main ways, which are summed up as follows. Firstly, it provides the inner dynamics of efficiency through the multifractality of US electricity indices i.e., Mass Hub, Mid C, Palo Verde, and PJM West. To do so, we employ MFDFA [36], which makes it possible to quantify the multiple scaling exponents within each financial time series. This method allows us to assess the long memory, persistency, predictability, and informational efficiency in non-linear financial markets. Secondly, we quantify the multifractality of electricity indices through the MLM measure to rank the efficiency and ensure the robustness of the results. Thirdly, this study provides the first evidence of a rolling MFDFA approach to investigate further the evolution of complexity parameters i.e., $q = 2$ and MLM of electricity indices. This approach has been extensively adopted in the literature, enabling us to capture the entire historical evolution of persistency and efficiency dynamics. For example, Guo et al. [52] used rolling MFDFA to analyze the dynamic efficiency of China's copper futures market, Zhu and Bao [20] used it for exchange-traded funds (ETFs) and Aloui et al. [35] employed it for European credit market sectors.

2. Data and Descriptive Statistics

This study is based on examining the efficiency through multifractality of US electricity indices. Currently, electricity products are being traded at over two dozen delivery points and hubs in North America. Intercontinental Exchange (ICE) is a leading over-the-counter (OTC) trading platform for "day ahead" or prompt electricity markets of North America. The electricity indices selected for study are Mass Hub, Mid C, Palo Verde, and PJM West, which are sourced from the website of US Energy Information Administration (EIA). These indices are derived directly from the transactions executed on the ICE platform and constructed using the following formula:

$$I = \sum(P{\cdot}V)/T \tag{1}$$

where I is the volumetric weighted average index price, P and V are the price and volume of a single transaction, while T denotes the total volume of every qualified transaction.

The indices are based on the largest electricity utilization areas of the US and have enough data availability i.e., between 2001 and 2021, which allows us to make a conclusive analysis of multifractal behavior. Mass hub distributes power all over New England, Mid C hub supplies electricity to Columbia for the Northwest region, and Palo Verde is supplied throughout the Southwest region. Lastly, PJM West provides electricity to Indiana, Illinois, Delaware, Maryland, Ohio, Kentucky, Michigan, North Carolina, New Jersey, Virginia, West Virginia, Tennessee, Pennsylvania, and the District of Columbia. Table 1 presents the names, symbols, time periods, and the total number of observations of the indices. For analysis, the daily returns r_t of prices are calculated by:

$$r_t = log(P_t) - log(P_{t-1}). \tag{2}$$

Table 1. Description of indices, symbols, and data range.

S. No.	Index	Symbol	Data Range	No. of Observations
1	Mass Hub	MASS	8 January 2001–18 May 2021	4787
2	Mid-C Hub	MIDC	29 March 2001–18 May 2021	4903
3	Palo Verde Hub	PALO	8 January 2001–18 May 2021	4933
4	PJM West Hub	PJM	3 January 2001–18 May 2021	5164

The daily logarithmic returns plots of these US electricity indices are presented in Figure 1. Table 2 indicates the descriptive statistics of returns and shows negative average returns in all series. The highest average loss (about 0.01%) is observed for PJM West, whereas the lowest average loss (around 0.15%) is noted for MIDC Hub. However, MIDC Hub has the highest maximum return of 439% in a single day, while Mass Hub has the highest loss of 110%. As shown in Figure 1, MIDC Hub is the most volatile index, followed by Mass Hub, PJM West, and Palo Verde. Skewness values are non-zero i.e., negative for all the sample series with the exception of Mass Hub. For kurtosis, all series demonstrate sharp peak characteristics, indicating the presence of fat tails. Hence, these electricity indices have significant deviations from normal distribution, implying the presence of multifractality, which could be created by non-Gaussian distribution such as fat tails [59].

Table 2. Descriptive statistics.

Statistics	MASS	MIDC	PALO	PJM
Mean	−0.0002	−0.0015	−0.0003	−0.0001
Median	−0.0044	−0.0029	−0.0029	−0.0033
Maximum	1.2667	4.3909	1.5718	1.1173
Minimum	−1.0956	−4.3160	−2.1324	−1.5302
Standard Deviation	0.1782	0.2964	0.1541	0.1772
Skewness	0.2485	−0.2305	−0.0597	−0.2077
Kurtosis	5.4886	50.9029	21.8383	8.1933

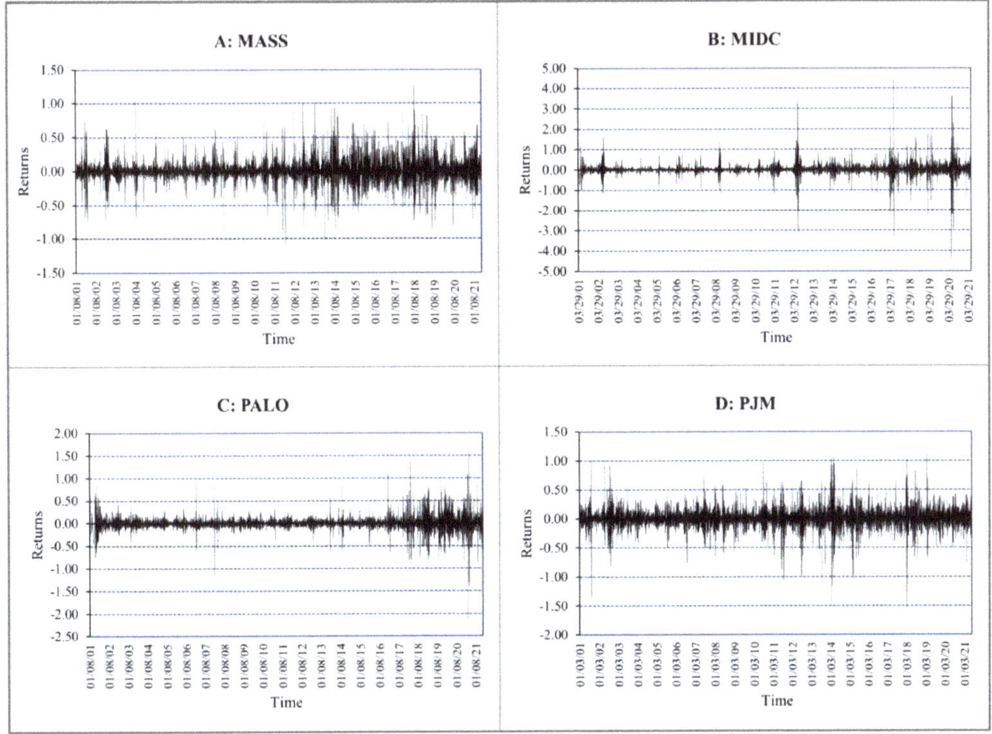

Figure 1. Daily fluctuations in log returns of electricity indices.

3. Methodology

For many years, MFDFA [36] has been frequently employed for multifractal non-stationary financial time series. Thus, a brief explanation of it has been provided here.

Assume a finite length financial time series of y_i, $i = 1, 2, \ldots, N$, where N is the number of observations. The five following steps are required to complete this method:

1. The profile value of $Y\,(i)$ is determined.

$$Y(i) = \sum_{t=1}^{i} (y_t - \bar{y}), \; i = 1, \ldots, N \tag{3}$$

where

$$\bar{y} = \frac{1}{N} \sum_{t=1}^{N} y_t. \tag{4}$$

2. The $Y(i)$ profile is divided into equal time scale s length of numerous non-overlapping components. As a result, the total number of components becomes $Ns = int(N/s)$. However, if N is not a multiple of the time scale s, a similar backward process is repeated to cover the full sample. After this, a total of $2N_s$ are obtained, which is followed by calculating the local trend s for each of the $2N_s$ segments by the kth-order polynomial fit.

3. Ordinary Least Square (OLS) is employed within each component to fit the sample appropriately, and then, the local trend is estimated for each component. In this study, the fitting polynomial for each component v is denoted as $y_v(i)$.

$$y_v(i) = \alpha_0 + \alpha_1 i + \cdots + \alpha_t i^t \tag{5}$$

with $i = 1, 2, \ldots, s; t = 1, 2, \ldots$

To estimate the variance, we apply:

$$F^2(s, v) = \begin{cases} \frac{1}{s} \sum_{i=1}^{s} \{Y[(v-1)s+i] - y_v(i)\}^2, & for \ v = 1, \ldots, N_s \\ \frac{1}{s} \sum_{i=1}^{s} \{Y[N - (v - N_s)s + i] - y_v(i)\}^2, & for \ v = N_s + 1, \ldots, 2N_s \end{cases} \quad (6)$$

The linear polynomial fitting is demonstrated here. In practice, linear $(m = 1)$, quadratic $(m = 2)$, and cubic $(m = 3)$ or those polynomials that have high orders can be applied when needed in order to fit the sample series. In this study, m is used as the order, and it should not be set too high to avoid overfitting the sample series. To obtain the optimal order of m, the findings for different ms should also be compared.

4. The fluctuation function $F_q(s)$ of q order is estimated for all components through:

$$F_q(s) = \left\{ \frac{1}{2N_s} \sum_{v=1}^{2N_s} \left[F^2(s, v) \right]^{\frac{q}{2}} \right\}^{\frac{1}{q}}. \quad (7)$$

The fluctuation function $F_0(s)$ is obtained for $q = 0$

$$F_0(s) = exp \left\{ \frac{1}{4N_s} \sum_{v=1}^{2N_s} \ln \left[F^2(s, v) \right] \right\}. \quad (8)$$

Steps 2 to 4 are repeated for various time scales s in order to examine how $F_q(s)$ is affected by different s on q.

5. The log–log plots of $F_q(s)$ vs. s are analyzed at different q levels. The generalized Hurst exponent [33] $h(q)$ is specified by Equation (7) if a long-range power law correlation is present in sample series.

$$F_q(s) \sim s^{h(q)} \quad (9)$$

The Hurst exponent identifies the features of multifractality of financial time series through the speed of local fluctuation growth with increasing scale s. A strong reliance of $h(q)$ on q is observed when series have multifractality, such as when large and small fluctuations scale differently. However, in the case of mono-fractal series, $h(q)$ is constant for every q. Since the scaling behavior of the variances $F^2(s, v)$ is the same for all the components, the averaging in Equation (7) produces the same scaling behavior for all values of q. There is a significant dependence of $h(q)$ on q if large and small fluctuations scale differently. In case of positive q, the components with larger variances $F^2(s, v)$ (i.e., large deviations from the corresponding fit) dominate the average $F_q(s)$. Hence, for positive q, $h(q)$ depicts the scaling behavior of the segments with larger fluctuations. On the contrary, the segments with small variance $F^2(s, v)$ dominate the average $F_q(s)$ for negative q. Thus, $h(q)$ describes the scaling behavior of the segments with small fluctuations for negative values of q.

The value of $\Delta h = q_{min} - q_{max}$ represents the range of $h(q)$ and indicates the degree of multifractality of a given time series. The higher the range of Δh, the stronger the multifractality [60] but lower the strength of market efficiency because of fat-tailed behavior and long-range autocorrelation properties. Since multifractal properties are negatively correlated with market efficiency, the wider the multiple spectrum, the less efficient the market will be [61]. The fluctuation associated with q shows a random walk at $h(q) = 0.5$, which is persistent at $h(q) > 0.5$ and anti-persistent at $h(q) < 0.5$.

The scaling exponent $\tau(q)$ is defined as follows:

$$\tau(q) = qh(q) - 1. \quad (10)$$

The singularity strength α and the singularity spectrum $f(\alpha)$ are examined through Legendre transform and are calculated by the following equations:

$$\alpha = \frac{d\tau(q)}{dq} = h(q) + qh'(q) \tag{11}$$

$$f(\alpha) = q\alpha - \tau(q) = 1 + q[\alpha - h(q)]. \tag{12}$$

Here, α is also known as the Holder exponent and is used to characterize the singularity of the time series, where $h'(q)$ represents the derivative of $h(q)$ with respect to q. Whereas $f(\alpha)$ defines the fractal dimension generated by all points with the same singularity exponent α and $f(\alpha) \sim \alpha$ is a single, peaked, bell-shaped fractal spectrum. In other words, the singularity spectrum describes the multifractal measure in terms of interlaced sets with singularity force α, while $f(\alpha)$ is the dimension of the contour subset characterized by α. For mono-fractal series, the uniqueness of the spectrum generates a single point, while for multifractal series, the uniqueness of the spectrum is generated by a downward concave function, whose degree of multifractality is evaluated by $f(\alpha)$. The width of the multifractal spectrum is calculated by taking the difference between maximum and minimum probability; i.e., α_{max} and α_{min}. If the width is small, the time series has higher efficiency and lower heterogeneity [62].

The selection of scale q is important when investigating multifractality. However, there is some uncertainty regarding the maximum and minimum values of q [63]. For instance, Zhang et al. [64] and Liu et al. [65] stated that the range $q = [-10, 0, 10]$ generally achieves the requirements. However, Kantelhardt and Koscielny-Bunde [66] indicated that a narrower range of q from -5 to 5 could be used to avoid a possible distortion of results by the so-called "freezing phenomenon" linked to the fat tails of time series. Therefore, in this study, we restrict q to $[-5, 0, 5]$ with a step size of 1.

Lastly, the methods of Wang et al. [56] and Wang et al. [67] are employed to quantify the level of inefficiency by calculating the market deficiency measure or the index of Magnitude of Long Memory (MLM). According to MLM, a market is efficient if all of its fluctuations follow a random walk behavior. This means that $h(q)$s related to different qs are equal to 0.5. The MLM measure is defined as:

$$\mathrm{MLM} = \frac{1}{2}(|h(-5) - 0.5| + |h(5) - 0.5|) = \frac{1}{2}\Delta h. \tag{13}$$

If large fluctuations ($q = 5$) and small fluctuations ($q = -5$) follow a random walk process, the market is qualified as efficient, and the MLM measure reaches zero. Whereas greater MLM values show weaker market efficiency and smaller MLM values show stronger market efficiency.

4. Results and Discussion

The MFDFA's empirical results for the US electricity market indices are presented in this section. For all series, the log-log graphs of the fluctuation function $F_q(s)$ vs. time scale s, the slopes of generalized Hurst exponent, Renyi exponent $\tau(q)$, and the multifractal spectrum $f(\alpha)$ are presented in Figure 2. The fitted lines for all series are observed in the log-log graphs for the fluctuation functions and scales, which is picked up for $q = [-5, 0, 5]$. As for slopes of $h(q)$ in Figure 2, the fitted lines that correspond to the generalized Hurst exponent $h(q)$ evidently depend on q. This declining pattern of q and its dependency for scaling exponent imply the presence of multifractal structures. The plots of the Renyi exponent $\tau(q)$ presented in Figure 2 are non-linear for all electricity indices. Finally, the presence of multifractality is proven by the multifractal spectrum $f(\alpha)$, which is a single humped shape.

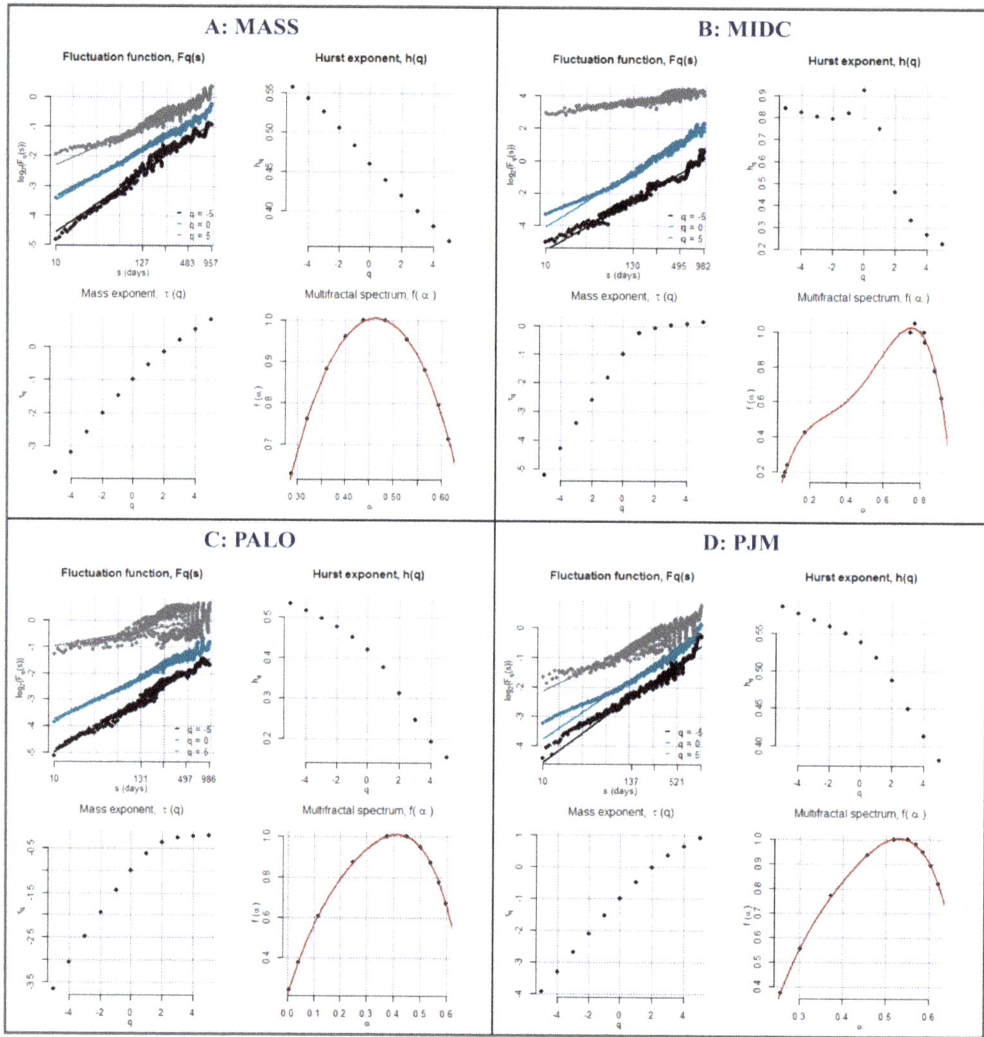

Figure 2. The Multifractal detrended fluctuation analysis (MFDFA) findings of electricity indices. The Fluctuation functions for $q = 5$, $q = 0$, *and* $q = 5$ are displayed at the top left. The top right shows the Generalized Hurst exponent for each q. The Mass exponent, $\tau(q)$, is presented at the bottom left, and the bottom right shows the Multifractal spectrum. The market's codes are presented in Table 1.

Table 3 presents the results of $h(q)$ for the range of $q = -5$ to $q = 5$ where a negative q signifies small price fluctuations and a positive q relates to large price fluctuations. It is well acknowledged that a market is said to be multifractal if $h(q)$ fluctuates with q from -5 to 5; otherwise, it is mono fractal. As shown in Table 3, the $h(q)$ values for the returns vary significantly with q from -5 to 5, signaling that the electricity market indices are multifractal. For example, the findings for Mass Hub show that $h(q)$ achieves a maximum of 0.40 at $q = -5$; then, it falls to 0.24 at $q = 0$ and finally to 0.10 at $q = 5$. The fact that the generalized Hurst exponent $h(q)$ is decreasing supports its dependency on q, implying the presence of multifractality in Mass Hub's time fluctuations. Similar patterns and findings are found for the remaining electricity indices.

Table 3. Generalized Hurst exponents ranging from $q = -5$ to $q = 5$.

Order q	MASS	MIDC	PALO	PJM
−5	0.3971	0.5249	0.4339	0.3100
−4	0.3766	0.4912	0.4154	0.2902
−3	0.3510	0.4500	0.3943	0.2686
−2	0.3194	0.4015	0.3695	0.2454
−1	0.2816	0.3462	0.3384	0.2204
0	0.2406	0.2831	0.2963	0.1930
1	0.2020	0.2134	0.2399	0.1622
2	0.1691	0.1449	0.1743	0.1272
3	0.1416	0.0872	0.1119	0.0893
4	0.1185	0.0430	0.0611	0.0517
5	0.0989	0.0104	0.0224	0.0174

The range or width of $h(q)$ is examined through Δh over the range $q \, \varepsilon \, [-5,\, 5\,]$, which reveals the strength of multifractality. Larger values of Δh are associated with higher multifractal patterns and lower efficiency levels shown by the sample series under analysis. Table 4 shows the findings of the width of Δh. The greatest width of the generalized Hurst exponent is noted for PJM West ($\Delta h = 0.67$) followed by Palo Verde ($\Delta h = 0.54$) and MIDC Hub ($\Delta h = 0.51$). Mass hub, on the other hand, has the lowest level of multifractality, with a Δh of 0.50. Hence, the findings reveal Mass Hub to be highly efficient, while PJM West is the least efficient index of them all. Similar results are confirmed in Figure 2, which presents the plots of the multifractal spectrum for all electricity market indices. PJM West has a very large width ($\Delta \alpha$), suggesting high multifractality levels compared to all electricity indices. The rationale behind the higher complexity of PJM is its bigger size than Mass Hub and other US electricity markets [68]. It serves more than 65 million people in 13 mid-Atlantic states and is the world's largest competitive wholesale power market by load [69]. On other hand, Mass Hub only provides electricity to around 7 million consumers in six US states known as New England.

Table 4. Results of efficiency.

	Hurst Average	Delta h	Delta Alpha	Fractal Dimension	MLM	Ranking
MASS	0.2451	0.5040	0.4586	1.7549	0.2520	1
MIDC	0.2723	0.5145	0.7797	1.7277	0.2573	2
PALO	0.2598	0.5437	0.6403	1.7402	0.2719	3
PJM	0.1796	0.6726	0.5090	1.8204	0.3363	4

Now, the classical Hurst exponent $H(q)$ at $q = 2$ is employed to examine the persistence of sample series, which is a key indicator of multifractal characteristics. For all electricity indices, the classical Hurst exponent values are less than 0.5, indicating anti-persistent behavior or negative autocorrelation. This means that any negative or positive change in one period will probably be followed by an opposite positive or negative change in the next period. Electricity prices have been found in the literature to be neither persistent nor random, but rather a mean-reverting (anti-persistent) process [70,71]. More recently, Kristjanpoller and Minutolo [5] found anti-persistent behavior while examining the multifractal cross-correlation through R/S of US electricity indices with WTI and natural gas.

For the robustness of results, the efficiency of electricity indices is further ranked according to the MLM measure. Higher MLM values are associated with greater market deficiency. In Table 4, the MLM results show PJM West to be most inefficient, with an MLM value of 0.34, whereas Mass Hub is the least inefficient, as indicated by the lowest MLM value of 0.25.

We now use a rolling window approach to examine the dynamic Hurst exponent ($q = 2$) and MLM to look at time-varying changes in the market efficiency of electricity indices. This method has been widely used to investigate the dynamic features of financial time series and to identify the potential repercussions of exogenous events [72]. Previous literature has discussed the importance of choosing the right length of rolling window. Zhang et al. (2018) argue that if the length of the rolling window is too large, the scaling exponents' evolution gets smoother, and the significant trends are easier to spot. However, the impact of such events on short-term market dynamics is covered. The estimated parameters due to economic cycling and seasonal factors might lose their locality and be unable to illustrate the evolution of short-term events. Conversely, in the case of too short a length, some short-term parameters may experience significant fluctuations, making it difficult to find the trend [73]. Therefore, we use Zhao and Cui [74]'s rolling window setting of 1500 days as the window length because of the large sample size. To begin, the first 1500 days from the start of returns are taken to obtain the Hurst exponent ($q = 2$) and MLM. Then, the window is moved forward by dropping the first observation and adding another at the end. This process is repeated until the series is completed. A sequence of Hurst exponent ($q = 2$) and MLM is obtained by looping over a range of dates across the entire data set, as shown in Figures 3 and 4.

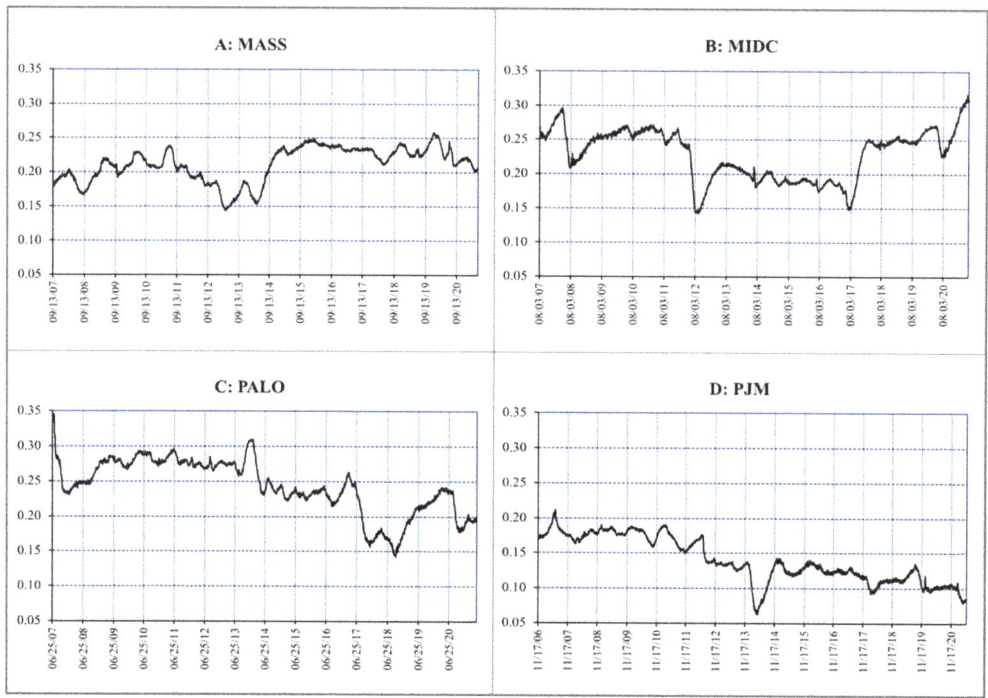

Figure 3. Dynamic Hurst exponents evolution for electricity indices ($q = 2$, and window = 1500).

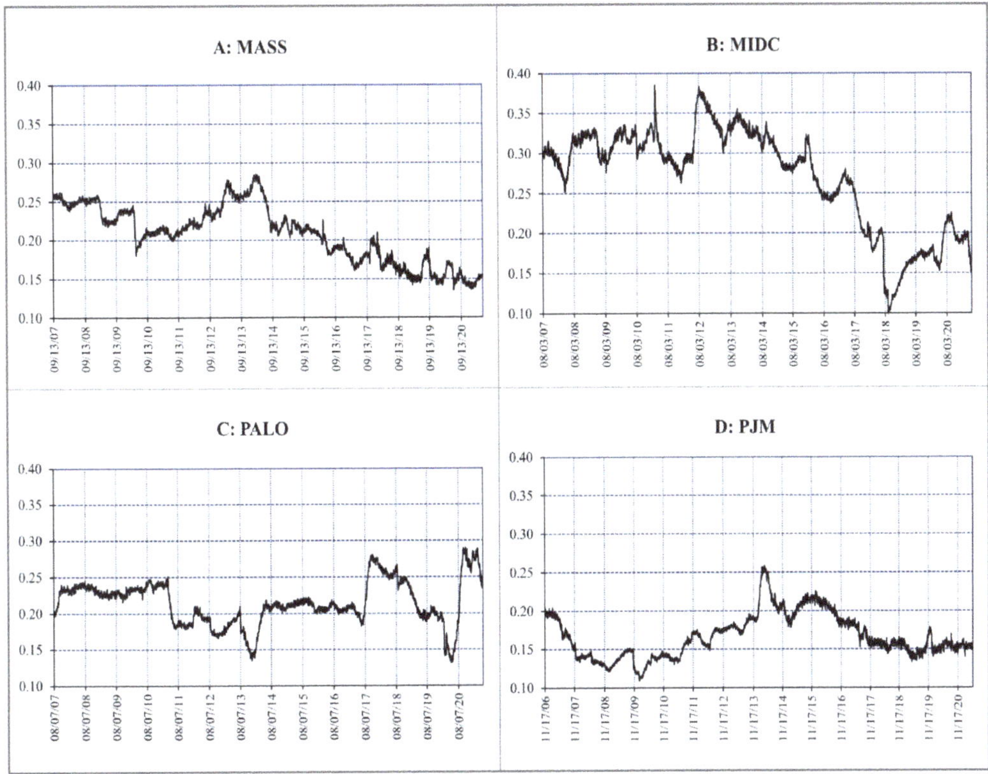

Figure 4. Dynamic evolution of multifractality degree for electricity indices (MLM, and *window* = 1500).

Figure 3 shows that the exponent lines never move up from 0.5 for all electricity indices throughout the time period. This indicates anti-persistent behavior (negative autocorrelation) during the whole sample. Figure 4 plots the results of the indices' market inefficiency using the MLM measure. Looking at the graphs, we can see that all return series show rich multifractal degrees, but the highest is in Palo Verde. After the first quarter of 2014, MLM shows a downward trend for Mass Hub (New England) and PJM West, indicating an upward trend of market efficiency. According to the 2021 Regional Electricity Outlook report by ISO New England, a winter reliability program was designed in 2014 to keep the power grid reliable during periods of fuel insecurity. It happened because the region's fuel delivery and energy security issues were highlighted by a brutal winter cold snap. For MIDC Hub, a significant improvement in efficiency is observed after the fourth quarter of 2012 with a maximum in 2018. The highest efficiency around the third quarter of 2018 could be due to the fact that electricity prices in the western US reached their highest levels since 2008. According to EIA [75], these high prices are the result of record high temperatures, which led to a relatively high demand for electricity. However, Palo Verde shows different behavior regarding market inefficiency, where the highest MLM values are found in 2017 and 2020, indicating the lowest efficiency. Electricity sales in the US declined by 80 billion kilowatt hours (kWh) in 2017, which is the largest drop since the economic recession in 2009. According to EIA [76], this major drop in sales is due to weather variations, which may have resulted in the highest inefficiency of Palo Verde in 2017. However, we found no significant impact of the recent COVID-19 pandemic on the degree of market efficiency for all electricity indices. On 11th March 2011, the Tohoku earthquake and tsunami struck Japan and sparked the Fukushima nuclear crisis.

Woo et al. [77] document that the impact of that crisis is international and could have a significant effect on electricity markets, at least in the short term, as shutting down a nuclear plant reduces supply in the electricity market, which leads to a high market price. Figure 3 indicates a significant decrease in efficiency in electricity markets overall, especially Mass Hub and PJM West after 11 March 2011. Moreover, the overall electricity prices were high from 2012 to 2013, which were driven largely by an increase in spot natural gas prices [78]. However, due to regional supply and demand issues, the percentage increases in electricity prices were higher in MIDC and Mass Hub. MIDC is often among the least expensive in the regions because of the regional concentration of hydroelectric generation. According to EIA, the spring in MIDC region in 2013 was drier than the prior two springs, which kept prices higher. This could be attributed to the significant decrease in the inefficiency of MIDC from 2012 to 2013, as depicted in Figure 3. The cold weather put a strain on the already stressed natural gas pipeline infrastructure in New England. This resulted in electricity prices spikes in 2013, which might be the reason for the inefficiency spikes of Mass Hub during this period.

5. Conclusions

The electricity market in the United States has undergone major deregulatory reforms since the 1990s, aiming to increase competition and benefit consumers. This rising competitiveness altered the price dynamics, resulting in an unavoidable increase in pricing volatility [7,79]. At the same time, electricity prices incorporate all the supply/demand shocks, which leads to highly complex characteristics i.e., autocorrelation, heteroscedasticity, nonlinearity, etc. [8]. In this context, the focus is on examining the efficiency dynamics of US electricity markets, using multifractality to test these complex characteristics. To do so, we used MFDFA [36] to find the generalized Hurst exponent $h(q)$ and the Renyi exponent $\tau(q)$. The four major electricity indices i.e., Mass Hub, Mid C, Palo Verde, and PJM West were studied. The roughly 20 years of data from 2001 to 2021 is a sufficient period to test its multifractality.

The findings of this study confirm a significance presence of multifractal behavior in all the US electricity markets studied. However, the strength of multifractality (Δh) varies, with PJM West having the highest and Mass Hub having the lowest. As multifractality is the indicator of market efficiency, the most efficient market is Mass Hub, while PJM is seen to be least efficient. The MLM measure, which is a useful tool to quantify the efficiency of financial markets, further confirms the findings. Moreover, all electricity indices at $q = 2$ exhibit anti-persistent (mean reverting) behavior. We employ an MFDFA rolling window approach to investigate the dynamic feature of persistency and efficiency through the classical Hurst exponent and MLM measure. The findings confirm anti-persistent behavior for all series over time as well. All electricity market indices show significant multifractal patterns over time with Palo Verde's being the most volatile of all. Specifically, we found an upward trend in the efficiency of Mass Hub and PJM West after the first quarter of 2014.

Several factors and variables may have an impact on the multifractality and efficiency of financial markets. For example, Rizvi et al. [80] found liquidity and speculative bubble problems, Chung and Hrazdil [81] discovered that a positive link between liquidity and arbitrage activity leads to increased efficiency, and more recently, Al-Yahyaee et al. [82] found efficiency to be positively linked with liquidity and negatively with volatility. Furthermore, herding behavior [60], temporal correlation [83], volatility predictability [84], crash predictability [84,85], complexity of markets [86], and inefficient market structures [67,72,87] may also impact the multifractality of financial markets. Most importantly, government reforms [56], financial liberalization [60], economic freedom, and competitive financial intermediaries [88] are also positively linked with market efficiency and decrease multifractality.

This study provides novel findings for regulatory authorities, policymakers, and decision makers at the government and corporate levels. The significant multifractal behavior of the US electricity market implies the existence of dependency and inefficiency. These inefficiencies could be linked to the market's predictability and imperfections, which in return lead to attaining abnormal returns [89]. Hence, regulators and policymakers should confirm transparency requirements to improve public access to information, which leads to reducing inefficiency. These findings may also be used by governments to decide on future deregulation of other possible natural monopolies, such as the electricity transmission sector. Additionally, the findings are useful for institutional investors who invest or trade in electricity market activities or energy stocks. This will help them to develop portfolios for better decision-making processes and risk management strategies, as according to previous studies, the width of the multiple spectrum is an indicator of future price fluctuations [84,90]. Therefore, these findings may also potentially help consumers who could see better electricity prices as producers and retailers are more able to match forecasts with demand. Future research should compare the efficiency and multifractality of the US electricity market with countries that have deregulated their electricity market and those that are still monopolistic. This research can also be extended by looking at the possible factors that may affect the multifractality of electricity markets.

Author Contributions: Conceptualization, H.A., F.A. and P.F.; Data curation, H.A., F.A. and P.F.; Formal analysis, H.A. and P.F.; Funding acquisition, F.A.; Writing—original draft, H.A., F.A. and P.F.; Writing—review and editing, H.A., F.A. and P.F. All authors have read and agreed to the published version of the manuscript.

Funding: Paulo Ferreira acknowledges the financial support of Fundação para a Ciência e a Tecnologia (grants UIDB/05064/2020 and UIDB/04007/2020).

Institutional Review Board Statement: Not applicable.

Informed Consent Statement: Not applicable.

Data Availability Statement: The data presented in this study are openly available at the website of US Energy Information Administration (EIA).

Conflicts of Interest: The authors declare no conflict of interest.

References

1. Brehm, P.A.; Zhang, Y. The efficiency and environmental impacts of market organization: Evidence from the Texas electricity market. *Energy Econ.* **2021**, *101*, 105359. [CrossRef]
2. Dertinger, A.; Hirth, L. Reforming the electric power industry in developing economies evidence on efficiency and electricity access outcomes. *Energy Policy* **2020**, *139*, 111348. [CrossRef]
3. Borenstein, S.; Bushnell, J. The US Electricity Industry after 20 Years of Restructuring. *Annu. Rev. Econ.* **2015**, *7*, 437–463. [CrossRef]
4. Rosellón, J.; Myslíková, Z.; Zenón, E. Incentives for transmission investment in the PJM electricity market: FTRs or regulation (or both?). *Util. Policy* **2011**, *19*, 3–13. [CrossRef]
5. Kristjanpoller, W.; Minutolo, M.C. Asymmetric multi-fractal cross-correlations of the price of electricity in the US with crude oil and the natural gas. *Phys. A Stat. Mech. Appl.* **2021**, *572*, 125830. [CrossRef]
6. Joskow, P.L. Lessons Learned from Electricity Market Liberalization. *Energy J.* **2008**, *29*. [CrossRef]
7. Dias, J.G.; Ramos, S.B. Heterogeneous price dynamics in U.S. regional electricity markets. *Energy Econ.* **2014**, *46*, 453–463. [CrossRef]
8. Heydari, A.; Nezhad, M.M.; Pirshayan, E.; Garcia, D.A.; Keynia, F.; De Santoli, L. Short-term electricity price and load forecasting in isolated power grids based on composite neural network and gravitational search optimization algorithm. *Appl. Energy* **2020**, *277*, 115503. [CrossRef]
9. Uniejewski, B.; Weron, R. Efficient Forecasting of Electricity Spot Prices with Expert and LASSO Models. *Energies* **2018**, *11*, 2039. [CrossRef]
10. Hong, Y.-Y.; Wu, C.-P. Day-Ahead Electricity Price Forecasting Using a Hybrid Principal Component Analysis Network. *Energies* **2012**, *5*, 4711–4725. [CrossRef]
11. Lin, X.; Yu, H.; Wang, M.; Li, C.; Wang, Z.; Tang, Y. Electricity Consumption Forecast of High-Rise Office Buildings Based on the Long Short-Term Memory Method. *Energies* **2021**, *14*, 4785. [CrossRef]
12. Weron, R. Electricity price forecasting: A review of the state-of-the-art with a look into the future. *Int. J. Forecast.* **2014**, *30*, 1030–1081. [CrossRef]
13. Lin, W.-M.; Gow, H.-J.; Tsai, M.-T. An enhanced radial basis function network for short-term electricity price forecasting. *Appl. Energy* **2010**, *87*, 3226–3234. [CrossRef]

14. Keles, D.; Scelle, J.; Paraschiv, F.; Fichtner, W. Extended forecast methods for day-ahead electricity spot prices applying artificial neural networks. *Appl. Energy* **2016**, *162*, 218–230. [CrossRef]
15. Agrawal, R.K.; Muchahary, F.; Tripathi, M.M. Ensemble of relevance vector machines and boosted trees for electricity price forecasting. *Appl. Energy* **2019**, *250*, 540–548. [CrossRef]
16. Luo, S.; Weng, Y. A two-stage supervised learning approach for electricity price forecasting by leveraging different data sources. *Appl. Energy* **2019**, *242*, 1497–1512. [CrossRef]
17. Matsumoto, T.; Yamada, Y. Simultaneous hedging strategy for price and volume risks in electricity businesses using energy and weather derivatives. *Energy Econ.* **2021**, *95*, 105101. [CrossRef]
18. Nakajima, T.; Toyoshima, Y. Examination of the Spillover Effects among Natural Gas and Wholesale Electricity Markets Using Their Futures with Different Maturities and Spot Prices. *Energies* **2020**, *13*, 1533. [CrossRef]
19. Fama, E. Efficient market hypothesis: A review of theory and empirical work. *J. Financ.* **1970**, *25*, 28–30. [CrossRef]
20. Zhu, X.; Bao, S. Multifractality, efficiency and cross-correlations analysis of the American ETF market: Evidence from SPY, DIA and QQQ. *Phys. A Stat. Mech. Appl.* **2019**, *533*, 121942. [CrossRef]
21. Mandelbrot, B.B.; Mandelbrot, B.B. *The Fractal Geometry of Nature*; WH Freeman: New York, NY, USA, 1982; Volume 1.
22. Lahmiri, S.; Uddin, G.S.; Bekiros, S. Nonlinear dynamics of equity, currency and commodity markets in the aftermath of the global financial crisis. *Chaos Solitons Fractals* **2017**, *103*, 342–346. [CrossRef]
23. Yang, Y.-H.; Shao, Y.-H.; Shao, H.-L.; Stanley, H.E. Revisiting the weak-form efficiency of the EUR/CHF exchange rate market: Evidence from episodes of different Swiss franc regimes. *Phys. A Stat. Mech. Appl.* **2019**, *523*, 734–746. [CrossRef]
24. Aloui, C.; Mabrouk, S. Value-at-risk estimations of energy commodities via long-memory, asymmetry and fat-tailed GARCH models. *Energy Policy* **2010**, *38*, 2326–2339. [CrossRef]
25. Herrera, R.; Rodriguez, A.; Pino, G. Modeling and forecasting extreme commodity prices: A Markov-Switching based extreme value model. *Energy Econ.* **2017**, *63*, 129–143. [CrossRef]
26. Oh, G.; Kim, S.; Eom, C. Long-term memory and volatility clustering in high-frequency price changes. *Phys. A Stat. Mech. Appl.* **2008**, *387*, 1247–1254. [CrossRef]
27. Adrangi, B.; Chatrath, A.; Dhanda, K.K.; Raffiee, K. Chaos in oil prices? Evidence from futures markets. *Energy Econ.* **2001**, *23*, 405–425. [CrossRef]
28. Naeem, M.A.; Bouri, E.; Peng, Z.; Shahzad, S.J.H.; Vo, X.V. Asymmetric efficiency of cryptocurrencies during COVID19. *Phys. A Stat. Mech. Appl.* **2021**, *565*, 125562. [CrossRef]
29. He, L.-Y.; Chen, S.-P. Multifractal Detrended Cross-Correlation Analysis of agricultural futures markets. *Chaos Solitons Fractals* **2011**, *44*, 355–361. [CrossRef]
30. Mandelbrot, B. The Variation of Some Other Speculative Prices. *J. Bus.* **1967**, *40*, 393. [CrossRef]
31. Peters, E.E. *Fractal Market. Analysis: Applying Chaos Theory to Investment and Economics*; John Wiley & Sons: Hoboken, NJ, USA, 1994; Volume 24.
32. Mensi, W.; Tiwari, A.K.; Al-Yahyaee, K.H. An analysis of the weak form efficiency, multifractality and long memory of global, regional and European stock markets. *Q. Rev. Econ. Financ.* **2019**, *72*, 168–177. [CrossRef]
33. Hurst, H.E. Long-Term Storage Capacity of Reservoirs. *Trans. Am. Soc. Civ. Eng.* **1951**, *116*, 770–799. [CrossRef]
34. Lo, A.W. Long-Term Memory in Stock Market Prices. *Econom. J. Econom. Soc.* **1991**, *59*, 1279. [CrossRef]
35. Aloui, C.; Shahzad, S.J.H.; Jammazi, R. Dynamic efficiency of European credit sectors: A rolling-window multifractal detrended fluctuation analysis. *Phys. A Stat. Mech. Appl.* **2018**, *506*, 337–349. [CrossRef]
36. Kantelhardt, J.W.; Zschiegner, S.A.; Koscielny-Bunde, E.; Havlin, S.; Bunde, A.; Stanley, H. Multifractal detrended fluctuation analysis of nonstationary time series. *Phys. A Stat. Mech. Appl.* **2002**, *316*, 87–114. [CrossRef]
37. Peng, C.-K.; Buldyrev, S.; Havlin, S.; Simons, M.; Stanley, H.E.; Goldberger, A.L. Mosaic organization of DNA nucleotides. *Phys. Rev. E* **1994**, *49*, 1685–1689. [CrossRef]
38. Fisher, A.J.; Calvet, L.E.; Mandelbrot, B.B. *Multifractality of Deutschemark/US Dollar Exchange Rates*; Elsevier: Amsterdam, The Netherlands, 1997.
39. Pasquini, M.; Serva, M. Multiscale behaviour of volatility autocorrelations in a financial market. *Econ. Lett.* **1999**, *65*, 275–279. [CrossRef]
40. Kwapien, J.; Oswiecimka, P.; Drożdż, S. Components of multifractality in high-frequency stock returns. *Phys. A Stat. Mech. Appl.* **2005**, *350*, 466–474. [CrossRef]
41. Oświe, P.; Kwapień, J.; Drożdż, S. Multifractality in the stock market: Price increments versus waiting times. *Phys. A Stat. Mech. Appl.* **2005**, *347*, 626–638. [CrossRef]
42. Aslam, F.; Mohti, W.; Ferreira, P. Evidence of Intraday Multifractality in European Stock Markets during the Recent Coronavirus (COVID-19) Outbreak. *Int. J. Financ. Stud.* **2020**, *8*, 31. [CrossRef]
43. Aslam, F.; Latif, S.; Ferreira, P. Investigating Long-Range Dependence of Emerging Asian Stock Markets Using Multifractal Detrended Fluctuation Analysis. *Symmetry* **2020**, *12*, 1157. [CrossRef]
44. Aslam, F.; Nogueiro, F.; Brasil, M.; Ferreira, P.; Mughal, K.S.; Bashir, B.; Latif, S. The footprints of COVID-19 on Central Eastern European stock markets: An intraday analysis. *Post-Communist Econ.* **2020**, *33*, 751–769. [CrossRef]
45. Maganini, N.D.; Filho, A.C.S.; Lima, F.G. Investigation of multifractality in the Brazilian stock market. *Phys. A Stat. Mech. Appl.* **2018**, *497*, 258–271. [CrossRef]

46. Aslam, F.; Aziz, S.; Nguyen, D.K.; Mughal, K.S.; Khan, M. On the efficiency of foreign exchange markets in times of the COVID-19 pandemic. *Technol. Forecast. Soc. Chang.* **2020**, *161*, 120261. [CrossRef] [PubMed]

47. Diniz-Maganini, N.; Rasheed, A.A.; Sheng, H.H. Exchange rate regimes and price efficiency: Empirical examination of the impact of financial crisis. *J. Int. Financ. Mark. Inst. Money* **2021**, *73*, 101361. [CrossRef]

48. Kakinaka, S.; Umeno, K. Cryptocurrency market efficiency in short- and long-term horizons during COVID-19: An asymmetric multifractal analysis approach. *Financ. Res. Lett.* **2021**, 102319. [CrossRef]

49. Mnif, E.; Jarboui, A.; Mouakhar, K. How the cryptocurrency market has performed during COVID 19? A multifractal analysis. *Financ. Res. Lett.* **2020**, *36*, 101647. [CrossRef]

50. Telli, Ş.; Chen, H. Multifractal behavior in return and volatility series of Bitcoin and gold in comparison. *Chaos Solitons Fractals* **2020**, *139*, 109994. [CrossRef]

51. Fernandes, L.H.; de Araújo, F.H.; Silva, I.E. The (in)efficiency of NYMEX energy futures: A multifractal analysis. *Phys. A Stat. Mech. Appl.* **2020**, *556*, 124783. [CrossRef]

52. Guo, Y.; Yao, S.; Cheng, H.; Zhu, W. China's copper futures market efficiency analysis: Based on nonlinear Granger causality and multifractal methods. *Resour. Policy* **2020**, *68*, 101716. [CrossRef]

53. Mensi, W.; Vo, X.V.; Kang, S.H. Upside-Downside Multifractality and Efficiency of Green Bonds: The Roles of Global Factors and COVID-19. *Financ. Res. Lett.* **2021**, 101995. [CrossRef]

54. Lee, Y.-J.; Kim, N.-W.; Choi, K.-H.; Yoon, S.-M. Analysis of the Informational Efficiency of the EU Carbon Emission Trading Market: Asymmetric MF-DFA Approach. *Energies* **2020**, *13*, 2171. [CrossRef]

55. Choi, S.-Y. Analysis of stock market efficiency during crisis periods in the US stock market: Differences between the global financial crisis and COVID-19 pandemic. *Phys. A Stat. Mech. Appl.* **2021**, *574*, 125988. [CrossRef]

56. Wang, Y.; Liu, L.; Gu, R. Analysis of efficiency for Shenzhen stock market based on multifractal detrended fluctuation analysis. *Int. Rev. Financ. Anal.* **2009**, *18*, 271–276. [CrossRef]

57. Li, J.; Lu, X.; Jiang, W.; Petrova, V.S. Multifractal Cross-correlations between foreign exchange rates and interest rate spreads. *Phys. A Stat. Mech. Appl.* **2021**, *574*, 125983. [CrossRef]

58. Shahzad, S.J.H.; Bouri, E.; Kayani, G.M.; Nasir, R.M.; Kristoufek, L. Are clean energy stocks efficient? Asymmetric multifractal scaling behaviour. *Phys. A Stat. Mech. Appl.* **2020**, *550*, 124519. [CrossRef]

59. He, L.-Y.; Chen, S.-P. Are crude oil markets multifractal? Evidence from MF-DFA and MF-SSA perspectives. *Phys. A Stat. Mech. Appl.* **2010**, *389*, 3218–3229. [CrossRef]

60. Cajueiro, D.; Gogas, P.; Tabak, B.M. Does financial market liberalization increase the degree of market efficiency? The case of the Athens stock exchange. *Int. Rev. Financ. Anal.* **2009**, *18*, 50–57. [CrossRef]

61. Caraiani, P. Evidence of Multifractality from Emerging European Stock Markets. *PLoS ONE* **2012**, *7*, e40693. [CrossRef]

62. Zunino, L.; Tabak, B.; Figliola, A.; Pérez, D.; Garavaglia, M.; Rosso, O. A multifractal approach for stock market inefficiency. *Phys. A Stat. Mech. Appl.* **2008**, *387*, 6558–6566. [CrossRef]

63. Zhan, C.; Liang, C.; Zhao, L.; Zhang, Y.; Cheng, L.; Jiang, S.; Xing, L. Multifractal characteristics analysis of daily reference evapotranspiration in different climate zones of China. *Phys. A Stat. Mech. Appl.* **2021**, *583*, 126273. [CrossRef]

64. Zhang, X.; Zhang, G.; Qiu, L.; Zhang, B.; Sun, Y.; Gui, Z.; Zhang, Q. A modified multifractal detrended fluctuation analysis (MFDFA) approach for multifractal analysis of precipitation in dongting lake basin, China. *Water* **2019**, *11*, 891. [CrossRef]

65. Liu, H.; Zhang, X.; Zhang, X. Multiscale multifractal analysis on air traffic flow time series: A single airport departure flight case. *Phys. A Stat. Mech. Appl.* **2020**, *545*, 123585. [CrossRef]

66. Kantelhardt, J.W.; Koscielny-Bunde, E.; Rybski, D.; Braun, P.; Bunde, A.; Havlin, S. Long-term persistence and multifractality of precipitation and river runoff records. *J. Geophys. Res. Atmos.* **2006**, *111*. [CrossRef]

67. Wang, Y.; Wei, Y.; Wu, C. Cross-correlations between Chinese A-share and B-share markets. *Phys. A Stat. Mech. Appl.* **2010**, *389*, 5468–5478. [CrossRef]

68. Mujeeb, S.; Javaid, N. ESAENARX and DE-RELM: Novel schemes for big data predictive analytics of electricity load and price. *Sustain. Cities Soc.* **2019**, *51*, 101642. [CrossRef]

69. Lavin, L.; Murphy, S.; Sergi, B.; Apt, J. Dynamic operating reserve procurement improves scarcity pricing in PJM. *Energy Policy* **2020**, *147*, 111857. [CrossRef]

70. Weron, R. Energy price risk management. *Phys. A Stat. Mech. Appl.* **2000**, *285*, 127–134. [CrossRef]

71. Weron, R.; Przybyłowicz, B. Hurst analysis of electricity price dynamics. *Phys. A Stat. Mech. Appl.* **2000**, *283*, 462–468. [CrossRef]

72. Cajueiro, D.O.; Tabak, B.M. Testing for time-varying long-range dependence in volatility for emerging markets. *Phys. A Stat. Mech. Appl.* **2005**, *346*, 577–588. [CrossRef]

73. Liu, L.; Wan, J. A study of correlations between crude oil spot and futures markets: A rolling sample test. *Phys. A Stat. Mech. Appl.* **2011**, *390*, 3754–3766. [CrossRef]

74. Zhao, R.; Cui, Y. Dynamic Cross-Correlations Analysis on Economic Policy Uncertainty and US Dollar Exchange Rate: AMF-DCCA Perspective. *Discret. Dyn. Nat. Soc.* **2021**, *2021*, 1–9. [CrossRef]

75. EIA. Summer Average Wholesale Electricity Prices in Western U.S. Were Highest since 2008. 2018. Available online: https://www.eia.gov/todayinenergy/detail.php?id=37112 (accessed on 10 September 2021).

76. EIA. In 2017, U.S. Electricity Sales Fell by the Greatest Amount since the Recession. 2018. Available online: https://www.eia.gov/todayinenergy/detail.php?id=35612 (accessed on 10 September 2021).

77. Woo, C.K.; Ho, T.; Zarnikau, J.; Olson, A.; Jones, R.; Chait, M.; Horowitz, I.; Wang, J. Electricity-market price and nuclear power plant shutdown: Evidence from California. *Energy Policy* **2014**, *73*, 234–244. [CrossRef]
78. EIA. New England and Pacific Northwest Had Largest Power Price Increases in 2013. 2014. Available online: https://www.eia.gov/todayinenergy/detail.php?id=14511 (accessed on 10 September 2021).
79. Assereto, M.; Byrne, J. The Implications of Policy Uncertainty on Solar Photovoltaic Investment. *Energies* **2020**, *13*, 6233. [CrossRef]
80. Rizvi, S.A.R.; Arshad, S.; Alam, N. A tripartite inquiry into volatility-efficiency-integration nexus—Case of emerging markets. *Emerg. Mark. Rev.* **2018**, *34*, 143–161. [CrossRef]
81. Chung, D.; Hrazdil, K. Liquidity and market efficiency: A large sample study. *J. Bank. Financ.* **2010**, *34*, 2346–2357. [CrossRef]
82. Al-Yahyaee, K.H.; Mensi, W.; Ko, H.-U.; Yoon, S.-M.; Kang, S.H. Why cryptocurrency markets are inefficient: The impact of liquidity and volatility. *N. Am. J. Econ. Financ.* **2020**, *52*, 101168. [CrossRef]
83. Green, E.; Hanan, W.; Heffernan, D. The origins of multifractality in financial time series and the effect of extreme events. *Eur. Phys. J. B* **2014**, *87*, 1–9. [CrossRef]
84. Wei, Y.; Huang, D. Multifractal analysis of SSEC in Chinese stock market: A different empirical result from Heng Seng index. *Phys. A Stat. Mech. Appl.* **2005**, *355*, 497–508. [CrossRef]
85. Grech, D.; Pamuła, G. The local Hurst exponent of the financial time series in the vicinity of crashes on the Polish stock exchange market. *Phys. A Stat. Mech. Appl.* **2008**, *387*, 4299–4308. [CrossRef]
86. Kumar, S.; Deo, N. Multifractal properties of the Indian financial market. *Phys. A Stat. Mech. Appl.* **2009**, *388*, 1593–1602. [CrossRef]
87. Tabak, B.; Cajueiro, D. Are the crude oil markets becoming weakly efficient over time? A test for time-varying long-range dependence in prices and volatility. *Energy Econ.* **2007**, *29*, 28–36. [CrossRef]
88. Dewandaru, G.; Rizvi, S.A.R.; Bacha, O.I.; Masih, M. What factors explain stock market retardation in Islamic Countries. *Emerg. Mark. Rev.* **2014**, *19*, 106–127. [CrossRef]
89. Dragotă, V.; Ţilică, E.V. Market efficiency of the Post Communist East European stock markets. *Cent. Eur. J. Oper. Res.* **2014**, *22*, 307–337. [CrossRef]
90. Su, Z.-Y.; Wang, Y.-T. An Investigation into the Multifractal Characteristics of the TAIEX Stock Exchange Index in Taiwan. *J. Korean Phys. Soc.* **2009**, *54*, 1385–1394. [CrossRef]

Review

Review of Renewable Energy Potentials in Indonesia and Their Contribution to a 100% Renewable Electricity System

Jannis Langer *, Jaco Quist and Kornelis Blok

Faculty of Technology, Policy and Management, Department of Engineering Systems and Services, Delft University of Technology, Jaffalaan 5, 2628BX Delft, The Netherlands; j.n.quist@tudelft.nl (J.Q.); k.blok@tudelft.nl (K.B.)
* Correspondence: j.k.a.langer@tudelft.nl

Abstract: Indonesia has an increasing electricity demand that is mostly met with fossil fuels. Although Indonesia plans to ramp up Renewable Energy Technologies (RET), implementation has been slow. This is unfortunate, as the RET potential in Indonesia might be higher than currently assumed given the archipelago's size. However, there is no literature overview of RET potentials in Indonesia and to what extent they can meet current and future electricity demand coverage. This paper reviews contemporary literature on the potential of nine RET in Indonesia and analyses their impact in terms of area and demand coverage. The study concludes that Indonesia hosts massive amounts of renewable energy resources on both land and sea. The potentials in the academic and industrial literature tend to be considerably larger than the ones from the Indonesian Energy Ministry on which current energy policies are based. Moreover, these potentials could enable a 100% renewables electricity system and meet future demand with limited impact on land availability. Nonetheless, the review showed that the research topic is still under-researched with three detected knowledge gaps, namely the lack of (i) economic RET potentials, (ii) research on the integrated spatial potential mapping of several RET and (iii) empirical data on natural resources. Lastly, this study provides research and policy recommendations to promote RET in Indonesia.

Keywords: renewable energy; potential; Indonesia; literature review; 100% renewables; scenario

Citation: Langer, J.; Quist, J.; Blok, K. Review of Renewable Energy Potentials in Indonesia and Their Contribution to a 100% Renewable Electricity System. *Energies* **2021**, *14*, 7033. https://doi.org/10.3390/en14217033

Academic Editors: Sergio Ulgiati, Hans Schnitzer and Remo Santagata

Received: 5 October 2021
Accepted: 22 October 2021
Published: 27 October 2021

Publisher's Note: MDPI stays neutral with regard to jurisdictional claims in published maps and institutional affiliations.

1. Introduction

Indonesia is a strongly growing country and could become the world's 4th largest economy by 2050 [1]. This development is reflected by Indonesia's rapidly increasing electricity demand of more than 6% p.a. since 2000 [2,3]. Until now, the archipelago mostly depends on its abundant domestic resources of coal and natural gas to meet this demand [4]. Nevertheless, Indonesia has committed to the energy transition via the national energy plan (Rencana Umum Energi Nasional—RUEN) and targets a share of Renewable Energy Technologies (RET) in the energy mix of 23% and 31% by 2025 and 2050, respectively [5].

Large hydropower, geothermal and biomass are already substantial parts of Indonesia's electricity mix with 17.3% in 2018 [4]. In contrast, the shares of alternatives like solar photovoltaics (PV) and wind power are considerably lower, while ocean energy has not been implemented at all. The reasons for the stagnant development of the latter technologies are manifold, including lack of experience, limited grid flexibility to balance intermittent power production [6–9] as well as opaque and incomplete pricing schemes, investment-repelling regulation and complicated, time-consuming licensing processes [5]. Notwithstanding, the implementation of RET might benefit from a more comprehensive and accurate overview of their potential in Indonesia. With such an overview, it would be possible to assess how much current and future electricity demand could be covered with RET. Moreover, energy scenarios like a 100% renewable electricity system and its requirements like land area could be deduced. With these insights, it would be possible to evaluate whether current RET implementation goals are in line with the potentials and whether adjustments are needed. To our knowledge, such an overview does not exist yet in literature. Therefore, this paper aims to address the following research question:

What is the state of contemporary literature on RET potentials for electricity production in Indonesia and to what extent could these potentials meet current and future electricity demands?

To answer the research question, existing academic, industrial, and governmental literature on the potentials of nine RET in Indonesia is reviewed. The focus is set on the provincial and national level and distinctions are made between the theoretical, technical, practical, and economic potential as shown in Table 1. Moreover, this study critically analyses what is necessary to activate these potentials in terms of required land areas, and indicates the impact of the potentials on current and future electricity demand. Light is also shed on how implementation proceeded compared to the plans expressed in the RUEN.

The scientific contribution of this work is not only to provide an overview of existing literature on RET potentials in Indonesia but also to critically put them into perspective in terms of impact and realisation requirements. Moreover, this study aims to raise awareness to researchers, policymakers, and investors about Indonesia as a country that not only hosts a diverse set of renewable resources but also has a large and rising energy demand to match these resources. By discovering current knowledge gaps in the literature, future research directed towards these gaps might contribute to knowledge on both Indonesia's energy transition and climate change mitigation with benefits beyond national borders. Therefore, the results also have significant policy relevance.

The paper is organised as follows. Section 2 elaborates on the methods to search and select literature as well as defining the boundaries of the review. Section 3 presents the results of the literature review, followed by a critical discussion in Section 4. The paper ends with a conclusion and recommendations in Sections 5 and 6, respectively.

2. Materials and Methods

An overview of the literature review is depicted in Figure 1. Backwards snowballing was used to trace primary literature with a maximum of two iteration cycles. Regarding language and grammar, studies were left out if the main message of the reviewed publication could not be unequivocally reconstructed. In case a study was filtered out on abstract scan, it was still fully read if its content was helpful to convey the storyline of this paper. Thus, the elaborations in the following sections are not only based on the 38 extracted studies in Figure 1. Out of the 38 reviewed studies, 4 come from the Ministry of Energy and Natural Resources (Kementerian Energi dan Sumber Daya Mineral—ESDM), 5 from industrial sources and 29 from academic literature. 22 publications focus on the national, 6 on the global level, and 5 studies each on the provincial and inter-provincial, regional level. Regarding the technologies, 7 studies each were about a set of RET and solar PV, 6 studies were about biomass, 5 studies each were about wave energy and tidal current and 2 studies each were about hydropower, OTEC, offshore wind and geothermal. 34 studies are in English, 4 in Indonesian (Bahasa Indonesia).

Figure 1 shows that 182 studies were filtered out due to being secondary literature or too regional scope. This study aims to draw insights from potential studies that can be scaled to the global or at least provincial level. Local case studies might not be scalable to such an extent, especially for locally sensitive technologies like wind power, which is why they are excluded here. Nonetheless, it is acknowledged that localised RET research is highly important, as decentralised RET can be a gateway for community empowerment and local socio-economic development [10,11].

The nine reviewed technologies comprise geothermal, large and small hydropower, biomass, solar PV, wind power as well as tidal power, wave energy and Ocean Thermal Energy Conversion (OTEC). In literature, the potential of these technologies is studied on different levels based on different definitions as Table 1 shows. To maintain consistency, this study uses the definitions found in academic and industrial literature, as these are broken down more distinctly compared to those of the ESDM. The ministry's potentials are adjusted where necessary to make them comparable. ESDM's technical and practical potentials are summarised as technical potentials and the acceptable potential becomes the practical potential. ESDM's theoretical and economic potentials are assumed to remain unchanged. The values drawn from the Global Energy Resource Database by Royal Dutch Shell are based on "[…] an estimate of the *realistic*, or constrained, technical potential, which accounts for technical as well as non-technical limitations […]" [12] (p. 240). Hence, the realistic potentials listed in the database are labelled as practical potentials here.

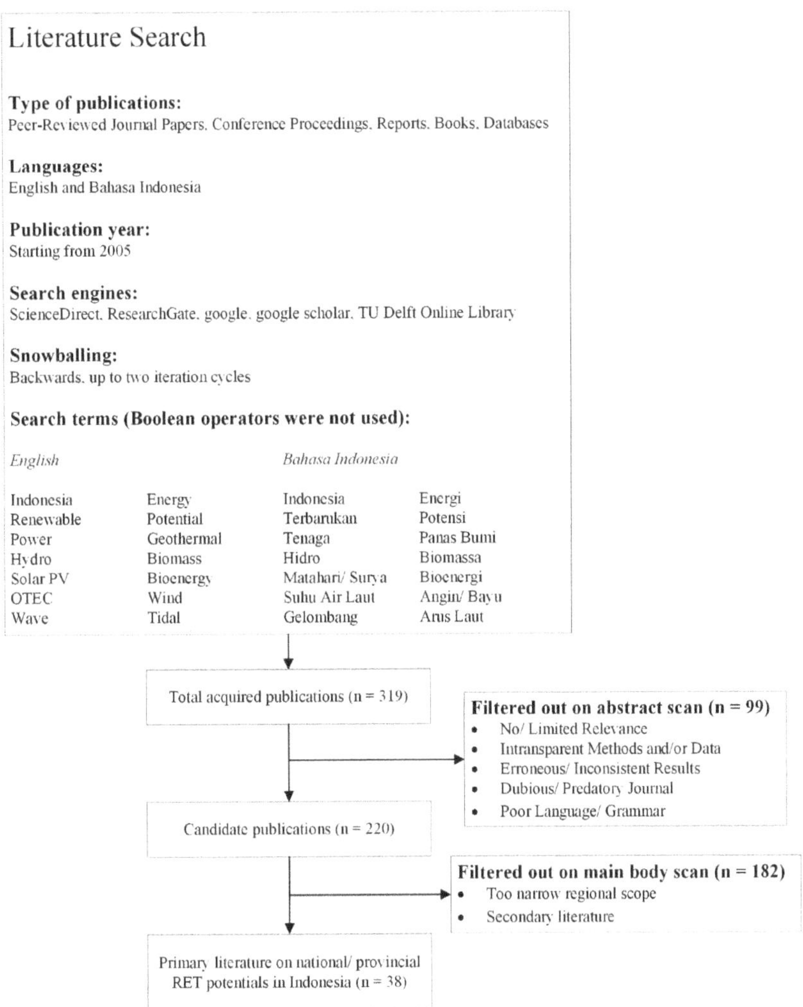

Figure 1. Methods used for the systematic literature review on RET potentials in Indonesia.

In this paper, the potentials found in literature are shown in their original physical units and converted to GW_e to make them comparable. In the case of units of energy, the potential is converted to GW_e using average generation efficiencies (electricity output divided by primary energy input) and capacity factors (generated electricity divided by installed capacity and 8760 h/year) of Indonesian power plants based on the statistics provided by ESDM [2].

Unless stated otherwise, this literature review focuses on RET for electricity production, while other applications such as heat, cold and transportation are excluded. Moreover, the state of the art of individual technologies and power plants is not reviewed as such works already exist as pointed out in the respective sub-sections. The review of energy statistics is limited to the context of RET since general statistics for the whole Indonesian energy system were covered recently [6,9,13,14].

Table 1. Different definitions of potentials found in literature.

	Academic and Industrial Literature [15–17] (Terminology Used in This Study)	Reports by ESDM [18]
Theoretical Potential	Potential restricted by physical limits (e.g., Carnot efficiency for thermal power plants, Betz limit for wind turbines, etc.)	Potential based on field data via a modelling system
Technical Potential	Potential restricted by technical limits (e.g., geo- and oceanographic restrictions, electrical and mechanical efficiencies, etc.)	Identified potential that can be implemented at a certain location
Practical Potential	Potential restricted by non-technical limits (e.g., protection zones and tourist areas)	Identified potential that can be implemented at a certain location based on long-term data
Acceptable Potential	-	Potential that considers demand, infrastructure, and communal approval
Economic Potential	Potential with unit costs equal to or lower than benchmark (e.g., wholesale electricity price)	Potential that can be actually utilised

3. Results

3.1. RET in Indonesia and Development Plans

In 2018, Indonesia's share of RET in the electricity mix was 17.4% as shown in Figure 2. The Levelized Cost of Electricity (LCOE) of renewables and their competitiveness against fossil-fuelled generators in Indonesia are shown in Table 2. What the most prominent RET in Indonesia, namely large hydropower, geothermal and biomass, have in common is their non-intermittent power production. In contrast, fluctuating RET like solar PV and wind power are still at an early stage of implementation in Indonesia [2]. But this might change with the government's current capacity development targets. Indonesia plans to ramp up the total installed capacity from 65 GW in 2018 to 443 GW until 2050, 168 GW of which from RET, as shown in Figure 3a. Moreover, Indonesia's electricity demand is expected to rise from 258 TWh in 2018 [4] to 2046 TWh in 2050 [5,19]. Despite the ambition to develop RET in the Indonesian electricity system, the dominance of fossil fuels would not end with the RUEN but get stronger as illustrated in Figure 3a. So far, both fossil and renewable capacity are not implemented as planned in the RUEN as seen in Figure 3b, at least in absolute terms. In relative terms, fossil capacity was developed at the planned annual rate of roughly 6%, while RET only grew by 5% per year instead of the planned 9%. This suggests that implementation targets might generally be set too high and that the development of fossil capacity proceeds smoother compared to renewable capacity.

Table 2. The levelized cost of electricity of renewables and fossil-fuelled generators in Indonesia. Values for OTEC based on [20], values for all other technologies based on [21].

Technology	Levelized Cost of Electricity [US¢/kWh]
Open-Cycle Gas Turbine	9.2–12.94
Combined-Cycle Gas Turbine	6.69–8.93
Coal Mine Mouth	5.01–7.31
Coal Sub Critical	6.11–8.41
Coal Super Critical	5.77–8.05
Coal Ultra Super Critical	5.83–8.38
Onshore Wind	7.39–16.1
Utility Scale Solar	5.84–10.28
Geothermal	4.56–8.7
Biomass	4.68–11.4
Ocean Thermal Energy Converson (Commecial Large-scale plant after 30 years of modelled upscaling.)	6.2–16.8

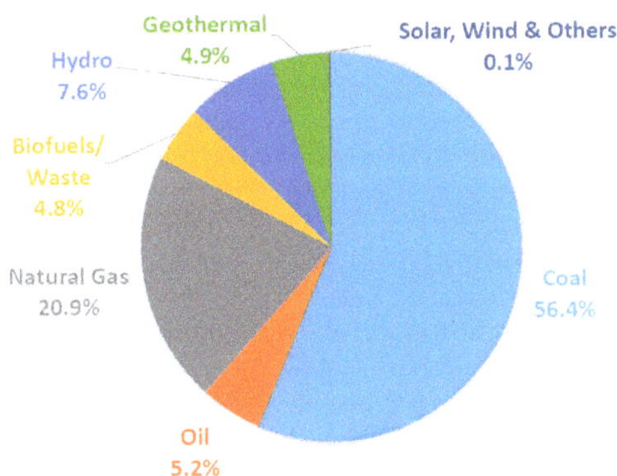

Figure 2. Total electricity supply of Indonesia in 2018 [4].

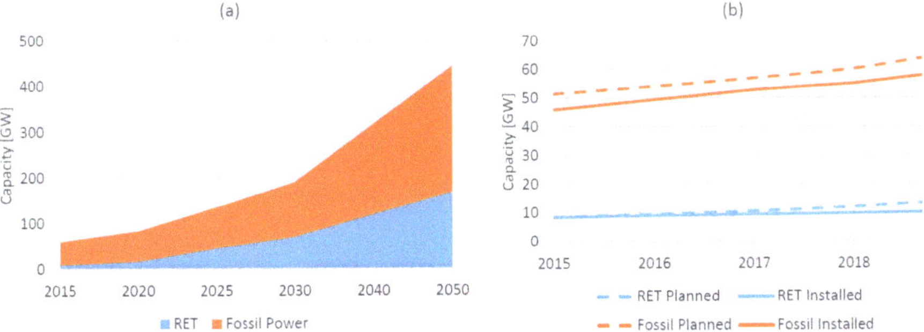

Figure 3. (**a**) Planned installed capacity based on RUEN [5] and (**b**) installed vs. planned capacity of fossil and RET until 2019 [2,5].

The following sub-sections show the results of the literature review on RET potentials in Indonesia and the impact of these potentials on demand coverage and area usage if possible. Furthermore, the current developments and barriers of each technology are discussed as well as their roles in the RUEN. Based on these insights, it might be possible to explain why RET implementation does not progress as planned.

3.2. Geothermal

Geothermal power plants produce electricity by extracting the heat generated and stored within the Earth at depths of around 1 km and below. According to current estimates, Indonesia hosts around 40% of global geothermal resources due to its location on the ring of fire, an area known for seismic and volcanic activity [22,23]. As of 2019, Indonesia deployed 2.1 GW_e or 9% of estimated geothermal resources which produced 14 TWh_e. With such a capacity, the country ranks 2nd in global geothermal implementation behind the USA [24].

Estimates on geothermal potentials in Indonesia mainly come from the Geology Agency of ESDM. In contrast to other RET, geothermal potentials are clustered in two categories, namely resources and reserves. Resources are rough estimations of geothermal heat, which might be exploitable if technical and economic prerequisites are met. Reserves on the other hand only include technically and economically recoverable heat based on geoscience survey tools and empirical data like reservoir temperature and size [25]. The Geology Agency aggregates resources and reserves to get a total [2]. Resources can become reserves if they can be extracted economically and vice versa, reserves can become resources again if detrimental economic developments render their extraction unprofitable.

In 2019, geothermal resources and reserves were 9.3 GW_{th} and 14.6 GW_{th} [2], respectively. Outside ESDM's work, only one academic study was found that estimated Indonesia's geothermal potential. However, that study did not calculate potentials but proposed a new accounting methodology based on ESDM's values [25]. Additionally, recent literature comprises literature reviews of the geothermal industry in Indonesia [22,23,26,27]. From the industrial side, Royal Dutch Shell [28] indicates practical resources of 1009 PJ_e per year or 42 GW_e if a capacity factor of 76% is assumed [2]. With such resources, Indonesia's electricity demand in 2018 [4] and 2050 [5,19] could be covered to 108% and 14%, respectively.

Until 2050, an installed geothermal capacity of 17.5 GW_e is planned, which is 4% of the total planned RET capacity [5]. For an additional capacity of 15.4 GW_e, more than 300 new plants would have to be built with an average capacity of 50 MW_e [2]. This exceeds the current thermal reserves, which implies that some part of the resources must become reserves. To which extent this is possible depends on economic developments and technical availability, as not all thermal resources are suitable for electricity generation [25]. As of now, the installed capacity in 2019 is 15% lower than projected in the RUEN [2]. Current challenges include complications in obtaining land permits, inadequate electricity tariffs, opposition from local communities, limited data availability as well as long lead times of 7–8 years on average amongst others [29].

3.3. Hydropower

Hydropower plants convert the energy of moving water into electricity. Depending on the system size, there is large and small hydropower. Although an accepted consensus of 10 MW has emerged as an upper limit for small hydropower, there is no formal definition, leading to regionally variable thresholds [30]. Some works aggregate the potential of both technologies, including Hoes et al. [31] calculating a theoretical potential of 241 GW and Royal Dutch Shell [28] with a practical potential of 205 PJ per year or 15 GW, if a capacity factor of 43% is assumed [2]. In the following, the two technologies are reviewed separately.

3.3.1. Large Hydropower

ESDM currently estimates a theoretical large hydropower potential of 75 GW [32,33], a value obtained in 1983 [34]. From this potential, 30% and 29% are situated in Papua and Kalimantan, respectively [18]. The only reviewed industrial study estimates a practical potential of 26 GW and includes restrictions like protected areas, tourism zones, reservoir size and resettlement of residents [35]. In 2019, roughly 5.6 GW or 7.5% of the theoretical potential was installed resulting in an electricity production of around 21 TWh. Malaysian hydropower is the only type of electricity that is imported to Indonesia. Including 1.7 TWh of these imports, large hydropower's contribution to the total electricity supply was 7.6% in 2019 [2].

Large hydropower will be an integral part of the Indonesian energy transition according to the RUEN. Until 2050, an additional capacity of 32.5 GW is planned, which is higher than the technical and practical potentials mentioned above. With a resulting capacity of 38 GW in 2050, large hydropower would form 8.6% of total installed capacity, making it the second largest renewable generator in Indonesia in terms of capacity after solar PV [5]. Moreover, 38 GW of large hydropower could cover 55% and 7% of Indonesia's electricity demand in 2018 [4] and 2050 [5,19], respectively. In 2019, implementation exceeded plans by 2% [2].

3.3.2. Small Hydropower

Currently, ESDM estimates a theoretical small hydropower potential of around 19.4 GW [18,36]. The highest share of that potential is in East and Central Kalimantan with 18% and 17%, respectively [33]. In 2019, the installed capacity was around 418 MW or 2% of the theoretical potential [2]. In academic literature, small hydropower enjoys more attention than its large counterpart with individual case studies [37–40], a review [41] and a climate change impact study [42].

The rapid upscaling of small hydropower is endorsed in Indonesia due to low costs, local expertise and reliable power production, amongst others [43]. An installed capacity of 7 GW until 2050 is targeted in the RUEN, most of which are in Sumatera, Kalimantan, Java and East Nusa Tenggara [5]. With 7 GW of small hydropower, 10% and 1% of Indonesia's electricity demand in 2018 [4] and 2050 [5,19] could be covered, respectively. However, implementation lagged by roughly 44% in 2019 [2]. Small hydropower is considered key for rural electrification and community empowerment, while reported barriers include lack of foreign investment, access to finance, as well as limited infrastructure [29].

3.4. Biomass

In the field of energy, biomass encompasses all renewable plant- and animal-based materials for power and heat production. Figure 4 summarises the different types of biomass available in Indonesia and the options for power generation.

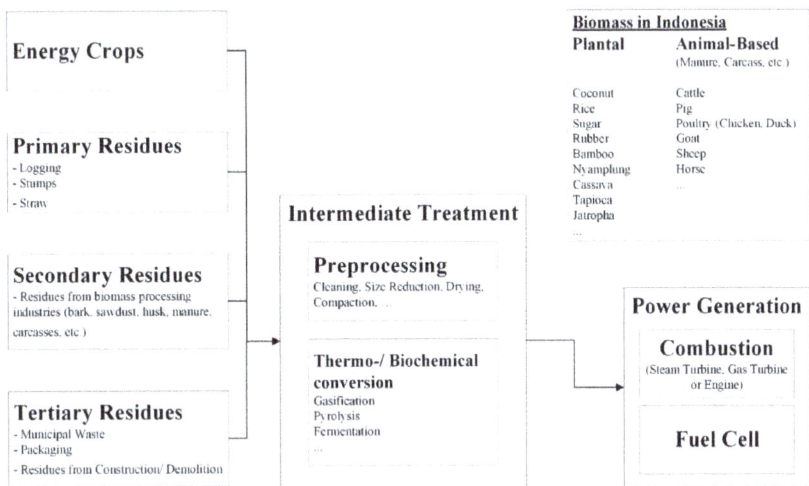

Figure 4. Biomass in Indonesia and options for power production (based on [44–49]).

The potential of biomass in Indonesia is studied widely by both governmental and academic research. Table 3 shows current literature on biomass potentials in both their original physical units as well as in terms of thermal energy and electrical capacity. ESDM estimated the potential of biofuels, residues from industrial agriculture and biogas for power generation. Elaborations on the methods and assumptions regarding the conversion from thermal to electrical energy could not be found. Currently, ESDM estimates a theoretical biomass potential of 32.7 GW_e [36], with most of the resources being located in Sumatera, the Java-Madura-Bali region and Kalimantan with roughly 48%, 28% and 16%, respectively. Palm oil, as well as rice husk, take the largest shares of the potential with 39% and 30%, respectively [18]. Out of the 32.7 GW_e, municipal waste and biogas from manure comprise potentials of 2.1 GW_e and 0.5 GW_e, respectively [18,32].

In academic literature, national theoretical and technical potentials were assessed for solid biomass [45,47,50], biogas [46,51] and bio-methanol for fuel cells [44]. A critical aspect of the sustainability of biomass is its origin. As mentioned above, biomass for energy conversion is primarily produced in palm oil plantations which often renders the local environment a degraded wasteland [52]. Therefore, an increased use of unsustainable biomass for electricity generation might exacerbate deforestation and undermine efforts to make Indonesia's energy system more environmentally friendly. One way to establish sustainable biomass supply chains is the renewed use of degraded land to cultivate plants like bamboo and nyamplung with additional benefits like soil recovery and non-interference with food production [47,50,53]. From a bottom-up perspective, challenges with this option are uncertain land tenure and local ownership as well conflicting interests between investors and local communities [54]. Although potentials in literature can vary considerably, it can be noted that the biomass potential tends to be the highest for energy crops, amongst others cultivated on degraded land. They could theoretically cover up to 28% of Indonesia's final energy demand and 32% of electricity demand in 2050 [5]. Compared to energy crops, the potential of biomass residues is less high, which is in line with the findings of ESDM.

Table 3. Biomass potentials in Indonesia. * (Co-)firing in steam plants with efficiency and capacity factor of 34.0% and 74.8% respectively. ** Combustion in gas plants with efficiency and capacity factor of 38.4% and 18.8%, respectively [2,4]. *** Density and heat value of methanol 792 kg/m^3 and 22.7 MJ/kg, respectively.

Ref.	Type of Biomass	Origin of Biomass	Type of Potential	Original Unit(s)	Potential Thermal Energy [PJ$_{th}$]	Potential Capacity [GW$_e$]
			Solid Biomass			
[32]	Primary & Secondary	Agriculture	Theoretical	28.0 GW$_e$	1940 *	28.0
[45]	Primary & Secondary	Industrial forestry	Technical	132.2 PJ$_{th}$	132.2	1.9 *
[47]	Energy Crops	Degraded land	Theoretical	1105 PJ$_{th}$	1105	15.9 *
[50]	Energy Crops	Degraded land	Theoretical	5000–7000 PJ$_{th}$	5000–7000	71.9–100.7 *
[28]	Energy Crops, Primary & Secondary	Industrial forestry and agriculture	Practical	1225 PJ$_{th}$	1225	17.6 *
			Biogas			
[32]	Secondary	Manure	Theoretical	535 MW$_e$	8.3 **	0.5
[46]	Secondary	Livestock farming	Theoretical Technical	9597.4 Mm^3/year 1.7 × 10^{10} kWh$_e$/year	159.4 **	10.3 **
			Waste-to-Energy			
[18]	Tertiary	Agriculture	Theoretical	2.1 GW$_e$	145.7 *	2.1
[51]	Tertiary	Households, industry, etc.	Theoretical Technical	2992 GWh$_{th}$/year 1172 GWh$_e$/year 343 MW$_e$	10.8	0.3
			Bio-Methanol			
[44]	Primary & Secondary	Natural and industrial forestry	Technical	40–169 × 10^9 L 42–176 Wh$_e$/year 10–42 GW$_e$	730–3040 ***	10–42

Recently, the use of biomass for power generation was increased significantly from 0.3% of total generation in 2017 [55] to 4.8% in 2018 [4]. Parts of that share come from the co-firing of biomass in coal plants, which is perceived as one of the cheaper options to promote the energy transition [48]. First tests have already been conducted by ESDM with positive results [56]. However, its feasibility for small-scale, rural application still needs to be addressed [54]. Moreover, there might be lock-in effects for coal-based power generation, as current co-firing plants are designed for a biomass rate of only 10–15% [57]. In the RUEN, a ramp-up to 26 GW until 2050 is projected, which would encompass 15.5% of the total planned renewable capacity [5]. As of 2019, implementation lags by 15% [2] due to barriers like insufficient tariffs, the resistance of local communities as well as lack of stakeholder coordination [29].

3.5. Solar PV

Solar energy can be converted to electricity in numerous ways, e.g., with PV modules on which this section will focus. ESDM estimates a theoretical solar PV potential of 3551 GW$_p$ [58] with forest areas and 1360 GW$_p$ [5,18] without forest areas, respectively. The theoretical potential is then multiplied with a uniform efficiency of 15%, resulting in a technical potential of 533 GW$_p$ [58] with forest areas and 208 GW$_p$ [5,18,33] without forest areas. Although ESDM only mentions forest areas as an exclusion criterion, their solar PV potential map indicates that conservation areas on land and sea are considered as well [5]. The largest shares of the photovoltaic power potentials are situated in the West and the North of the country, especially in Sumatera with 32% and Kalimantan with 25%, respectively [18].

However, if forest and conservation areas are the only exclusion criteria, the technical potentials are rather small as shown in Table 4. Assuming an installed capacity of 150 W_p/m^2 and using current statistics on total land, forest, water, and conservation areas, the technical solar PV potential would be 99 TW_p if the entire eligible area would be covered with solar panels. Only 0.21% of eligible land area and 0.07% of total land area would be needed for 208 GW_p which seems conservative. For instance, roughly 0.1% of Germany's total land area was already covered with solar PV in 2019. (Based on installed capacity of 49 GW_p [59], power production of 150 W/m^2, and total land area of 357,386 km^2.) It could be that ESDM used further exclusion criteria, but these were not confirmed by the reviewed material. Nonetheless, solar PV's prospect of becoming Indonesia's key energy technology is apparent even with ESDM's values. Assuming an annual solar electricity production of 1377 kWh/kW_p [60], the electricity production from 208 GW_p would be enough to cover 111% of Indonesia's electricity demand in 2018 [4] and 14% of the projected demand in 2050 [5,19].

Table 4. The technical potential of solar PV based on ESDM [5,18,33] and own estimations for maximum land coverage excluding forest, water, and conservation areas.

Region	BPS [61] Land Area (Excl. Forest, Water, and Conservation Area) [km²]	ESDM Tech. Potential [GWp]	Own Estimation Total Area Coverage for ESDM Potentials [%]	Own Estimation Tech. Potential with Land Area [GWp]
Sumatera	251,603	69	0.070	37,740
Java	96,312	32	0.032	14,447
Bali, East & West Nusa Tenggara	43,870	19	0.019	6581
Kalimantan	176,921	53	0.053	26,538
Sulawesi	53,422	23	0.023	8013
Maluku & North Maluku	14,547	5	0.005	2182
Papua & West Papua	20,991	8	0.008	3149
Total	657,666	208	0.21	98,650

The potentials found in academic and industrial work are much higher than the ones from ESDM and are summarised in Table 5. The Institute for Essential Services Reform (IESR) [62] found a technical potential of 20 TW_p while excluding protected and forest areas, water bodies, wetlands, airports, harbours and areas with slopes higher than 10°. With further exclusion criteria like agricultural and settlement areas, a practical potential of 3.4 TW_p was calculated, which would cover Indonesia's electricity demand in 2018 [4] 18 times and the projected demand in 2050 twice [5,19]. For such a capacity, 3.4% of Indonesia's suitable land area is necessary. Another industrial estimation on practical solar PV potentials comes from the Royal Dutch Shell Database [28] with 6569 PJ or 1.3 TW_p.

Solar PV is planned to be the most dominant technology in terms of installed capacity with 45 GW_p in 2050, which would be 10.1% of total and 26.8% of renewable capacity. For this, the roofs of up to 30% of government buildings and up to 25% of developed residential housing should be occupied by solar PV. Another plan is the development of a vertically integrated, domestic PV industry [5]. However, solar PV struggles to gain traction in Indonesia today and implementation trails behind by over 73% [2], as shown in Figure 5a.

Table 5. Overview of academic and industrial solar PV potential research.

Ref.	Publication Type	Regional Scope	Potential			
			Theoretical	Technical	Practical	Economic
[62]	Report	National	-	20,000	3400	-
[28]	Database	National	-	-	1300	-
[63]	Journal Paper	On-Grid National	-	1100	27	0.4
[64]	Journal Paper	Off-Grid National	-	-	0.8	-
[65]	Journal Paper	National	-	3200 (on-grid) 45,900 (off-grid)	73.3 (on-grid) 0.4 (off-grid)	-
[66]	Journal Paper	West Kalimantan	-	148	-	-
[67]	Journal Paper	West Kalimantan	-	2.0	-	-
[68,69]	Report	Bali	-	80	-	-

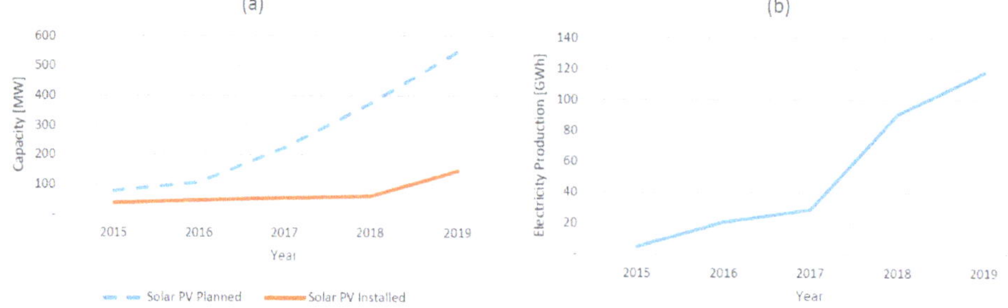

Figure 5. (**a**) Planned vs. installed solar PV capacity. (**b**) Electricity production from solar PV [2,5].

A closer look into ESDM's statistics reveals that the problems mostly come from grid-connected systems. Although off-grid PV systems only comprise 28% of the total installed capacity of 146 MW_p in 2019, they produced 54% of the total solar electricity production. Based on Figure 5, an average capacity factor as low as 2% underlines the operational problems of some solar PV systems documented in the literature [7,8]. Then again, statistical errors might also be responsible for the low factor, given that Figure 5a,b are not always aligned. On a positive note, solar PV already contributes to the electrification of rural communities. As part of a government programme, over 360,000 solar-powered lamps have been distributed across Indonesian communities [18,70]. These and other efforts seem to pay off and the recorded performance of the off-grid solar system should be an encouragement to promote even more solar systems in Indonesia both on- and off-grid. To do so, several barriers must be tackled, which for solar PV include complications in land ownership, unattractive tariffs and policy support, lack of local experience [29] as well as active resistance from the state-owned utility company Perusahaan Listrik Negara (PLN) [8].

3.6. Wind Power

The kinetic energy of moving air can be converted to electricity using wind turbines. ESDM estimated a theoretical and technical onshore wind potential with and without forest and conservation areas. At locations with average wind speeds above 6 m/s, 1 MW wind turbines were assumed with an area requirement of 1 km^2 per turbine. At locations with wind speeds between 4 and 6 m/s, 100 kW turbines with an area requirement of 0.25 km^2 were assumed [71]. The wind speeds were mapped at heights between 30–50 m at 120 locations [18]. Offshore locations were excluded in ESDM's assessments and the differences between potentials were not elaborated. The theoretical and technical potentials of onshore wind are 113.5 GW and 30.8 GW [71] with and 60.6 GW and 18.1 GW [5,18] without forest and conservation areas. Assuming a capacity factor of 36% [2], the latter technical potential would be enough to cover 22% of Indonesia's electricity demand in

2018 [4] and 3% of the projected demand in 2050 [5,19]. Most of the theoretical potential is in Java and East Nusa Tenggara with 38% and 17%, respectively. More comprehensive wind measurements and analyses are recommended to refine the potential [33]. As with solar PV, ESDM's wind potentials might be too conservative for three reasons. First, it is again not clear whether forest and conservation areas were the only spatial restriction areas on land, given that 18 GW of wind power would merely require 2.7% of Indonesia's total land area. Second, the assumed capacity densities might be too pessimistic, as current practice and studies suggest a density of 7 MW/km^2 [12]. Third, the omission of offshore wind removes a vast and otherwise eligible area for wind power deployment. Although there are good reasons to omit offshore wind in some areas, for example interfering shipping routes and high risk of natural catastrophes, no explanation for the exclusion could be found in ESDM's reports.

Rethinking the exclusion of offshore wind might be worthwhile, as academic and industrial sources suggest a far higher offshore than onshore wind potential. Bosch et al. [72] conducted a global offshore wind analysis and calculated a practical potential of 3.0 TW and 8318 TWh in Indonesia, using Economic Exclusive Zones (EEZ), conservation areas, vicinity of marine cables and water depth as exclusion criteria. This potential could cover Indonesia's electricity demand in 2018 [4] 36 times and the projected demand in 2050 [5,19] four times. To implement such a capacity, roughly 8% of the 5,568,600 km^2 [73] of the total available sea area of Indonesia would be required. In the database of Royal Dutch Shell [28], the practical on- and offshore wind resources are 69 and 14,174 PJ, or 6 and 1248 GW with a capacity factor of 36% [2], respectively. In the underlying study of the database [12], floating wind turbines were included up to a water depth of 1000 m. This opens up a new dimension of potential as mounted offshore turbines cannot be implemented at such depths today.

Gernaat et al. [74] estimate a technical offshore potential of 53 EJ, which translates to a capacity of 4668 GW or 260 times ESDM's potential. It is unclear why this potential is so high, given that the water depth and distance to shore were restricted to 80 m and 139 km, respectively, while the other two studies [28,72] above include depths of 1000 m for floating wind turbines and a distance to shore of more than 200 km. There might be differences in input data and limited accuracy due to low-resolution data. In contrast to Deng et al. [12] and Bosch et al. [72], who use wind speed data with a resolution of 19 km and 5 km, respectively, Gernaat et al. [74] do not mention the data resolution, so their estimation could not be checked. No other academic study on the national or provincial potential of wind power was found to validate these numbers. Instead, both international [75–77] and Indonesian [78–80] research tends to focus more on local case studies. Even if Gernaat et al.'s [74] potential would be technically possible, the practical hurdles are very high given that 11% of Indonesia's available sea area would be needed for such a capacity.

Until 2050, 28 GW of wind power are planned to be installed, but Figure 6a shows that implementation lagged by roughly 60% in 2019, notwithstanding a significant growth of electricity production from wind power since 2017 as Figure 6b illustrates [5]. In 2019, 154 MW or 0.25% of ESDM's technical potential were tapped. But as with solar PV, the unaligned development of capacity and electricity production in Figure 6a,b indicate that there might be statistical errors. Current barriers are unattractive tariffs as well as a lack of stakeholder coordination and experience [29]. Besides increasing the quantity and quality of wind resource assessments and feasibility studies, the RUEN calls for the development of wind turbines in isolated regions, outermost islands and at the country borders [5], which might imply wind power's vital role for future rural electrification.

3.7. Ocean Energy

Ocean energy is the least developed RET in Indonesia and no commercial plants are operating yet. However, being the largest archipelago worldwide, Indonesia has exceptional potentials to use the energy stored in the ocean, namely through the motion of tides and waves or the thermal energy of the water. In recent reports of ESDM and the RUEN, the collective theoretical and technical potentials of ocean energy are estimated as 288 GW and 18–72 GW, respectively, though without elaboration on methods, assumptions and distinction between individual technologies [33]. The further assessment and refinement of ocean energy technologies are explicitly encouraged, and their upscaling is currently projected to start in 2025 with a target capacity in 2050 of 6.1 GW [5]. Besides ESDM, the Indonesian Ocean Energy Association (Asosiasi Energi Laut Indonesia—ASELI) assessed the potentials of individual ocean technologies. However, despite being frequently cited in other papers [16,81–83], the underlying study or seminar protocols could not be found. The internet presence of ASELI was not accessible anymore in December 2020. Thus, the primary study from ASELI could unfortunately not be reviewed.

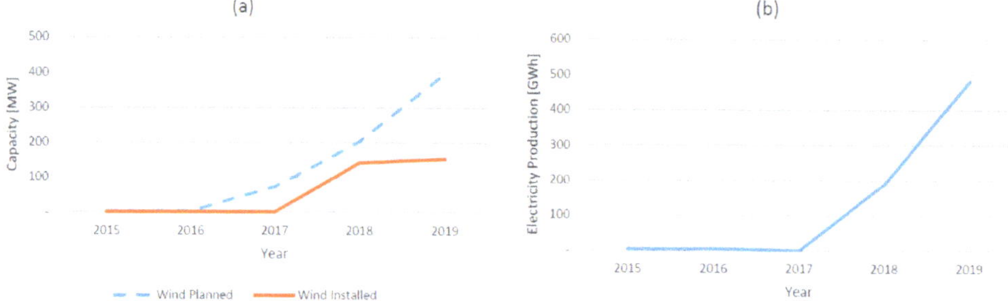

Figure 6. (**a**) Planned vs. installed wind capacity. (**b**) Electricity production from wind power [2,5].

3.7.1. OTEC

OTEC generates electricity using the temperature difference between warm surface and cold deep-sea water. As a tropical archipelago, Indonesia is a very interesting country for OTEC [84–86]. Recently, the practical and economic potential of moored OTEC in Indonesia has been estimated with and without upscaling and technological learning. There, a practical potential of 102 GW_e is estimated, which would span over 14% of the available marine area. Without upscaling and technological learning, the economic potential is refined to 0–2.0 GW_e [87] and increases to 6–41 GW_e if these two important mechanisms are included [20,88]. OTEC could cover up to 22% of Indonesia's electricity demand in 2050 [5,19]. Besides that, a nominal 100 MW_{nom} plant at 20 °C seawater temperature difference could produce around 1200 GWh of electricity annually [89] due to real average temperature differences far higher than 20 °C of up to 25.4 °C [90].

3.7.2. Tidal Power

The movement of water caused by the gravitational forces between the Earth, Moon and Sun can be exploited for electricity generation. The only estimation of tidal energy's theoretical potential in Indonesia originates from an IRENA report [91] in collaboration with ESDM and comprises 18 GW, which would be 6% of the total theoretical ocean energy potential above. Besides that, academic research focuses on local power densities [92–94] and regional potentials [95–99] of tidal current power, while studies on alternatives like tidal barrages could not be found. Among existing literature, the most researched sites are the straits in Bali, Lombok, Larantuka and Alas. In Alas, the technical potential could be as high as 2.3 GW, while Larantuka and Bali could have technical potentials between 0.2–0.3 GW and 0.5–1.0 GW, respectively [95,96]. These low potentials might be explained by suboptimal local tide properties and moderate flow velocities [98]. To the knowledge of the authors, no academic or industrial work has shed light on national tidal power potentials in Indonesia yet.

3.7.3. Wave Energy Conversion

Wave energy converters produce electricity from the kinetic energy of waves. Within the global wave energy research network, many concepts have been studied over the last decades. Many of these designs are limitedly comparable due to technical differences [100] and uncertain design parameters [101]. For Indonesia, the oscillating wave column emerged as the most frequently studied technology and the potential of wave converters have been assessed as parts of global studies [102,103] as well as country-specifically on a national [83,101], provincial [104], cross-provincial [105,106] and local levels [107,108]. In the field of wave energy, the specific potential is usually expressed in the unit of kW/m, which represents the power per wave crest width [100]. In Indonesia, South Java is considered to have promising wave energy resources of up to 30 kW/m [101–103]. Other interesting areas are the Arafuru Sea [83], South Sumatera coastline [106] and South Kuta Bali [109]. An aggregated potential in kW is only available for individual sites [83,104], but not aggregated over provincial or national boundaries.

3.8. Potential Overview and 100% RET Scenario

The national RET potentials found in literature are summarised in Table 6. Solar PV and offshore wind power have the highest technical potential in Indonesia with a capacity of 20 TW$_p$ and 4.7 TW and electricity production of 27,540 TWh and 14,722 TWh, respectively. This would be enough to cover the demand in 2018 and 2050 more than 163 and 20 times, respectively. However, these two technologies are also amongst the least developed ones in the Indonesian electricity system and less than 1% of each potential is currently tapped. Compared to more established RET like geothermal and large hydropower, less established RET like solar PV, wind power and small hydropower were implemented slower than projected in the RUEN. Table 6 summarises the potentials for both ESDM and other sources. It shows that ESDM's potentials do not go beyond the technical level and although definitions for an acceptable and economic potential exist, no publication could be found that reports these potentials for any RET.

Table 7 shows how a 100% RET electricity scenario in 2050 could be shaped with the reviewed potentials. Until 2050, large hydropower, geothermal and biomass can still be considerably scaled up. On a national level, they could comprise 6–14% of the electricity mix. Most of the electricity would have to be supplied with solar PV and wind power with a combined share of 66%. The area requirements for the necessary capacity would be limited, as only 0.5% of the marine area would be necessary for offshore wind farms, and only 0.5% of the suitable land area of solar PV parks. The conceptual feasibility of a 100% RE system is in line with recent studies on Indonesia [110–112], although there are differences in the roles of RET and land use. Compared to IESR's recent deep decarbonisation report [113], the major difference to our projection is that solar PV's role is more prominent in their work with a share of 88% in 2050. With the reviewed potentials, such a share could have been reproduced here as well, but we decided to diversify the electricity mix over a broader set of RET with 33% of solar PV, 33% of wind energy and 33% of other RET. Compared to Simaremare et al. [110], Günther [111], and Günther & Eichinger [112], our land use shares are much smaller which can be explained by differences in regional scope. All of the three studies look into the Java-Bali region, while our scenario spans across the entire country. This shows that most RET in our scenario would not be in the economic heart of Indonesia in the Java-Bali region but the economically less developed East. Therefore, large investments in transmission infrastructure are probably required to transport the electricity produced in the East to the demand centres in the West. Moreover, creating a RET hub in East Indonesia could boost socio-economic development there and empower local communities with clean, decentralised electricity. A significant share of baseload could be provided by OTEC without interfering with land use, which is an interesting insight. Although not included in Table 7, other ocean energies like wave and tidal energy could contribute locally as well. Note that our 100% RET scenario is just a rough projection and comes with several limitations. Besides the aforementioned necessary transmission capacity from the East to the West, the scenario does not consider the necessary storage capacity to cope with the short-term and seasonal fluctuations of solar and wind power production. Moreover, OTEC would have to be scaled up with an annual growth rate of 28% until 2050 [20,88]. The necessary growth rates for solar PV and offshore wind should be even higher. Moreover, the economic feasibility of this projection will require more attention in future research. Nonetheless, the scenario shows that current energy transition plans could be reshaped towards more ambitious targets.

Table 6. Potential of RET in Indonesia. For references, see respective sections.

| Technology | | National Potential [GW$_e$] | | | | | | | | Installed Capacity 2019 [GW$_e$] | Planned until 2050 [GW$_e$] | Demand Coverage in 2050 [%] (Pract. Potential) |
| | | Theoretical | | Technical | | Practical | | Economic | | | | |
		ESDM (Theo)	Rest	ESDM (Tech + Pract)	Rest	ESDM (Accep)	Rest	ESDM (Eco)	Rest			
Hydro	Large	75	241	-	-	-	26	-	-	5.6	38	3
	Small	19		-	-	-	15	-	-	0.4	7	
Biomass	Solid	28	16–101	-	2	-	18	-	-			6
	Waste	2.1	-	-	0.3	-	-	-	-	1.8	26	-
	Methanol	-	-	-	10–42	-	-	-	-			-
	Biogas	0.5	-	-	10	-	-	-	-			-
Solar PV		1360–3551	-	208–533	1100–19,835	-	28–3397	-	0.4	0.15	45	2–229
Wind		61–114	-	18–72	4668	-	1254–2976	-	-	0.15	28	193–406
Ocean	OTEC		-		-	-	102	-	6–41			40
	Tidal	288	-	18–72	-	-	-	-	-	-	6.1	-
	Wave	-		-	-	-	-	-	-			-

| | | Resources | | | Reserves | | | | | |
		Speculative	Hypothetical	Possible	Probable	Proven				
Geothermal	ESDM	6 GW$_{th}$	3 GW$_{th}$	10 GW$_{th}$	2 GW$_{th}$	3 GW$_{th}$	2.1	17.5	-
	Rest	-	-		42 GW$_e$				14

Table 7. 100% RET scenario until 2050 based on the reviewed potentials.

100% RET System in 2050 (with Electricity Demand of 2,046,000 GWh [5,19])						
RET	**Potential (Type) [GWe]**	**Potential Electricity Production [GWh/Year]**	**Share of Practical Potential [%]**	**Deployed Capacity [GWe]**	**Annual Electricity Production [GWh/Year]**	**Share of Electricity Generation [%]**
Geothermal	42 (pract)	279,619	100%	42	279,619	14%
Large Hydro	38 (RUEN)	143,138	100%	38	143,138	7%
Small Hydro	7 (RUEN)	26,368	100%	7	26,368	1%
Biomass	18 (pract)	115,324	100%	18	115,324	6%
Solar PV	3397 (pract)	4,677,669	14%	491	676,306	33%
Wind Energy	2976 (pract)	8,318,237	7%	214	676,306	33%
OTEC	102 (pract)	339,045	16%	16	128,940	6%
Total	6580	13,899,400	-	826	2,046,000	100%

4. Discussion

4.1. Limitations

Although the methods described in Section 2 yielded more than 300 publications, it cannot be guaranteed that all available literature was retrieved. The use of additional search engines, terms and techniques could have resulted in an even more comprehensive collection. Moreover, there can be a subjective bias in the classification of potentials, especially in the cases where studies did not specify the type of potential or definitions differed substantially across studies. Therefore, the differences in potentials throughout studies might stem from the underlying differences in assumptions. This was especially apparent for the reports from ESDM, where methods are not always elaborated or scattered across multiple reports. The potential definitions in Table 1 are not consistently used, which could be because different departments within ESDM use different definitions. Therefore, there are uncertainties involved about the potentials from ESDM, which this study can only point out, but not resolve. These limitations aside, this paper still provides the most comprehensive overview of the general state of research on Indonesia's RET potentials so far.

4.2. Knowledge Gaps

Three knowledge gaps can be identified. A first knowledge gap comprises the limited work on RET potentials in Indonesia beyond the technical level. Most potentials reviewed in this paper originate from reports by ESDM, which do not always elaborate on the used data, methods, and assumptions. Most academic literature covers localised case studies with limited applicability to provincial and national levels. Many of these case studies were excluded from this review due to conceptual and methodological inconsistencies. If national potentials are mentioned in journal papers, they are generally directly adopted from ESDM [9,13,14,46,114]. This is reasonable as the potentials from ESDM are not only useful for review papers and energy policy planning but also provide a foundation for energy scenarios in academic research [13,14,115]. However, this paper provides reasons to assume that ESDM's potentials are too conservative and therefore current strategies like the RUEN. Although potentials can vary considerably across academic publications, they tend to be significantly higher compared to ESDM's potentials. If these academic estimations hold, Indonesia's potential to implement RET might be much larger than currently assumed. Alternative development strategies might capture these updated potentials more adequately than the RUEN enabling more progressive implementation targets. But to consolidate these arguments, more in-depth research is required.

The second knowledge gap builds upon the first one, as there is not only limited work on the potential of individual technologies but also on how these potentials relate to each other. Outside the field of ocean energy, no study was found that assesses the potential of several RET in Indonesia simultaneously. If the applicability of RET across Indonesia was mapped, it was either done for individual technologies [87] and in the case of solar PV and wind power [63,64] solely onshore, thus excluding alternatives like floating PV and offshore wind. For ocean energy, collective potential maps exist [16,116], but they are qualitative and do not offer insights into their technical and economic performance. As a result, current literature does not offer a map of the collective potential of several RET across Indonesia and the interaction between individual technologies.

The third knowledge gap refers to the lack of thorough data on natural resources such as wind and ocean data. As mentioned in two biomass studies, datasets on the same metric could differ significantly between sources, thus affecting the results based on the choice of the dataset [44,117]. Regarding wind power, both ESDM and academia agree that thorough field data is needed for more refined potentials, although the costs of acquisition are a hurdle [5,114,118,119]. This might explain why current research focuses more on local case studies since these cases can be studied more cost-effectively via simulations [76,79] or local on-site measurements [78,80]. These complications also apply to ocean energy research, as there are only a few data observation stations [101] and research is currently predominantly performed locally. None of the reviewed wind and ocean energy studies used simulated resource data from reanalysis models like HYCOM or NASA MERRA-2 as a proxy for measured field data.

5. Conclusions

In this paper, contemporary literature was reviewed to show what the potential of Renewable Energy Technologies (RET) in Indonesia is and how they could contribute to meeting current and future electricity demand in 2050. This study concludes that Indonesia hosts massive renewable energy resources spread over a wide range of different technologies on land and sea. Moreover, a 100% RET system could be technically feasible to meet Indonesia's future electricity demand. However, the research field is still underdeveloped and could benefit from more attention, potentially targeting the three knowledge gaps discovered in this study. First, there is limited work on RET potentials beyond the technical level with most existing knowledge originating from the Indonesian Energy Ministry and its subdivisions. These potentials might be too conservative based on the methodological assumptions. Second, existing studies mostly assess individual technologies and do not offer insights on the aggregated potential of multiple technologies and their distribution across the country. Third, there is a lack of thorough empirical data on natural resources such as wind and ocean data, due to which contemporary literature focuses more on local case and feasibility studies with little applicability to larger regional scopes.

The implementation of most RET, especially of unestablished ones like small hydropower, solar PV, and wind power, has proceeded slower than planned in the RUEN. This and the lack of academic and industrial research oppose the potential that RET might possess in Indonesia. Potentials from non-governmental studies tend to be much higher than the ones from ESDM. For example, the technical potential of wind power might be 260 times higher than currently projected in the national energy plan. If these projections hold, Indonesia has the luxury to choose between multiple options to promote the energy transition beyond what is already planned in the RUEN. However, due to the limited body of academic and industrial studies, more research is required to make more solid estimations of these potentials.

The assessment of RET potentials in Indonesia is a promising and worthwhile pursuit. Indonesia is a strongly growing country with the outlook of becoming one of the largest economies in the world; a development that might precipitate an equally robust growth in electricity demand. At the same time, the recent exhaustion of domestic oil resources in Indonesia shows that fossil fuels are finite [9], with the currently abundant resources of coal and natural gas being no exception. Therefore, the archipelago has splendid prerequisites to move away from fossil fuels towards a more sustainable energy system with beneficial effects beyond national borders.

6. Recommendations

Based on the literature review and three knowledge gaps found in this study, the following research and policy recommendations are proposed. The research recommendations are not ordered by relevance, but by the knowledge gaps in Section 4.2.

1. *Assessment of RET Potentials Beyond the Technical Level*

As shown in Table 6, there is only limited work on potentials beyond the technical level for virtually all reviewed technologies. To consolidate the potentials found in literature, more research on the potentials under practical and economic constraints is needed. For example, Langer et al. [87] assessed the economic potential of OTEC considering marine protected areas, water depth, connection points from sea to shore and the local electricity tariff. The methodology proposed there might be adapted for other RET, as recently done for wind power as a master thesis project at TU Delft [120].

2. *Aggregation and Spatial Mapping of Potentials of Several RET*

The potentials of individual technologies do not provide insights into how these technologies interact with each other. For instance, OTEC plants could be complemented with floating solar energy modules [121,122], but not with offshore wind turbines due to potential harmful interference of the offshore structures. Therefore, it might be helpful to pursue an integrated approach and to map the potential of several technologies across Indonesia. If multiple non-combinable technologies overlap at one location, the one with the higher potential could be preferred. Such work could connect the existing work on individual technologies, e.g., visualising the potential of wave energy conversion in South Java, while highlighting solar PV potentials in East Nusa Tenggara.

3. *Utilisation of Simulation and Forecast Models for an Initial Potential Estimation*

In literature, the collection of thorough field data is mentioned to refine the potential analyses. This might not be necessary and instead, the collection of field data could be limited to high-potential areas based on grounded estimations. For example, a preliminary assessment with data from sources like HYCOM and the Global Wind Atlas could reveal interesting areas for further investigation. For example, Namrole on Buru Island emerged as an economically interesting location for OTEC based on simulated data from HYCOM [87]. Thus, field data could be collected there to validate the simulation data, methods, and potential of OTEC.

4. *Re-shape provincial and national targets for RET implementation until 2050*

A key insight of this study is that the potential of RET in Indonesia is far higher than currently assumed by ESDM. However, current energy policies are built around ESDM's work, so the RUEN does not consider these increased potentials or even leaves out entire technologies like offshore wind. Therefore, this study recommends to re-assess current energy transition strategies to consider the potential of RET more appropriately. An important step towards this was PLN's recent pledge to become carbon neutral by 2060 [123]. To achieve this ambitious goal, the role of solar PV and offshore wind should become far more prominent as well as storage technologies to deal with short-term fluctuations in power supply. The integrated potential map discussed above and the scenarios derived from it could serve as the conceptual baseline of an updated energy transition strategy.

Author Contributions: J.L.: Conceptualization; Data curation; Formal analysis; Investigation; original draft; J.Q.: Contributions to methodology; Supervision; Validation; Writing—review & editing; K.B.: Contributions to methodology; Supervision; Validation; Writing—review & editing. All authors have read and agreed to the published version of the manuscript.

Funding: The work reported in this paper is funded by a grant (grant number W 482.19.509) from the Dutch research council NWO for the project entitled "Regional Development Planning and Ideal Lifestyle of Future Indonesia", under the NWO Merian Fund call on collaboration with Indonesia.

Institutional Review Board Statement: Not applicable.

Informed Consent Statement: Not applicable.

Data Availability Statement: No new data were created or analysed in this study. Data sharing does not apply to this article.

Acknowledgments: We want to express our gratitude to Josef Sergio Simanjuntak and Femke Pragt for their valuable input on wind power and solar PV potentials, respectively.

Conflicts of Interest: The authors declare no conflict of interest.

Abbreviations

Abbreviation	Meaning
ASELI	Asosiasi Energi Laut Indonesia (Indonesian Ocean Energy Association)
EEZ	Exclusive Economic Zones
ESDM	Kementerian Energi dan Sumber Daya Mineral (Ministry of Energy and Mineral Resources)
IESR	Institute for Essential Services Reform
LCOE	Levelized Cost of Electricity
OTEC	Ocean Thermal Energy Conversion
PLN	Perusahaan Listrik Negara (State Electricity Company)
PV	Photovoltaic
RUEN	Rencana Umum Energi Nasional (National Energy Plan)

References

1. Hawksworth, J.; Audino, H.; Clarry, R. *The Long View: How Will the Global Economic Order Change by 2050?* Price Waterhouse and Coopers: London, UK, 2017; pp. 1–72.
2. ESDM. *Handbook of Energy & Economic Statistics of Indonesia 2019*; ESDM: Jakarta, Indonesia, 2020.
3. ESDM. *2010 Handbook of Energy & Economic Statistics of Indonesia*; ESDM: Jakarta, Indonesia, 2010.
4. International Energy Agency. *World Energy Balances 2020*; International Energy Agency: Paris, France, 2020.
5. Presiden Republik Indonesia. *Rencana Umum Energi Nasional*; Presiden Republik Indonesia: Jakarta, Indonesia, 2017.

6.   Burke, P.J.; Widnyana, J.; Anjum, Z.; Aisbett, E.; Resosudarmo, B.; Baldwin, K.G.H. Overcoming barriers to solar and wind energy adoption in two Asian giants: India and Indonesia. *Energy Policy* **2019**, *132*, 1216–1228. [CrossRef]

7.   Kennedy, S.F. Indonesia's energy transition and its contradictions: Emerging geographies of energy and finance. *Energy Res. Soc. Sci.* **2018**, *41*, 230–237. [CrossRef]

8.   Setyowati, A.B. Mitigating energy poverty: Mobilizing climate finance to manage the energy trilemma in Indonesia. *Sustainability* **2020**, *12*, 1603. [CrossRef]

9.   Maulidia, M.; Dargusch, P.; Ashworth, P.; Ardiansyah, F. Rethinking renewable energy targets and electricity sector reform in Indonesia: A private sector perspective. *Renew. Sustain. Energy Rev.* **2019**, *101*, 231–247. [CrossRef]

10.   Chaurey, A.; Ranganathan, M.; Mohanty, P. Electricity access for geographically disadvantaged rural communities-technology and policy insights. *Energy Policy* **2004**, *32*, 1693–1705. [CrossRef]

11.   Dóci, G.; Vasileiadou, E. "Let's do it ourselves" Individual motivations for investing in renewables at community level. *Renew. Sustain. Energy Rev.* **2015**, *49*, 41–50. [CrossRef]

12.   Deng, Y.Y.; Haigh, M.; Pouwels, W.; Ramaekers, L.; Brandsma, R.; Schimschar, S.; Grözinger, J.; de Jager, D. Quantifying a realistic, worldwide wind and solar electricity supply. *Glob. Environ. Chang.* **2015**, *31*, 239–252. [CrossRef]

13.   Handayani, K.; Krozer, Y.; Filatova, T. From fossil fuels to renewables: An analysis of long-term scenarios considering technological learning. *Energy Policy* **2019**, *127*, 134–146. [CrossRef]

14.   Purwanto, W.W.; Pratama, Y.W.; Nugroho, Y.S.; Warjito; Hertono, G.F.; Hartono, D.; Deendarlianto; Tezuka, T. Multi-objective optimization model for sustainable Indonesian electricity system: Analysis of economic, environment, and adequacy of energy sources. *Renew. Energy* **2015**, *81*, 308–318. [CrossRef]

15.   Blok, K.; Nieuwlaar, E. *Introduction to Energy Analysis*, 3rd ed.; Taylor and Francis: Abingdon, UK, 2021.

16.   Prabowo, H. Atlas Potensi Energi Laut. *Maj. Miner. Energi* **2012**, *10*, 65–71.

17.   Hoogwijk, M.; de Vries, B.; Turkenburg, W. Assessment of the global and regional geographical, technical and economic potential of onshore wind energy. *Energy Econ.* **2004**, *26*, 889–919. [CrossRef]

18.   Directorate General of Renewable Energy and Energy Conservation (EBTKE). *Rencana Strategis 2020–2024*; EBTKE: Jakarta, Indonesia, 2020.

19.   Dewan Energi Nasional. *Indonesia Energy Outlook 2019*; Dewan Energi Nasional: Jakarta, Indonesia, 2019.

20.   Langer, J.; Quist, J.; Blok, K. Upscaling scenarios for ocean thermal energy conversion with technological learning in Indo-nesia and their global relevance. *Renew. Sustain. Energy Rev.* **2021**. submitted.

21.   Institute for Essential Services Reform. *Levelized Cost of Electricity in Indonesia*; Institute for Essential Services Reform: Jakarta, Indonesia, 2019.

22.   Nasruddin, M.; Alhamid, M.I.; Daud, Y.; Surachman, A.; Sugiyono, A.; Aditya, H.B.; Mahlia, T.M.I. Potential of geothermal energy for electricity generation in Indonesia: A review. *Renew. Sustain. Energy Rev.* **2016**, *53*, 733–740. [CrossRef]

23.   Setiawan, H. Geothermal Energy Development in Indonesia: Progress, Challenges and Prospect. *Int. J. Adv. Sci. Eng. Inf. Technol.* **2014**, *4*, 224. [CrossRef]

24.   Richter, A. Global Geothermal Capacity Reaches 14,900 MW—New Top 10 Ranking of Geothermal Countries, 2019. ThinkGeoEnergy. Available online: https://www.thinkgeoenergy.com/global-geothermal-capacity-reaches-14900-mw-new-top10-ranking/ (accessed on 9 November 2020).

25.   Fauzi, A. Geothermal resources and reserves in Indonesia: An updated revision. *Geotherm. Energy Sci.* **2015**, *3*, 1–6. [CrossRef]

26.   Pambudi, N.A. Geothermal power generation in Indonesia, a country within the ring of fire: Current status, future development and policy. *Renew. Sustain. Energy Rev.* **2018**, *81*, 2893–2901. [CrossRef]

27.   Suharmanto, P.; Fitria, A.N.; Ghaliyah, S. Indonesian Geothermal Energy Potential as Source of Alternative Energy Power Plant. *KnE Energy* **2015**, *1*, 119. [CrossRef]

28.   Royal Dutch Shell Global Energy Resources Database. Available online: https://www.shell.com/energy-and-innovation/the-energy-future/scenarios/shell-scenarios-energy-models/energy-resource-database.html#iframe=L3dlYmFwcHMvRW5lcmd5UmVzb3VyY2VEYXRhYmFzZS8 (accessed on 2 December 2020).

29.   Price Waterhouse and Coopers (PwC) Indonesia. *Power in Indonesia*; Price Waterhouse and Coopers: Jakarta, Indonesia, 2018.

30.   Breeze, P. *Hydropower*; Academic Press: Cambridge, MA, USA, 2018; pp. 53–62, ISBN 9780128129067.

31.   Hoes, O.A.C.; Meijer, L.J.J.; Van Der Ent, R.J.; Van De Giesen, N.C. Systematic high-resolution assessment of global hydropower potential. *PLoS ONE* **2017**, *12*, e171844. [CrossRef]

32.   Directorate General of Renewable Energy and Energy Conservation (EBTKE). *Renstra (Rencana Strategis) Ditjen EBTKE 2015–2019*; EBTKE: Jakarta, Indonesia, 2016.

33.   Directorate General of Renewable Energy and Energy Conservation (EBTKE). *Statistik EBTKE 2016*; EBTKE: Jakarta, Indonesia, 2016.

34.   Persero (PLN). *Electric Power Supply Business Plan (2019–2028)*; PLN: Jakarta, Indonesia, 2019.

35.   Koei, N. *Project for the Master Plan Study of Hydropower Development in Indonesia*; Japan International Cooperation Agency Nippon Koei Co., Ltd.: Tokyo, Japan, 2011.

36.   Amin, A.Z.; Gielen, D.; Saygin, D.; Rigter, J. *Renewable Energy Prospects: Indonesia, a REmap Analysis*; IRENA: Abu Dhabi, UAE, 2017.

37. Pontoiyo, F.; Sulaiman, M.; Budiarto, R.; Novitasari, D. Sustainability Potential for Renewable Energy System in Isolated Area that Supports Nantu Boliyohuto Wildlife Reserve. *IOP Conf. Ser. Earth Environ. Sci.* **2020**, *520*, 012026. [CrossRef]

38. Ratnata, I.W.; Surya, S.W.; Somantri, M. Analisis Potensi Pembangkit Energi Listrik Tenaga Air Di Saluran Air Sekitar Universitas Pendidikan Indonesia. In Proceedings of the FPTK Expo—UPI., Bandung, Indonesia, 13–14 November 2013; pp. 254–261.

39. Subekti, R.A. Survey Potensi Pembangkit Listrik Tenaga Mikro Hidro di Kuta Malaka Kabupaten Aceh Besar Propinsi Nanggroe Aceh Darussalam. *J. Mechatron. Electr. Power Veh. Technol.* **2010**, *1*, 5–12. [CrossRef]

40. Sulaeman; Jaya, R.A. Perencanaan pembangunan sistem pembangkit listrik tenaga mikro hidro (pltmh) di kinali pasaman barat. *J. Tek. Mesin* **2014**, *4*, 90–96.

41. Darmawi; Sipahutar, R.; Bernas, S.M.; Imanuddin, M.S. Renewable energy and hydropower utilization tendency worldwide. *Renew. Sustain. Energy Rev.* **2013**, *17*, 213–215. [CrossRef]

42. Anugrah, P.; Setiawan, A.A.; Budiarto, R. Sihana Evaluating Micro Hydro Power Generation System under Climate Change Scenario in Bayang Catchment, Kabupaten Pesisir Selatan, West Sumatra. *Energy Procedia* **2015**, *65*, 257–263. [CrossRef]

43. Blum, N.U.; Sryantoro Wakeling, R.; Schmidt, T.S. Rural electrification through village grids—Assessing the cost competitiveness of isolated renewable energy technologies in Indonesia. *Renew. Sustain. Energy Rev.* **2013**, *22*, 482–496. [CrossRef]

44. Suntana, A.S.; Vogt, K.A.; Turnblom, E.C.; Upadhye, R. Bio-methanol potential in Indonesia: Forest biomass as a source of bio-energy that reduces carbon emissions. *Appl. Energy* **2009**, *86*, S215–S221. [CrossRef]

45. Simangunsong, B.C.H.; Sitanggang, V.J.; Manurung, E.G.T.; Rahmadi, A.; Moore, G.A.; Aye, L.; Tambunan, A.H. Potential forest biomass resource as feedstock for bioenergy and its economic value in Indonesia. *For. Policy Econ.* **2017**, *81*, 10–17. [CrossRef]

46. Khalil, M.; Berawi, M.A.; Heryanto, R.; Rizalie, A. Waste to energy technology: The potential of sustainable biogas production from animal waste in Indonesia. *Renew. Sustain. Energy Rev.* **2019**, *105*, 323–331. [CrossRef]

47. Jaung, W.; Wiraguna, E.; Okarda, B.; Artati, Y.; Goh, C.S.; Syahru, R.; Leksono, B.; Prasetyo, L.B.; Lee, S.M.; Baral, H. Spatial assessment of degraded lands for biofuel production in Indonesia. *Sustainability* **2018**, *10*, 4595. [CrossRef]

48. Sharma, R.; Wahono, J.; Baral, H. Bamboo as an alternative bioenergy crop and powerful ally for land restoration in Indonesia. *Sustainability* **2018**, *10*, 4367. [CrossRef]

49. De Jong, W.; Van Ommen, J.R. (Eds.) *Biomass as a Sustainable Energy Source for the Future: Fundamentals of Conversion Processes*; Wiley: Hoboken, NJ, USA, 2014; ISBN 9781118304914.

50. Nijsen, M.; Smeets, E.; Stehfest, E.; van Vuuren, D.P. An evaluation of the global potential of bioenergy production on degraded lands. *GCB Bioenergy* **2012**, *4*, 130–147. [CrossRef]

51. Anshar, M.; Ani, F.N.; Kader, A.S.; Mechanics, F.; Centre, M.T. The Energy Potential of Municipal Solid Waste for Power Generation in Indonesia. *J. Mek.* **2014**, *37*, 42–54.

52. Obidzinski, K.; Andriani, R.; Komarudin, H.; Andrianto, A. Environmental and social impacts of oil palm plantations and their implications for biofuel production in Indonesia. *Ecol. Soc.* **2012**, *17*. [CrossRef]

53. Milbrandt, A.; Overend, R.P. *Assessment of Biomass Resources from Marginal Lands in APEC Economies*; National Renewable Energy Lab. (NREL): Golden, CO, USA, 2009; p. 52.

54. Pirard, R.; Bär, S.; Dermawan, A. Challenges and opportunities of bioenergy development in Indonesia. In Proceedings of the A synthesis of the workshop co-organized by Ministry of National Development Planning/Bappenas and CIFOR, Jakarta, Indonesia, 31 May 2016.

55. International Energy Agency. *World Energy Outlook 2017*; International Energy Agency: Paris, France, 2017.

56. ESDM. Successful Co-firing at PLTU Ropa and PLTU Bolok. Available online: https://www.esdm.go.id/en/media-center/news-archives/terus-didorong-ini-hasil-uji-coba-co-firing-di-pltu-ropa-dan-pltu-bolok- (accessed on 1 December 2020).

57. Miller, B. *Fuel Considerations and Burner Design for Ultra-Supercritical Power Plants*; Woodhead Publishing Limited: Cambridge, UK, 2013; ISBN 9780857091161.

58. P3TKEBTKE. Peta Potensi Energi Surya Indonesia. Available online: https://twitter.com/p3tkebtke/status/857400607522422784?lang=sk (accessed on 5 February 2021).

59. Statista. Cumulative solar photovoltaic capacity in Germany from 2013 to 2019. Available online: https://www.statista.com/statistics/497448/connected-and-cumulated-photovoltaic-capacity-in-germany/ (accessed on 22 March 2021).

60. World Bank Group. *Solar Resource and Photovoltaic Power Potential of Indonesia*; World Bank Group: Washington, DC, USA, 2017.

61. Statistik, B.P. *Statistik Indonesia 2020*; Badan Pusat Statistik: Jakarta, Indonesia, 2020.

62. Institute for Essential Services Reform. *Beyond 207 Gigawatts: Unleashing Indonesia's Solar Potential*; Institute for Essential Services Reform: Jakarta, Indonesia, 2021.

63. Veldhuis, A.J.; Reinders, A.H.M.E. Reviewing the potential and cost-effectiveness of grid-connected solar PV in Indonesia on a provincial level. *Renew. Sustain. Energy Rev.* **2013**, *27*, 315–324. [CrossRef]

64. Veldhuis, A.J.; Reinders, A.H.M.E. Reviewing the potential and cost-effectiveness of off-grid PV systems in Indonesia on a provincial level. *Renew. Sustain. Energy Rev.* **2015**, *52*, 757–769. [CrossRef]

65. Kunaifi, K.; Veldhuis, A.J.; Reinders, A.H.M.E. *The Electricity Grid in Indonesia: The Experiences of End-Users and Their Attitudes toward Solar Photovoltaics*; Springer: Berlin, Germany, 2020; ISBN 9783030383411.

66. Sunarso, A.; Ibrahim-Bathis, K.; Murti, S.A.; Budiarto, I.; Ruiz, H.S. GIS-Based Assessment of the Technical and Economic Feasibility of Utility-Scale Solar PV Plants: Case Study in West Kalimantan Province. *Sustainability* **2020**, *12*, 6283. [CrossRef]

67. Ruiz Rondan, H.S.; Sunarso, A.; Ibrahim-Bathis, K.; Murti, S.A.; Budiarto, I. GIS-AHP Multi-Decision-Criteria-Analysis for the optimal location of solar energy plants at Indonesia. *Energy Rep.* **2020**, *6*, 3249–3263. [CrossRef]
68. Syanalia, A.; Winata, F. Decarbonizing Energy in Bali With Solar Photovoltaic: GIS-Based Evaluation on Grid-Connected System. *Indones. J. Energy* **2018**, *1*, 5–20. [CrossRef]
69. Sah, B.P.; Wijayatunga, P. *Geographic Information System-Based Decision Support System for Renewable Energy Development an Indonesian Case Study*; Asian Development Bank: Manila, Philippines, 2017.
70. Sambodo, M.T.; Novandra, R. The state of energy poverty in Indonesia and its impact on welfare. *Energy Policy* **2019**, *132*, 113–121. [CrossRef]
71. P3TKEBTKE. Peta Potensi Energi Angin Indonesia. Available online: https://twitter.com/P3TKEBTKE/status/857401162617634816/photo/1 (accessed on 5 February 2021).
72. Bosch, J.; Staffell, I.; Hawkes, A.D. Temporally explicit and spatially resolved global offshore wind energy potentials. *Energy* **2018**, *163*, 766–781. [CrossRef]
73. Pratama, O. Konservasi Perairan Sebagai Upaya Menjaga Potensi Kelautan dan Perikanan Indonesia. Available online: https://kkp.go.id/djprl/artikel/21045-konservasi-perairan-sebagai-upaya-menjaga-potensi-kelautan-dan-perikanan-indonesia (accessed on 22 March 2021).
74. Gernaat, D.E.H.J.; Van Vuuren, D.P.; Van Vliet, J.; Sullivan, P.; Arent, D.J. Global long-term cost dynamics of offshore wind electricity generation. *Energy* **2014**, *76*, 663–672. [CrossRef]
75. Fios, F. Mapping the potential of green energy to border societies of Indonesia and Timor Leste (a preliminary study). *MATEC Web Conf.* **2018**, *197*, 1–4. [CrossRef]
76. Putro, W.S.; Prahmana, R.A.; Yudistira, H.T.; Darmawan, M.Y.; Triyono, D.; Birastri, W. Estimation Wind Energy Potential Using Artificial Neural Network Model in West Lampung Area. *IOP Conf. Ser. Earth Environ. Sci.* **2019**, *258*, 012006. [CrossRef]
77. Sari, D.P.; Kusumaningrum, W.B. A technical review of building integrated wind turbine system and a sample simulation model in central java, Indonesia. *Energy Procedia* **2014**, *47*, 29–36. [CrossRef]
78. Kamal, S. Studi Potensi Energi Angin Daerah Pantai Purworejo Untuk Mendorong Penyediaan Listrik menggunakan Sumer Energi Terbarukan Yang Ramah Lingkungan. *J. Mns. Lingkung.* **2007**, *14*, 26–34.
79. Soeprino, M.; Ibrochim, M. Analisa Potensi Energi Angin dan Estimasi Energi Output Turbin Angin di Lebak Banten. *J. Teknol. Dirgantara* **2009**, *7*, 51–59.
80. Bachtiar, A.; Hayyatul, W. Analisis Potensi Pembangkit Listrik Tenaga Angin PT. Lentera Angin Nusantara (LAN) Ciheras. *J. Tek. Elektro ITP* **2018**, *7*, 34–45. [CrossRef]
81. Quirapas, M.A.J.R.; Lin, H.; Abundo, M.L.S.; Brahim, S.; Santos, D. Ocean renewable energy in Southeast Asia: A review. *Renew. Sustain. Energy Rev.* **2015**, *41*, 799–817. [CrossRef]
82. Aprilia, E.; Aini, A.; Frakusya, A.; Safril, A. Potensi Panas Laut Sebagai Energi Baru Terbarukan Di Perairan Papua Barat Dengan Metode Ocean Thermal Energy Conversion (Otec). *J. Meteorol. Klimatologi Geofis.* **2019**, *6*, 7–14. [CrossRef]
83. Anggraini, D.; Ihsan Al Hafiz, M.; Fathin Derian, A.; Alfi, Y. Quantitative Analysis of Indonesian Ocean Wave Energy Potential Using Oscillating Water Column Energy Converter. *MATTER Int. J. Sci. Technol.* **2015**, *1*, 228–239. [CrossRef]
84. Vega, L.A. Economics of Ocean Thermal Energy Conversion (OTEC): An Update. In Proceedings of the Offshore Technology Conference, Houston, TX, USA, 16–19 August 2010; pp. 3–6. [CrossRef]
85. IRENA. *Ocean Thermal Energy Conversion: Technology Brief*; IRENA: Abu Dhabi, UAE, 2014.
86. Bluerise Offshore OTEC. *Feasibility Study of a 10 MW Installation*; Bluerise Offshore OTEC: Delft, The Netherlands, 2014; p. 63.
87. Langer, J.; Cahyaningwidi, A.A.; Chalkiadakis, C.; Quist, J.; Hoes, O.; Blok, K. Plant siting and economic potential of ocean thermal energy conversion in Indonesia a novel GIS-based methodology. *Energy* **2021**, *224*, 120121. [CrossRef]
88. Langer, J.; Quist, J.; Blok, K. Harnessing the economic potential of ocean thermal energy conversion in upscaling scenarios: Methodological progress illustrated for Indonesia. In Proceedings of the 20th European Roundtable on Sustainable Consumption and Production, Graz, Austria, 8–10 September 2021.
89. Asian Development Bank. *Wave Energy Conversion and Ocean Thermal Energy Conversion Potential in Developing Member Countries*; Asian Development Bank: Manila, Philippines, 2014; ISBN 978-92-9254-530-7.
90. Langer, J.; Cahyaningwidi, A.A.; Chalkiadakis, C.; Quist, J.; Hoes, O.A.C.; Blok, K. *Practical Sites for OTEC Deployment in Indonesia*; 4TU.ResearchData: Delft, The Netherlands, 2021.
91. IRENA. *Renewable Energy Prospects: Indonesia*; IRENA: Abu Dhabi, UAE, 2017.
92. Rachmayani, R.; Atma, G.; Suprijo, T.; Sari, N. Marine Current Potential Energy for Environmental Friendly Electricity Generation in Bali, Lombok and Makassar Straits. In Proceedings of the Environmental Technology Management Conference, Zurich, Switzerland, 29–30 July 2006.
93. Ajiwibowo, H.; Lodiwa, K.S.; Pratama, M.B.; Wurjanto, A. Field measurement and numerical modeling of tidal current in larantuka strait for renewable energy utilization. *Int. J. Geomate* **2017**, *13*, 124–131. [CrossRef]
94. Ribal, A.; Amir, A.K.; Toaha, S.; Kusuma, J.; Khaeruddin, K. Tidal current energy resource assessment around Buton Island, Southeast Sulawesi, Indonesia. *Int. J. Renew. Energy Res.* **2017**, *7*, 857–865.
95. Orhan, K.; Mayerle, R. Assessment of the tidal stream power potential and impacts of tidal current turbines in the Strait of Larantuka, Indonesia. *Energy Procedia* **2017**, *125*, 230–239. [CrossRef]

96. Orhan, K.; Mayerle, R.; Narayanan, R.; Pandoe, W. Investigation of the Energy Potential From Tidal Stream Currents in Indonesia. *Coast. Eng. Proc.* **2017**, 10. [CrossRef]
97. Yuningsih, A. Potensi Arus Laut Untuk Pembangkit Energi Baru Terbarukan Di Selat Pantar, Nusa Tenggara Timur. *Maj. Miner. Energi* **2011**, *9*, 61–72.
98. Blunden, L.S.; Bahaj, A.S.; Aziz, N.S. Tidal current power for Indonesia? An initial resource estimation for the Alas Strait. *Renew. Energy* **2013**, *49*, 137–142. [CrossRef]
99. Orhan, K.; Mayerle, R.; Pandoe, W.W. Assesment of Energy Production Potential from Tidal Stream Currents in Indonesia. *Energy Procedia* **2015**, *76*, 7–16. [CrossRef]
100. Babarit, A. A database of capture width ratio of wave energy converters. *Renew. Energy* **2015**, *80*, 610–628. [CrossRef]
101. Ribal, A.; Babanin, A.V.; Zieger, S.; Liu, Q. A high-resolution wave energy resource assessment of Indonesia. *Renew. Energy* **2020**, *160*, 1349–1363. [CrossRef]
102. Cornett, A.M. A Global Wave Energy Resource Assessment. In Proceedings of the Proceedings of the 18th International Offshore and Polar Engineering Conference, Vancouver, BC, Canada, 6–11 July 2008.
103. Mørk, G.; Barstow, S.; Kabuth, A.; Pontes, M.T. Assessing the global wave energy potential. *Proc. Int. Conf. Offshore Mech. Arct. Eng.-OMAE* **2010**, *3*, 447–454. [CrossRef]
104. Safitri, L.E.; Jumarang, M.I.; Apriansyah, A. Studi Potensi Energi Listrik Tenaga Gelombang Laut Sistem Oscillating Water Column (OWC) di Perairan Pesisir Kalimantan Barat. *Positron* **2016**, *6*, 8–16. [CrossRef]
105. Zikra, M. Preliminary Assessment of Wave Energy Potential around Indonesia Sea. *Appl. Mech. Mater.* **2017**, *862*, 55–60. [CrossRef]
106. Rizal, A.M.; Ningsih, N.S. Ocean wave energy potential along the west coast of the Sumatra island, Indonesia. *J. Ocean Eng. Mar. Energy* **2020**, *6*, 137–154. [CrossRef]
107. Wijaya, I.W.A. Teknologi Oscilating Water Column Di Perairan Bali. *Teknol. Elektro* **2010**, *9*, 165–174.
108. Sugianto, D.N.; Purwanto, P.; Handoyo, G.; Prasetyawan, I.B.; Hariyadi, H.; Alifdini, I. Identification of wave energy potential in Sungai Suci Beach Bengkulu Indonesia. *ARPN J. Eng. Appl. Sci.* **2017**, *12*, 4877–4886.
109. Alifdini, I.; Iskandar, N.A.P.; Nugraha, A.W.; Sugianto, D.N.; Wirasatriya, A.; Widodo, A.B. Analysis of ocean waves in 3 sites potential areas for renewable energy development in Indonesia. *Ocean Eng.* **2018**, *165*, 34–42. [CrossRef]
110. Simaremare, A.A.; Bruce, A.; Macgill, I. Least Cost High Renewable Energy Penetration Scenarios in the Java Bali Grid Least Cost High Renewable Energy Penetration Scenarios in the Java Bali Grid System. In Proceedings of the Asia Pacific Solar Research Conference, Melbourne, Australia, 5–7 December 2017.
111. Günther, M. Challenges of a 100% renewable energy supply in the Java-Bali grid. *Int. J. Technol.* **2018**, *9*, 257–266. [CrossRef]
112. Günther, M.; Eichinger, M. Cost optimization for the 100% renewable electricity scenario for the Java-Bali grid. *Int. J. Renew. Energy Dev.* **2018**, *7*, 269–276. [CrossRef]
113. IESR; Agora Energiewende; LUT University. *Deep Decarbonization of Indonesia's Energy System Deep Decarbonization of Indonesia's Energy System: A Pathway to Zero Emissions*; IESR: Jakarta Selatan, Indonesia, 2021.
114. Hasan, M.H.; Mahlia, T.M.I.; Nur, H. A review on energy scenario and sustainable energy in Indonesia. *Renew. Sustain. Energy Rev.* **2012**, *16*, 2316–2328. [CrossRef]
115. Kumar, S. Assessment of renewables for energy security and carbon mitigation in Southeast Asia: The case of Indonesia and Thailand. *Appl. Energy* **2016**, *163*, 63–70. [CrossRef]
116. Purba, N.P.; Kelvin, J.; Sandro, R.; Gibran, S.; Permata, R.A.I.; Maulida, F.; Martasuganda, M.K. Suitable Locations of Ocean Renewable Energy (ORE) in Indonesia Region-GIS Approached. *Energy Procedia* **2015**, *65*, 230–238. [CrossRef]
117. Danish Energy Agency. *Kalimantan Regional Energy Outlook*; Danish Energy Agency: København, Denmark, 2019.
118. Martosaputro, S.; Murti, N. Blowing the wind energy in Indonesia. *Energy Procedia* **2014**, *47*, 273–282. [CrossRef]
119. Rumbayan, M.; Nagasaka, K. Assessment of Wind Energy Potential in Indonesia Using Weibull Distribution Function. *Int. J. Electr. Power Eng.* **2011**, *5*, 229–235. [CrossRef]
120. Simanjuntak, J.S. *Techno-Economic and Institutional Assessment of Wind Energy in Indonesia*; Delft University of Technology: Delft, The Netherlands, 2021.
121. Straatman, P.J.T.; van Sark, W.G.J.H.M. A new hybrid ocean thermal energy conversion-Offshore solar pond (OTEC-OSP) design: A cost optimization approach. *Sol. Energy* **2008**, *82*, 520–527. [CrossRef]
122. Malik, M.Z.; Musharavati, F.; Khanmohammadi, S.; Baseri, M.M.; Ahmadi, P.; Nguyen, D.D. Ocean thermal energy conversion (OTEC) system boosted with solar energy and TEG based on exergy and exergo-environment analysis and multi-objective optimization. *Sol. Energy* **2020**, *208*, 559–572. [CrossRef]
123. Reuters. Indonesian State Utility to Retire Coal Plants Gradually. Available online: https://www.reuters.com/article/indonesia-coal-idUSL3N2NE3FM (accessed on 7 July 2021).

MDPI

St. Alban-Anlage 66

4052 Basel

Switzerland

Tel. +41 61 683 77 34

Fax +41 61 302 89 18

www.mdpi.com

Energies Editorial Office

E-mail: energies@mdpi.com

www.mdpi.com/journal/energies